The ASQ Certified Quality Inspector Handbook

The ASQ Certified Quality Inspector Handbook

Fourth Edition

H. Fred Walker, Ahmad K. Elshennawy, and Mary McShane-Vaughn, Editors

ASQExcellence
Milwaukee, Wisconsin

Published by ASQExcellence, Milwaukee, WI

Produced and distributed by Quality Press, ASQ, Milwaukee, WI

Publisher's Cataloging-in-Publication data

Names: Walker, Harvey Fred, 1963–, editor. | Elshennawy, Ahmad K., 1952–, editor. | McShane-Vaughn, Mary A., 1963–, editor.
Title: The ASQ certified quality inspector handbook , fourth edition / H. Fred Walker; Ahmad K. Elshennawy; Mary McShane-Vaughn, Editors.
Description: Includes bibliographical reference and index. | Milwaukee, WI: ASQExcellence, 2025.
Identifiers: LCCN: 2025942114 | ISBN: 9781636941646 (hardcover) | 9781636941653 (pdf) | 9781636941660 (epub)
Subjects: LCSH Production management. | Quality control—Handbooks, manuals, etc. | Quality control inspectors—Certification. | BISAC BUSINESS & ECONOMICS / Industries / Manufacturing | BUSINESS & ECONOMICS / Quality Control | TECHNOLOGY & ENGINEERING / Measurement
Classification: LCC TS156 .A77 2025 | DDC 658.5/62—dc23

ASQ and ASQExcellence advance individual, organizational, and community excellence worldwide through learning, quality improvement, and knowledge exchange.

Bookstores, wholesalers, schools, libraries, businesses, and organizations: Quality Press and ASQExcellence books are available at quantity discounts for bulk purchases for business, trade, or educational uses. For more information, please contact Quality Press at 800-248-1946 or books@asq.org.

To place orders or browse the selection of all Quality Press titles, visit our website at: http://www.asq.org/quality-press.

30 29 28 27 26 GP 7 6 5 4 3 2 1

ASQ

Quality Press
600 N. Plankinton Ave.
Milwaukee, WI 53203-2914
Email: books@asq.org
Excellence Through Quality™

ASQE™
Inspire Excellence

In Memoriam

We dedicate this new edition of *The ASQ Certified Quality Inspector Handbook* to our colleague and friend Bhisham C. Gupta, who passed away on December 14, 2022.

Dr. Gupta had fifty-two years of experience consulting for the semiconductor, paper, and medical/healthcare industries. Bhisham was a Professor Emeritus in the Department of Mathematics and Statistics at the University of Southern Maine, where he served in many leadership roles and authored or co-authored more than thirty-five refereed research papers; two books published by UFRJ-Press, Brazil; three books published by ASQ Quality Press; and two books published by John Wiley & Sons.

Dr. Gupta earned his PhD in Statistics, an MA in Mathematics, and an MS in Mathematics and Statistics.

Bhisham once said in a book acknowledgment that he was indebted to his wife Swarn, daughters Anita and Anjali, son Shiva, sons-in-law Prajay and Mark, daughter-in-law Aditi, and several grandchildren for their deep love and devotion.

In loving memory of my father, Carl Ellsworth Walker.
—Fred

In loving memory of my parents, Mohammed Elshennawy and Ikram Ismail.
—Ahmad

To Ila and Ian, the next generation of Vaughns.
—Mary

Table of Contents

List of Figures and Tables

Acknowledgments

Ahmad would like to acknowledge all of God's blessings and guidance, without which none of this would be possible. He also thanks his family for their support and encouragement throughout his career, which has enabled him to contribute to the world of quality as an academician, author, and researcher.

Mary thanks ASQ Quality Press for the opportunity to contribute to this latest edition of the CQI Handbook, and her husband, Jim, for providing hot cups of coffee during the editing process.

CONTRIBUTORS' ACKNOWLEDGMENTS

The American Society for Quality and Quality Press would like to thank the Quality Press Peer Review Committee for its invaluable volunteer participation and contributions to this work. Without our volunteers' subject matter expertise, time, and passion for creating content, none of our efforts would be possible.

QUALITY PRESS PEER REVIEW COMMITTEE MEMBERS

Scott A. Laman, Chair
Peter Pylipow, Vice Chair
Melvin Alexander
Lance Coleman
Ahmad Elshennawy
Marc Hamilton
Gary Jing
Trevor Jordan
Jane Keathley
Betsy Macht
Mary McShane-Vaughn
Jayet Moon
G.S. Sureshchandar
Tiea Theurer

Technical Mathematics

Section Title

Chapter 1

Basic Shop Math

Solve basic shop math problems using addition, subtraction, multiplication, division of fractions and decimals, squares, and square roots. Use methods such as truncating and rounding to obtain significant digits for positive and negative numbers. Demonstrate when and how to use percent change calculation (new value—original value / original value x 100). (Apply)

Body of Knowledge I.A

PROPERTIES OF REAL NUMBERS

When performing arithmetic operations on numbers, keep in mind the following fundamental properties shown in Table 1.1.

Let a, b, and c represent real numbers.

POSITIVE AND NEGATIVE NUMBERS

A positive number is one that is greater than zero; a negative number is less than zero. For example, the number 6 is a positive number, while -2 is a negative number.

Adding and Subtracting with Positive and Negative Numbers

We can add and subtract positive and negative numbers, as the examples show:

$$4+3=7$$
$$4+-3=1$$
$$-4+3=-1$$
$$-4+-3=-7$$
$$8-2=6$$
$$8-(-2)=10$$
$$-8-2=-10$$
$$-8-(-2)=-6$$

Table 1.1 Properties of real numbers.

Commutative property for addition	$a+b=b+a$
Commutative property for multiplication	$a\times b=b\times a$
Associative property for addition	$(a+b)+c=a+(b+c)$
Associative property for multiplication	$(a\times b)\times c=a\times(b\times c)$
Distributive property	$a(b+c)=ab+ac$

Multiplying and Dividing with Positive and Negative Numbers

We can also multiply and divide positive and negative numbers, as the examples show:

$$4\times 3=12$$
$$4\times -3=-12$$
$$-4\times 3=-12$$
$$-4\times -3=12$$
$$8\div 2=4$$
$$8\div(-2)=-4$$
$$-8\div 2=-4$$
$$-8\div(-4)=2$$

FRACTIONS

A fraction relates a number of parts to the whole and is usually written in the form $\frac{a}{b}$.

The number a is the numerator and the number b is the ***denominator***. For example, the fraction $\frac{3}{8}$ refers to 3 parts from a total of 8 parts. Here, 3 is the numerator and 8 is the denominator.

Equivalent Fractions

Equivalent fractions are those that express the same proportion of parts to the whole. For example, the fractions $\frac{1}{2}$, $\frac{2}{4}$, and $\frac{40}{80}$ are equivalent.

Simplified Fractions

A simplified fraction refers to a fraction in which numerator and denominator do not share any factors in common. A simplified fraction is found by dividing both numerator and denominator by the greatest common divisor (GCD). We can find the GCD by factoring both numerator and denominator and choosing the largest factor in common.

For example, given the fraction $\frac{27}{45}$ we can factor numerator and denominator $\frac{3\times9}{5\times9}$ and determine that the GCD is 9. Dividing numerator and denominator each by 9 yields the simplified fraction $\frac{3}{5}$.

Proper and Improper Fractions

A *proper fraction* is one in which the numerator is less than the denominator. The fractions $\frac{2}{3}$, $\frac{14}{17}$, and $\frac{75}{100}$ are all examples of proper fractions. An *improper fraction* is one in which the numerator is larger than the denominator. Improper fractions can be converted into a ***mixed number***, or a whole number and a proper fraction, by dividing the denominator into the numerator. For example:

$$\frac{10}{7} = 7\sqrt{10} = 1\frac{3}{7}.$$

Adding and Subtracting Fractions with Like Denominators

To add or subtract fractions with equal denominators:

1. Add or subtract the numerators and leave the denominator the same.
2. Simplify the resulting fraction, if necessary.

$$\frac{3}{8} + \frac{4}{8} - \frac{1}{8} = \frac{6}{8} = \frac{3}{4}$$

Adding and Subtracting Fractions with Unequal Denominators

Many times, the denominators of the fractions we wish to add or subtract are not equal. In this case:

1. Find the ***least common denominator*** (LCD) using ***prime factorization.***
2. Find the equivalent fractions using the LCD.
3. Perform the addition or subtraction.
4. Simply the resulting fraction, if necessary.

Prime Factorization

A ***prime number*** is an integer greater than one that is a multiple of only itself and the number 1. For example, the number 7 is a prime number, since it can only be divided evenly by the numbers 1 and 7, which are called the ***factors*** of 7. Conversely, 8 is not a prime number, since its factors are 1, 2, 4, and 8. A list of the first few prime numbers follows: 1, 2, 3, 5, 7, 11, 13, 17, 19, 23, 29, . . .

A ***prime factorization*** expresses a number in terms of prime number factors. For example, the prime factorization of the number 12 is $(2\times2\times3)$. In order to find the prime factorization of a given value, we can draw a number tree in which

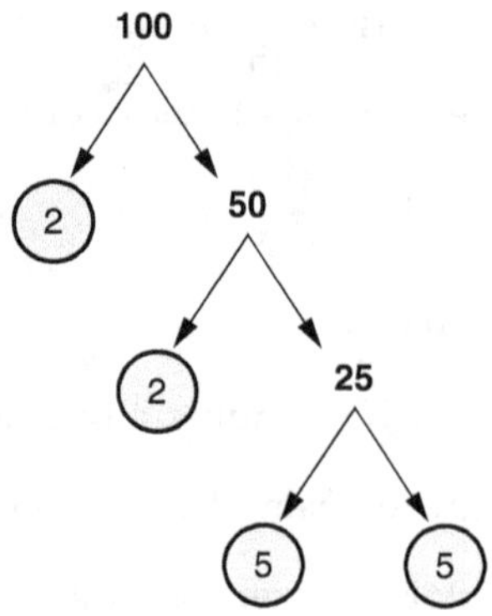

Figure 1.1 Example of prime factorization tree.

we successively break down the factors until we find the prime components. For example, the prime factorization of the number 100 is $(2\times2\times5\times5)$, which can be found using the tree in Figure 1.1.

Least Common Denominator

To find the LCD, express each denominator in terms of its prime factors. For example, if asked to solve

$$\frac{1}{6}+\frac{12}{18}-\frac{1}{4}=?,$$

we begin with expressing the denominators in terms of their prime factors:

$$6=(2\times3)$$
$$18=(2\times3\times3)$$
$$4=(2\times2)$$

Record the maximum number of times each factor is used in any one expression. This will indicate how many times each factor must be multiplied to find the LCD. The factor 2 appears at most two times in the expressions above, and the factor 3 appears at most 2 times. Therefore, the LCD is calculated as:

$$\text{LCD}=(2\times2)\times(3\times3)=36$$

Next, find the equivalent fractions using the LCD. The numerators can be found by cross multiplying:

$$\frac{\text{old numerator}}{\text{old denominator}}=\frac{\text{new numerator}}{\text{LCD}}$$

$$\frac{\text{old numerator}\times\text{LCD}}{\text{old denominator}}=\text{new numerator}$$

$$\frac{1\times 36}{6}=6$$

$$\frac{12\times 36}{18}=24$$

$$\frac{1\times 36}{4}=9$$

Now add the equivalent fractions and simplify the result if needed:

$$\left(\frac{1}{6}+\frac{12}{18}-\frac{1}{4}\right)=\frac{6+24-9}{36}=\frac{21}{36}=\frac{7}{12}$$

Adding or Subtracting Mixed Numbers

To add or subtract mixed numbers, we can either convert the mixed numbers back into improper fractions and then proceed, or we can first add and subtract the whole number parts and then add and subtract the fractions. For example:

$$\left(1\frac{1}{2}+2\frac{3}{4}\right)=\left(\frac{3}{2}+\frac{11}{4}\right)=\left(\frac{6+11}{4}\right)=\frac{17}{4}=4\frac{1}{4}$$

$$\left(1\frac{1}{2}+2\frac{3}{4}\right)=3+\left(\frac{1}{2}+\frac{3}{4}\right)=3+\left(\frac{2+3}{4}\right)=\left(3+\frac{5}{4}\right)=4\frac{1}{4}$$

Multiplying and Dividing Fractions

To multiply fractions:

1. Multiply the numerators to obtain the numerator of the result.
2. Multiply the denominators to obtain the resulting denominator.
3. Simplify the resulting fraction if necessary.

For example:

$$\frac{3}{8}\times\frac{4}{5}\times\frac{1}{2}=\frac{12}{80}=\frac{3}{20}$$

To divide fractions, we convert the problem into a multiplication problem:

1. Switch numerator and denominator of the second fraction.
2. Multiply the numerators to obtain the numerator of the result.
3. Multiply the denominators to obtain the resulting denominator.
4. Simplify the resulting fraction if necessary.

For example:

$$\frac{2}{9}\div\frac{4}{6}=\frac{2}{9}\times\frac{6}{4}=\frac{12}{36}=\frac{1}{3}$$

DECIMALS

Given a number of the form:

ABCDEFG.HIJKLM

we can define the place values for each letter-position, as shown in Table 1.2.

Decimal and Fraction Equivalents

Decimal and fraction equivalents are shown in Table 1.3.

Table 1.2 Place values for ABCDEFG.HIJKLM.

A	millions
B	hundred thousands
C	ten thousands
D	thousands
E	hundreds
F	tens
G	ones
H	tenths
I	hundredths
J	thousandths
K	ten-thousandths
L	hundred-thousandths
M	millionths

Table 1.3 Decimal and fraction equivalents.

One tenth	0.1	$\frac{1}{10}$
One hundredth	0.01	$\frac{1}{100}$
One thousandth	0.001	$\frac{1}{1000}$
One ten-thousandth	0.0001	$\frac{1}{10000}$
One hundred-thousandth	0.00001	$\frac{1}{100000}$
One millionth	0.000001	$\frac{1}{1000000}$

Converting Fractions to Decimals

Fractions can be converted to decimals by dividing denominator into the numerator. For example:

$$\frac{3}{8} = 8\sqrt{3} = 0.375$$

We read this decimal as *375 thousandths.*

Rational and Irrational Numbers

A ***rational number*** can be represented in fractional form. It either has a finite number of decimal places, or an infinite (never-ending) number of repeating decimal places. For example, the numbers $0.37, 0.6894$, 1.33562, $4.\overline{3}$, and $8.\overline{519}$ are all examples of rational numbers.

An ***irrational number*** is a number that has an infinite number of decimal places that do not repeat. An example of an irrational number is $\pi = 3.14159\ldots$ since its digits are non-repeating and infinite.

Converting Decimals to Percentages

To convert a decimal into a percentage, multiply by 100%, as shown:

$$0.375 \times 100\% = 37.5\%$$

Adding and Subtracting Decimals

To add or subtract decimals, line up the numbers at their decimal places. For example, the numbers 3.475, 11.55, and 2.2 can be added as shown:

$$\begin{array}{r} 3.475 \\ 11.55 \\ +2.2 \\ \hline 17.225 \end{array}$$

Multiplying and Dividing Decimals

To multiply or divide decimals, count the total number of decimal places in the problem. The final answer will have that number of decimal places.

$$\begin{array}{rl} 3.75 & \textit{2 decimal places} \\ \times 2.3 & \textit{1 decimal place} \\ \hline 8.625 & \textit{3 decimal places} \end{array}$$

SQUARES AND SQUARE ROOTS

The square of a number is simply that number multiplied by itself. The square of a number can also be written as that number to the second power. For example, the square of 7 is calculated as:

$$7 \times 7 = 7^2 = 49$$

A square root is denoted as follows:

$$\sqrt{25} = 25^{1/2} = 5, \text{ since } 25 = (5 \times 5) = 5^2$$

Every positive number has a square root. For example, using a calculator, we can determine:

$$\sqrt{54.3} = 7.369$$

Imaginary Numbers

The square root of a negative number does not exist on the real number line. Rather, the ***imaginary number i*** is defined as the square root of negative one. The result is written in terms of ***i***. For example:

$$\sqrt{-25} = \sqrt{-1 \times 25} = \sqrt{-1} \times \sqrt{25} = 5i$$

Exponents

A whole number exponent indicates how many times a number is multiplied by itself. For example:

$$7 \times 7 = 7^2$$
$$7 \times 7 \times 7 = 7^3$$

ORDER OF OPERATIONS

The order of operations can be remembered by using the mnemonic device of **PEMDAS**, or *Please **E**xcuse **M**y **D**ear **A**unt **S**ally*, which stands for Parentheses, Exponents, Multiplication, Division, Addition, and Subtraction. As an illustration, the expression $6 \times 4 + (9-1) \div 2 - 5^2$ is correctly evaluated as follows:

$$\begin{aligned} 6 \times 4 + (9-1) \div 2 - 5^2 &= 6 \times 4 + \mathbf{8} \div 2 - 5^2 && \textbf{P} \\ &= 6 \times 4 + 8 \div 2 - \mathbf{25} && \textbf{E} \\ &= \mathbf{24} + 8 \div 2 - 25 && \textbf{M} \\ &= 24 + \mathbf{4} - 25 && \textbf{D} \\ &= \mathbf{28} - 25 && \textbf{A} \\ &= \mathbf{3} && \textbf{S} \end{aligned}$$

FACTORIALS

A factorial is a mathematical operation denoted by an exclamation point (!) and is evaluated in the following manner:

$$5! = 5 \times 4 \times 3 \times 2 \times 1$$

By definition, $0! = 1$.

Factorials are used to calculate probabilities from the binomial, hypergeometric, and Poisson distributions, among others.

TRUNCATING, ROUNDING, AND SIGNIFICANT DIGITS

To display a result with a certain number of decimal places, we can choose to ***truncate*** the number. For example, we can truncate the number 3.527 to two decimal places by writing 3.52.

We can ***round*** a result to a certain number of digits by looking at the value of the digit to the right of the decimal place of interest. For example, if we want to round the number 3.527 to the hundredths place, we will look at the third decimal place (in this case 7) as our decision point. The rule for rounding is this: If the decision number is less than five, do not round up. If it is 5 or greater, round up. Therefore, 3.527 would be rounded up to 3.53. However, if we rounded to the tenths place, 3.527 would be displayed as 3.5, since 2 would be our decision number.

When dealing with the precision of a measurement, we must consider ***significant digits***. For example, instrument measurements have a certain inherent precision, which can be expressed in terms of a certain number of significant digits. We must take care to express instrument measurements using the proper number of significant digits.

Adding and Subtracting Measurements

When adding or subtracting measurements, the final answer will only be as precise as the *least* precise reading. Therefore, the number of decimal places of the answer should match those of the least precise reading. In the example below, the reading of 5.2 is the least precise, and the final answer is rounded to the tenths place:

$$\begin{array}{r} 4.573 \\ 3.77 \\ 5.2 \\ +\,6.1299 \\ \hline 19.6729 \end{array} \cong 19.7$$

Multiplying and Dividing Using Significant Digits

When multiplying measurements, display the final result with the same number of significant digits as the *least* accurate reading. For example, the result below will be rounded to 3 significant digits since the value 3.21 has three significant digits and 20.45 has four:

$$\begin{array}{r} 20.45 \\ \times\,3.21 \\ \hline 65.6445 \end{array} \cong 65.6$$

PERCENT CHANGE CALCULATIONS

To report a percentage change, we use the formula:

$$\%Change = \frac{(New\ value - Original\ value)}{Original\ value} \times 100\%$$

A positive result refers to a percentage increase in the original metric, and a negative result refers to a percentage decrease.

For example, consider a production line that currently runs at 300 pieces per minute. After a new machine is introduced, the line now produces 350 pieces per minute. The percent change in output is calculated:

$$\begin{aligned}\%\,Change &= \frac{(New\ value - Original\ value)}{Original\ value} \times 100\% \\ &= \frac{(375-300)}{300} \times 100\% \\ &= \frac{75}{300} \times 100\% \\ &= 25\%\end{aligned}$$

The new machine has increased the production rate by 25%.

If a process improvement results in 12 scrapped items per week, versus the original rate of 45 units, the percent change in scrap rate is calculated:

$$\begin{aligned}\%\,Change &= \frac{(New\ value - Original\ value)}{Original\ value} \times 100\% \\ &= \frac{(12-45)}{45} \times 100\% \\ &= \frac{-33}{45} \times 100\% \\ &= -73.3\%\end{aligned}$$

The process improvement has resulted in a 73.3% reduction in the scrap rate.

Chapter 2

Basic Algebra

> Solve or simplify first-degree and single-variable equations. (Apply)
>
> **Body of Knowledge I.B**

SOLVING ALGEBRAIC EQUATIONS

When we solve an algebraic equation, we are finding the value of the unknown variable, which is denoted by a letter. To illustrate, the following equations are all examples of algebraic expressions:

$$x+3=5$$
$$6a-2=24$$
$$4s+4=6s-9$$
$$\frac{7}{8}y+30\geq 51$$

In each, the goal is to solve for the value or values of the variable that make the expression true. We accomplish this by collecting terms and then *isolating* the unknown variable on one side of the equation by performing inverse operations.

Properties of Real Numbers

In addition to the commutative, associative, and distributive properties of real numbers presented in Chapter 1, real numbers have certain fundamental properties, as shown in Table 2.1. Let a, b, and c be real numbers.

Inverse Operations

To isolate the variable on one side of the equation, we perform a series of inverse operations on both sides of the equation. Table 2.2 shows a list of inverse operations.

Table 2.1 Properties of real numbers.

Reflexive property	$a=a$
Symmetric property	*If* $a=b$, *then* $b=a$
Transitive property	*If* $a=b$ *and* $b=c$, *then* $a=c$
Additive identity	$a+0=a$
Multiplicative identity	$a\times 1=a$

Table 2.2 Inverse operations.

Operation	Inverse Operation	Example
Addition	Subtraction	$a+(-a)=0$
Subtraction	Addition	$-a+a=0$
Multiplication	Division	$a\times\frac{1}{a}=1$
Division	Multiplication	$\frac{1}{a}\times a=1$
Exponent r	rth root	$\sqrt[r]{a^r}=a$, *for* $a\ge 0$
rth root	Exponent r	$\left(\sqrt[r]{a}\right)^r=a$, *for* $a\ge 0$
Power of 10	Log base 10	$\log_{10}(10^a)=a$
Log base 10	Power of 10	$10^{\log_{10}(a)}=a$

Solving Equations Using One Inverse Operation

To isolate *x*, we perform the appropriate inverse operation on *both* sides of the equation. In the example below, note that the inverse of adding 3 is subtracting 3:

$$x+3=5$$
$$x+3-3=5-3$$
$$x=2$$

Always perform a *check of your work* by substituting the value of the variable into the original equation:

$$x+3=5$$
$$2+3=5$$

☑

Solving Equations Using Two or More Inverse Operations

To solve a problem such as $ax+b=c$, we must do a series of inverse operations on both sides of the equation to isolate the variable *a*. In the following example, addition and division inverse operations are employed in succession to find the solution:

$$
\begin{aligned}
6a-2&=22\\
6a-2+2&=22+2\\
6a&=24\\
6a\div 6&=24\div 6\\
a&=4
\end{aligned}
$$

Next, check the solution:

$$
\begin{aligned}
6a-2&=22\\
6\times 4-2&=22\\
24-2&=22\\
22&=22
\end{aligned}
$$

Solving Equations by Collecting Terms

To solve a problem where the variable and/or constants appear in more than two terms, we must first collect terms and then perform the inverse operations. For example, to solve the equation $8z+5+4z+-1=16$, we must collect the variable and the constant terms before proceeding with the solution:

$$
\begin{aligned}
8z+5+4z-1&=16\\
12z+4&=16\\
12z+4-4&=16-4\\
12z&=12\\
12z\div 12&=12\div 12\\
z&=1
\end{aligned}
$$

Check the solution:

$$
\begin{aligned}
8z+5+4z-1&=16\\
(8\times 1)+5+(4\times 1)-1&=16\\
8+5+4-1&=16\\
16&=16
\end{aligned}
$$

To solve a problem such as $4s+4=6s-9$, we must collect terms that appear on both sides of the equation:

$$
\begin{aligned}
4s+4&=6s-9\\
4s+4-4s&=6s-4s-9\\
4&=2s-9\\
4+9&=2s-9+9\\
13&=2s\\
13\div 2&=2s\div 2\\
\frac{13}{2}&=s\\
6\frac{1}{2}&=s
\end{aligned}
$$

Check the solution:

$$\begin{aligned}4s+4&=6s-9\\4\times6\frac{1}{2}+4&=6\times6\frac{1}{2}-9\\(24+2)+4&=(36+3)-9\\30&=30\end{aligned}$$

☑

Solving Equations with Parentheses

To solve equations with parentheses, follow the order of operations rule PEMDAS, in which parentheses, exponents, multiplication, division, addition, and subtraction operations are performed left to right, in that order. Using the rule, operations included in parentheses are evaluated first. For example:

$$\begin{aligned}2(10x-2x)+50&=14\\2\times8x+50&=14\\16x+50-50&=14-50\\16x&=-36\\16x\div16&=-36\div16\\x&=-2\tfrac{1}{4}\end{aligned}$$

Converting $-2\frac{1}{4}$ to $-\frac{9}{4}$, we can check the solution:

$$\begin{aligned}2\left(10\left(-\frac{9}{4}\right)-2\left(-\frac{9}{4}\right)\right)+50&=14\\2\times\frac{-36}{2}+50&=14\\-36+50&=14\\14&=14\end{aligned}$$

☑

Solving Inequalities

To solve an inequality, remember that the solution will be a *range* of values. In the following example, the equation is true for any value of y equal to or greater than 24:

$$\begin{aligned}\frac{7}{8}y+30&\geq51\\\frac{7}{8}y+30-30&\geq51-30\end{aligned}$$

$$\frac{7}{8}y \geq 21$$
$$\frac{8}{7} \times \frac{7}{8}y \geq 21 \times \frac{8}{7}$$
$$y \geq 21 \times \frac{8}{7}$$
$$y \geq 24$$

To check your work, you must first check to see if the equality holds true. Then choose a number that is in the range of solutions to make sure the inequality is also correct, as shown below:

Check for the equality condition:

$$\frac{7}{8}y + 30 \geq 51$$

$$\frac{7}{8}24 + 30 = 51$$
$$21 + 30 = 51$$
$$51 = 51$$

☑

To check for the inequality condition, use a number in the solution range. Recall the solution to the example was $y \geq 24$. Here, we arbitrarily choose 40 to perform the check:

$$\frac{7}{8}y + 30 \geq 51$$

$$\frac{7}{8}40 + 30 \geq 51$$
$$35 + 30 \geq 51$$
$$65 \geq 51.$$

☑

Solving Inequalities with Negatively Signed Values

Be careful when solving an inequality if you are multiplying or dividing both sides by a negative number—the inequality sign reverses! Checking your work after solving an inequality is very important to make sure that the sign of your answer is correct. For example:

$$-7x + 4 \leq 25$$
$$-7x + 4 - 4 \leq 25 - 4$$
$$-7x \leq 21$$
$$-7x \div -7 \geq 21 \div -7$$
$$x \geq -3$$

Check for the equality condition:

$$-7x + 4 \le 25$$

$$\begin{aligned} (-7 \times -3) + 4 &= 25 \\ 21 + 4 &= 25 \\ 25 &= 25 \end{aligned}$$

Check for the inequality condition. Here, the number 2 is chosen for the check:

$$-7x + 4 \le 25$$

$$\begin{aligned} (-7 \times 2) + 4 &\le 25 \\ -14 + 4 &\le 25 \\ -10 &\le 25 \end{aligned}$$

Chapter 3

Basic Geometry

> Calculate general parameters for basic geometric shapes, such as area, circumference, diameter, radius, perimeter, and volume. Calculate complementary and supplementary angles. Select the minimum number of coordinate points for shapes (e.g., lines and circles). (Apply)
>
> Body of Knowledge I.C

AREAS OF BASIC GEOMETRIC SHAPES

The area of a shape is expressed in terms of squared units, such as cm^2 or ft^2. Table 3.1 shows the calculations for finding the area of basic geometric shapes.

Examples of Area Calculations

A rectangle has length 6 cm and width 4 cm. Its area is $6\text{ cm} \times 4\text{ cm} = 24\text{ cm}^2$.

A square has sides of length 10 inches. Its area is $10\text{ in.} \times 10\text{ in.} = 100\text{ in}^2$.

A parallelogram has a height equal to 32 mm and a base of 40 mm. Its area is $32\text{ mm} \times 40\text{ mm} = 1280\text{ mm}^2$.

A trapezoid has a length of 12 inches at the top and 24 inches at the base. Its height is 15 inches. Its area is $[(12\text{ in.} + 24\text{ in.}) \times 15\text{ in.}]\ /2 = 270\text{ in}^2$.

A regular hexagon (meaning one with all sides the same length) has sides of length 8 cm. Its area is $2.597 \times (8\text{ cm})^2 = 2.597 \times 64\text{ cm}^2 = 166.208\text{ cm}^2$.

A regular hexagon has a radius of 3.5 inches. Its area is $3.464 \times (3.5\text{ in.})^2 = 3.464 \times 12.25\text{ in}^2 = 42.434\text{ in}^2$.

A right triangle (meaning that the triangle has a 90 degree angle) has a base of 65 mm and a height of 120 mm. Its area is $(1/2) \times 65\text{ mm} \times 120\text{ mm} = 3900\text{ mm}^2$.

A triangle has a base of 34 cm and a height of 20 cm. Its area is $(1/2) \times 34\text{ cm} \times 20\text{ cm} = 340\text{ cm}^2$.

A circle has a radius of 25 inches. Its area is $\pi \times (25\text{ in.})^2 = 1963.495\text{ in}^2$. When possible, use a calculator to find areas that use the number π (pi) to avoid rounding errors.

A circular ring has outer radius of 10.5 cm and an inner radius of 3.5 cm. Its area is $\pi \times [(10.5\text{ cm})^2 - (3.5\text{ cm})^2] = \pi \times (110.25\text{ cm}^2 - 12.25\text{ cm}^2) = 307.8761\text{ cm}^2$.

Table 3.1 Calculation of area for basic geometric shapes.

Rectangle	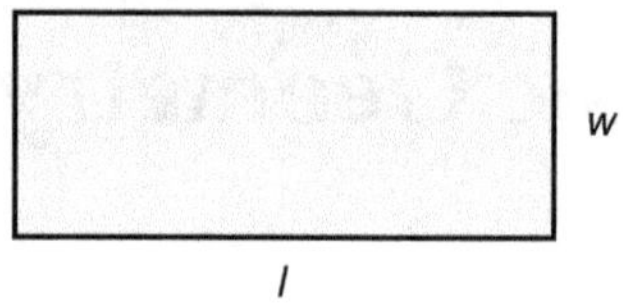	$A = l \times w$
Square	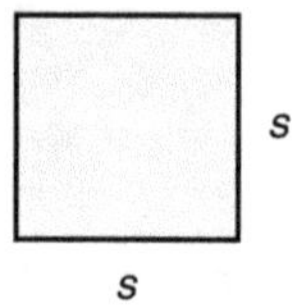	$A = s^2$
Parallelogram	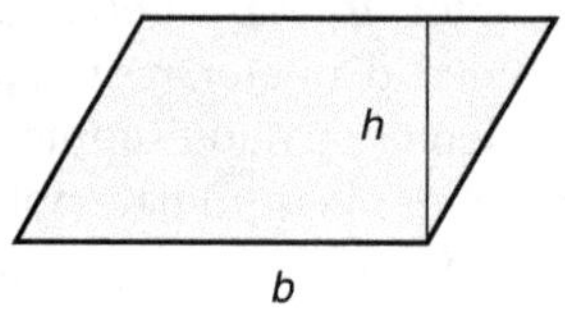	$A = b \times h$ *where b is base and h is height*
Trapezoid	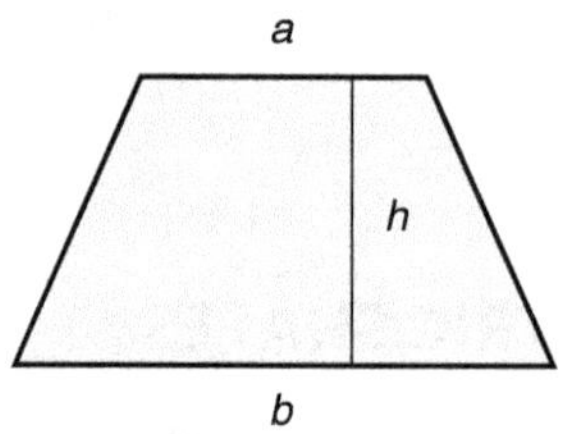	$A = \frac{(a+b)h}{2}$
Regular hexagon	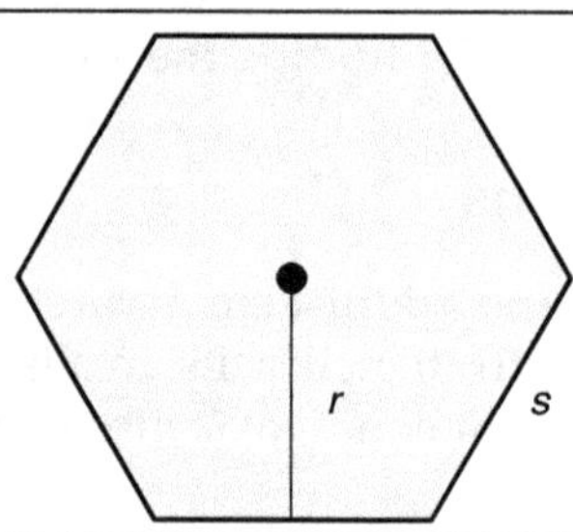	$A = 2.597s^2$ $A = 3.464r^2$
Right triangle	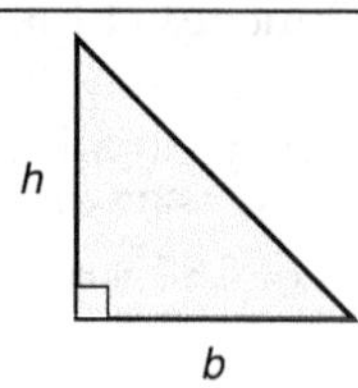	$A = \frac{1}{2} b \times h$
Triangle	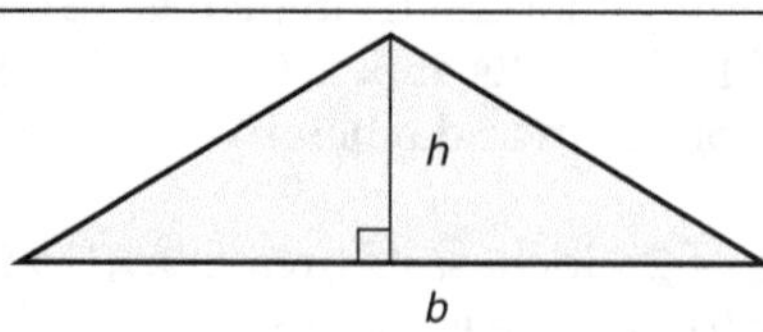	$A = \frac{1}{2} b \times h$

(continued)

Table 3.1 Calculation of area for basic geometric shapes. (continued)

Circle	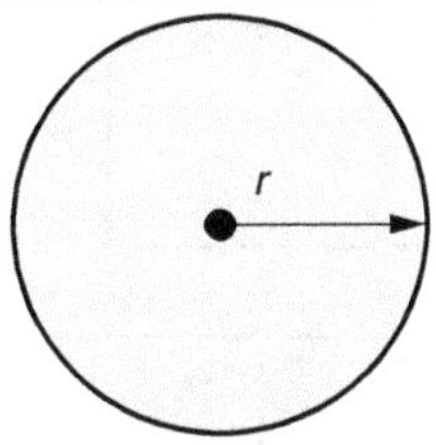	$A = \pi r^2$ *where r is radius and* $\pi = pi = 3.14159\ldots$
Circular ring	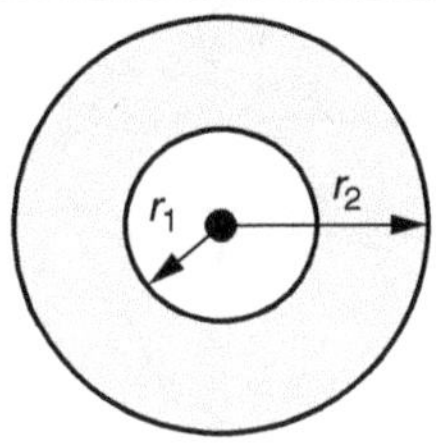	$A = \pi\left(r_2^2 - r_1^2\right)$

PERIMETER, DIAMETER, RADIUS, AND CIRCUMFERENCE OF BASIC GEOMETRIC SHAPES

Perimeter and circumference are expressed in terms of linear units, such as cm or ft. The term perimeter is used for the linear distance around a shape that has angles; the term circumference is reserved for the linear distance around a circle. Table 3.2 shows formulas for perimeter and circumference.

Examples of Perimeter and Circumference Calculations

A rectangle has length 6 cm and width 4 cm. Its perimeter is (2×6 cm)+(2×4 cm) =20 cm.

A square has sides of length 10 inches. Its perimeter is 4×10 in.=40 in.

A parallelogram has two sides equal to 26 mm and a top and base each equal to 40 mm. Its perimeter is (2×26 mm)+(2×40 mm)=132 mm.

A trapezoid has sides of length 12, 15, 24, and 27 cm. Its perimeter is (12 cm+ 15 cm+24 cm+27 cm)=78 cm.

A regular hexagon (meaning one with all sides the same length) has sides of length 8 cm. Its perimeter is 6×8 cm=48 cm.

A right triangle (meaning that the triangle has a 90 degree angle) has a base of 65 mm, a height of 120 mm and a hypotenuse of 136.47 mm. Its perimeter is (65 mm +120 mm+136.47 mm)=321.47 mm.

A triangle has sides of length 20 cm, 34 cm, and 40 cm. Its perimeter is (20 cm+ 34 cm+40 cm)=94 cm.

A circle has a radius of 10 cm. Its diameter is (2×10)=20 cm.

A circle has a radius of 25 inches, and hence a diameter of 50 inches. Its circumference is (2×π×25 in.)=(π×50 in.)=157.08 in. Since the formula involves pi, when possible, use a calculator to find circumference to avoid rounding errors.

Table 3.2 Calculation of perimeter and circumference for basic geometric shapes.

Rectangle	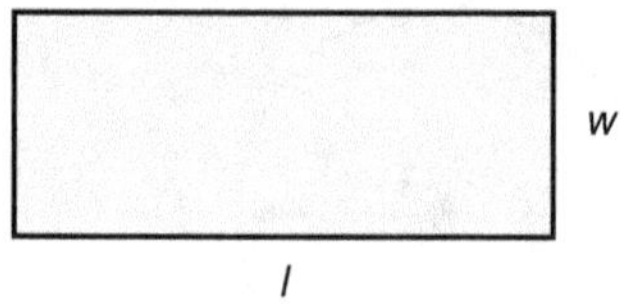	$P=2l+2w$
Square	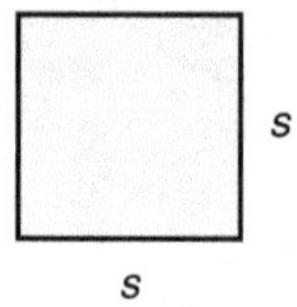	$P=4s$
Parallelogram	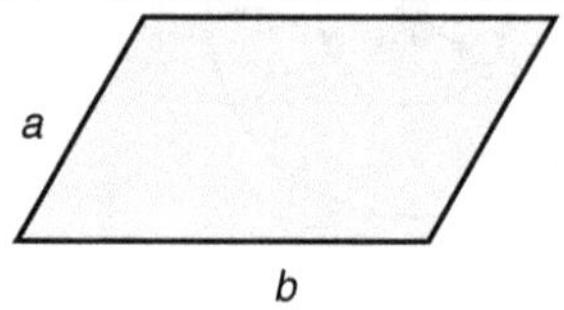	$P=2a+2b$
Trapezoid	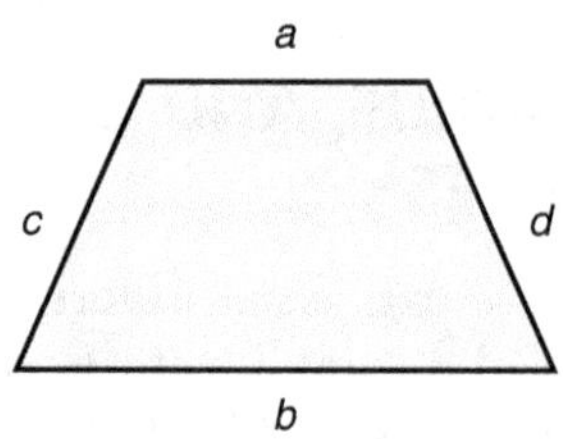	$P=a+b+c+d$
Regular hexagon (all sides equal length)	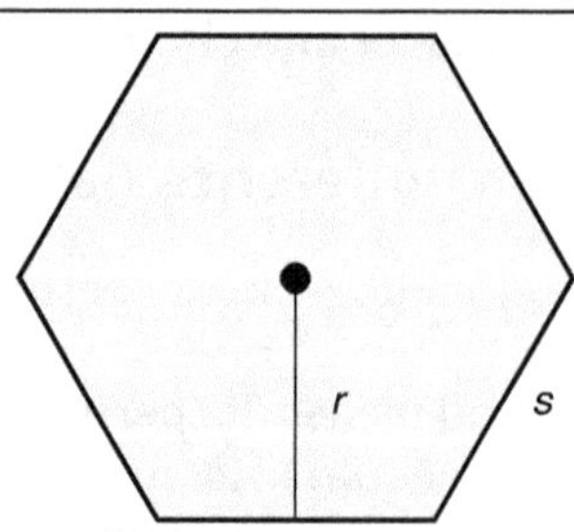	$P=6s$
Right triangle	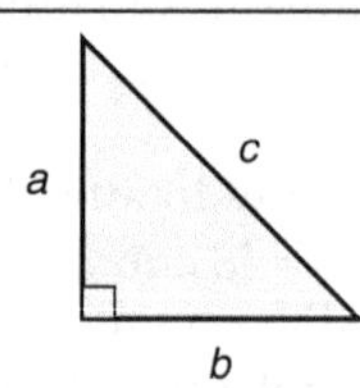	$P=a+b+c$ Note: $a^2+b^2=c^2$ (Pythagorean theorem)
Triangle	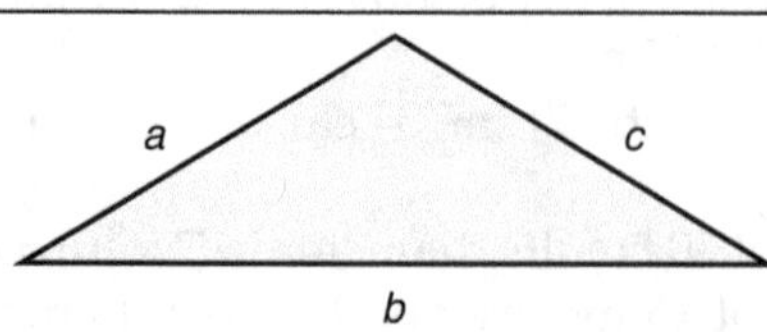	$P=a+b+c$

(continued)

Table 3.2 Calculation of perimeter and circumference for basic geometric shapes. (continued)

Circle	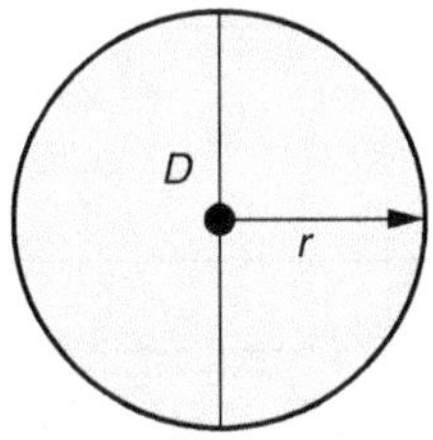	*Circumference* $= 2\pi r = \pi D$ *where r is radius and D is diameter* $D = 2r$

VOLUME OF BASIC GEOMETRIC SHAPES

Volume is expressed in terms of cubic units, such as cm^3 or ft^3. Table 3.3 shows formulas for calculating volume.

Examples of Volume Calculations

A square prism has length, width, and height equal to 23 cm, 12 cm, and 30 cm, respectively. Its volume is (23 cm × 12 cm × 30 cm) = 8280 cm^3.

A cube has sides of length 13 inches. Its volume is $(13 \text{ in.})^3 = (13 \text{ in.} \times 13 \text{ in.} \times 13 \text{ in.}) = 2197 \text{ in}^3$.

A cylinder has radius 120 mm and height 250 mm. Its volume is $\pi \times (120 \text{ mm})^2 \times (250 \text{ mm}) = 11{,}309{,}734 \text{ mm}^3$.

A cone has a radius of 3.5 inches and a height of 12 inches. Its volume is $(1/3) \times \pi \times (3.5 \text{ in.})^2 \times 12 \text{ in.} = 153.938 \text{ in}^3$.

A sphere has radius equal to 2.05 cm. Its volume is $(4/3) \times \pi \times (2.05 \text{ cm})^3 = 36.087 \text{ cm}^3$.

Note, since the formulas involve pi, when possible, use a calculator to find the volume of a cylinder, cone, and sphere to avoid round off errors.

SURFACE AREA OF BASIC GEOMETRIC SHAPES

Surface area is expressed in terms of squared units, such as cm^2 or ft^2. It is the area of the outside of a shape. Table 3.4 shows calculations for surface area.

Examples of Surface Area Calculations

A square prism has length, width, and height equal to 23 cm, 12 cm, and 30 cm, respectively. Its surface area is $2 \times [(23 \text{ cm} \times 12 \text{ cm}) + (12 \text{ cm} \times 30 \text{ cm}) + (23 \text{ cm} \times 30 \text{ cm})] = 2 \times (276 \text{ cm}^2 + 360 \text{ cm}^2 + 690 \text{ cm}^2) = 2652 \text{ cm}^2$.

A cube has sides of length 13 inches. Its surface area is $6 \times (13 \text{ in.})^2 = (6 \times 13 \text{ in.} \times 13 \text{ in.}) = 1014 \text{ in}^2$.

Table 3.3 Calculation of volume for basic geometric shapes.

Square prism	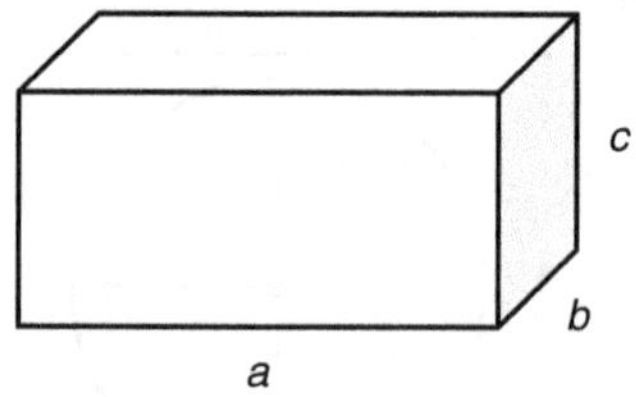	$V = abc$
Cube (all sides equal length)	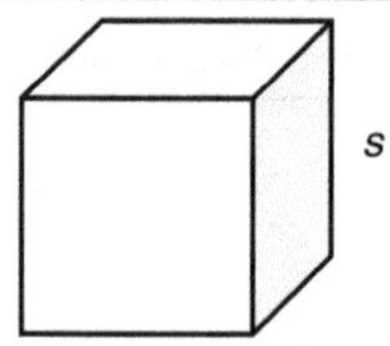	$V = s^3$
Cylinder	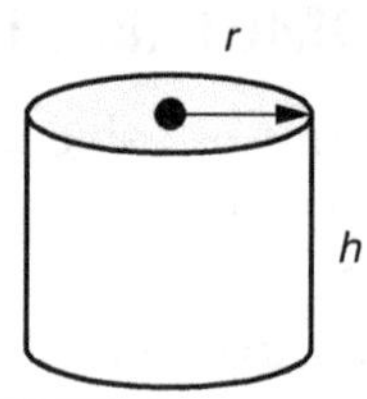	$V = \pi r^2 h$
Cone	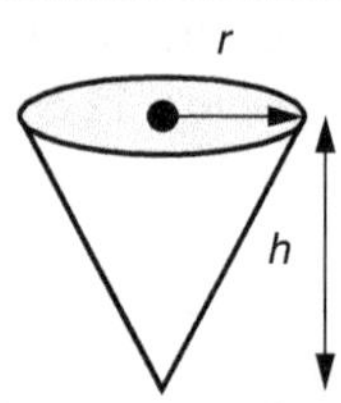	$V = \frac{1}{3}\pi r^2 h$
Sphere	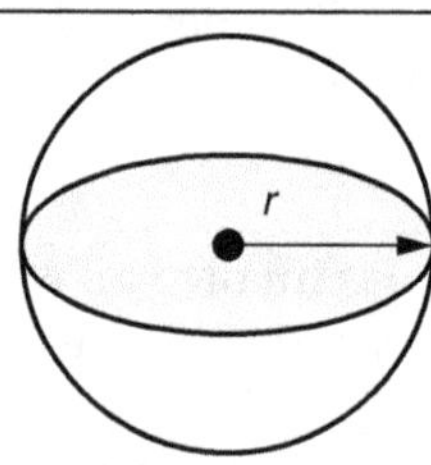	$V = \frac{4}{3}\pi r^3$

A cylinder has radius 120 mm and height 250 mm. Its surface area is $2 \times \pi \times (120\text{ mm})^2 + (2 \times \pi \times 120\text{ mm} \times 250\text{ mm}) = 278{,}973.4\text{ mm}^2$.

A cone has a radius of 3.5 inches and a height of 12 inches. Its surface area is $\pi \times (3.5\text{ in.})^2 + \pi \times (3.5\text{ in.}) \times \text{sqrt}\,((3.5\text{ in.})^2 + (12\text{ in.})^2) = 175.93\text{ in}^2$.

A sphere has radius equal to 2.05 cm. Its surface area is $4 \times \pi \times (2.05\text{ cm})^2 = 52.81\text{ cm}^2$.

Since the formulas involve pi, when possible, use a calculator to find surface area of a cylinder, cone, and sphere to avoid rounding errors.

Table 3.4 Calculation of surface area for basic geometric shapes.

Shape	Figure	Formula
Square prism	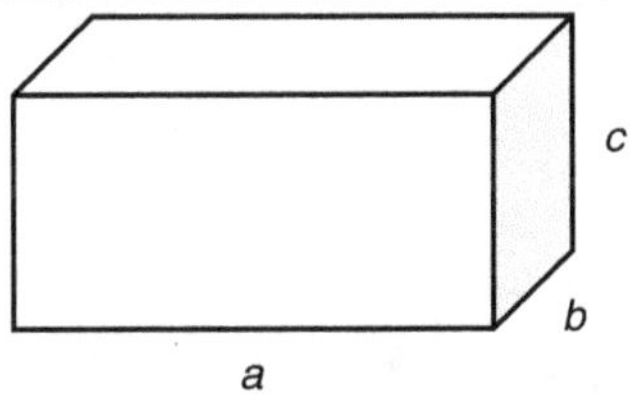	$SA = 2(ab + bc + ac)$
Cube (all sides equal length)	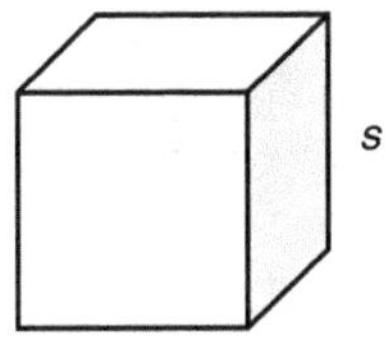	$SA = 6s^2$
Cylinder	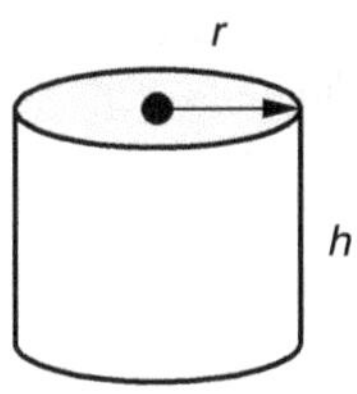	$SA = 2\pi r^2 + 2\pi rh$
Cone (includes the area of the base)	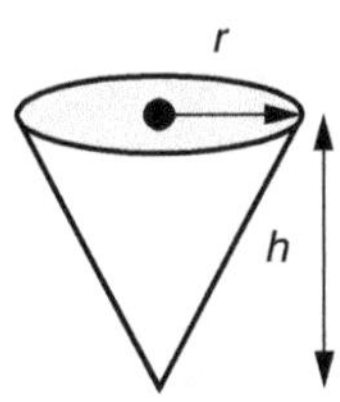	$SA = \pi r^2 + \pi r\sqrt{r^2 + h^2}$
Sphere	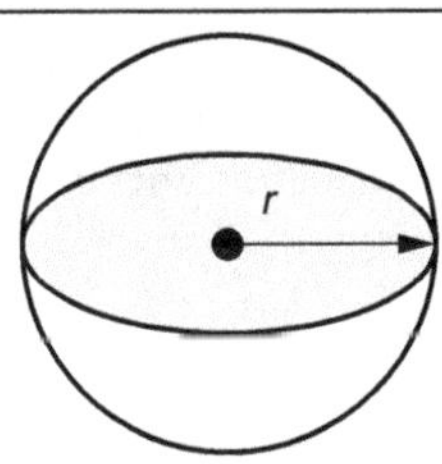	$SA = 4\pi r^2$

COMPLEMENTARY AND SUPPLEMENTARY ANGLES

Two angles that add to 90° 90°, or a right angle, are called *complementary angles*. For example, the angles 30° and 60° are complementary angles.

Two angles that add to 180°, which is a straight line, are called *supplementary angles*. For example, the angles 45° and 135° are supplementary angles.

MINIMUM NUMBER OF COORDINATE POINTS FOR SHAPES

A line can be fully described by two coordinates, or (x, y), points. For example, given the points (3, 4) and (5, 10), we can draw the line, as shown in Figure 3.1.

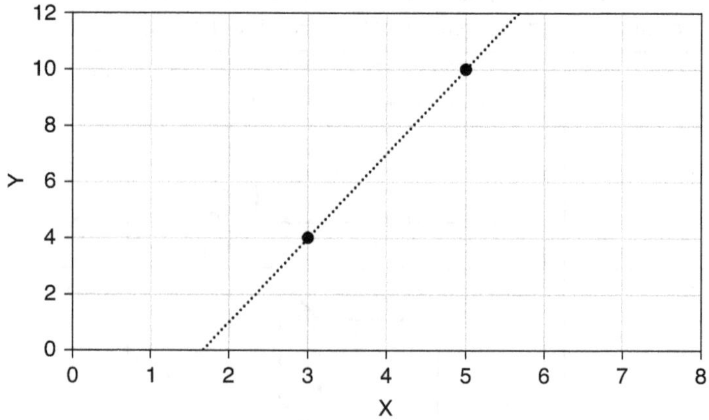

Figure 3.1 Line described by two coordinate points.

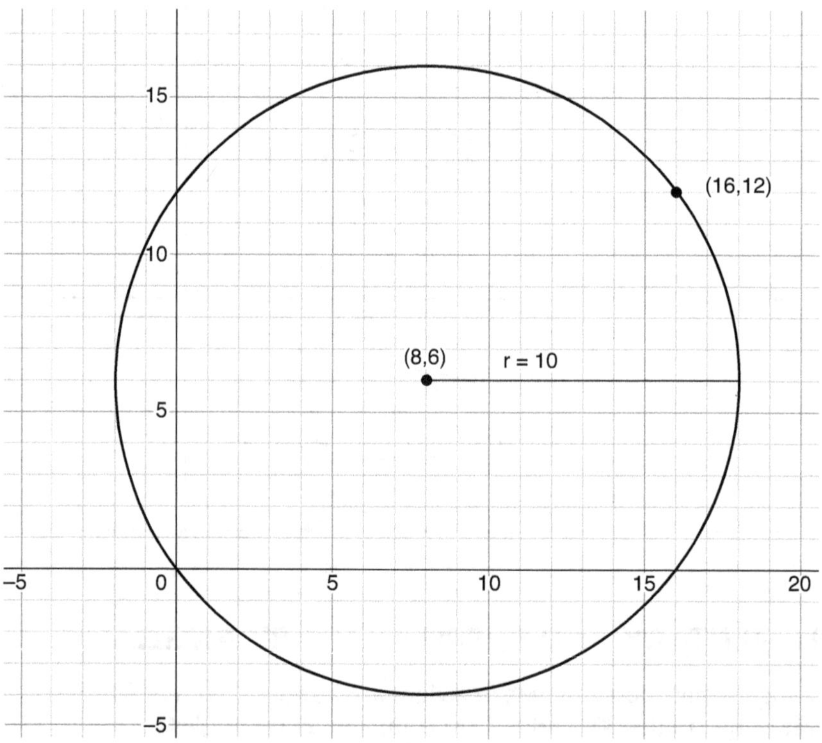

Figure 3.2 Circle described by a center point and a point on the circumference.

To fully describe a two-dimensional plane, three points are needed.

To define a circle in a two-dimensional plane, we require two points: the center point of the circle (h, k) and one point on the circumference of the circle (x, y). We can use the equation for a circle to calculate the radius:

$$r = \sqrt{(x-h)^2 + (y-k)^2}$$

With these two pieces of information, the circle is fully defined.

For example, we can draw a circle given the center point (h, k) = (8, 6) and a point on the circumference (x, y) = (16, 12). The radius is calculated:

$$\begin{aligned} r &= \sqrt{(x-h)^2 + (y-k)^2} \\ &= \sqrt{(16-8)^2 + (12-6)^2} \\ &= \sqrt{64+36} \\ &= \sqrt{100} \\ &= 10 \end{aligned}$$

Starting at the point (8.6), we can draw a circle with a radius equal to 10, as shown in Figure 3.2.

Chapter 4

Basic Trigonometry

> Solve for angles and lengths using trigonometric functions, such as sine, cosine, tangent, and the Pythagorean Theorem. (Apply)
>
> **Body of Knowledge I.D**

THE RIGHT TRIANGLE

As shown in Figure 4.1, a right triangle has sides a (the height), b (the base), and c (the hypotenuse). There is one 90°, or right, angle. Since the angles of a triangle add to 180°, the sum of the remaining two angles will equal 90°, making them *complementary* angles. Here, angles $\angle A$ and $\angle B$ will sum to 90°. Because each of these angles will be less than 90°, they are also classified as *acute* angles. Angles that are greater than 90° are called *obtuse* angles. Note that there are no obtuse angles in a right triangle.

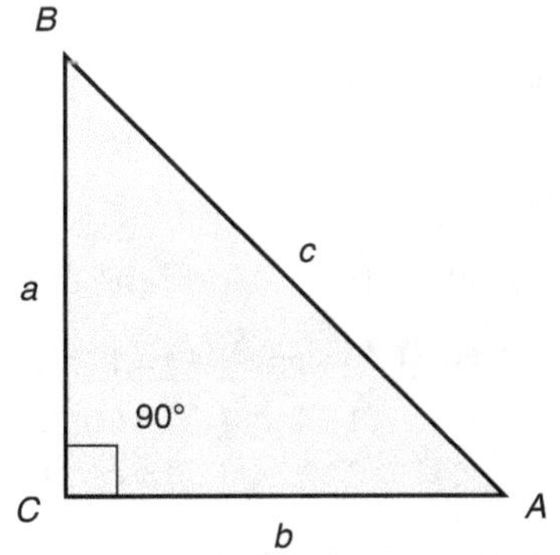

Figure 4.1 The right triangle.

Pythagorean Theorem

The sides of a right triangle are related according to the *Pythagorean theorem*, which states that the sum of the squared sides of a right triangle equals the hypotenuse squared:

$$a^2 + b^2 = c^2$$

The relationship can also be written as:

$$\sqrt{a^2 + b^2} = c$$

Using the theorem, the length of a third side can be found if two sides are known. For example, if side $a = 3$ cm and $b = 4$ cm, the length of hypotenuse c can be found:

$$\begin{aligned} \sqrt{3^2 + 4^2} &= c \\ \sqrt{25} &= c \\ 5 &= c \end{aligned}$$

Trigonometric Functions

There are six main trigonometric functions, which are ratios of the sides of a right triangle. These functions are: sine (sin), cosine (cos), tangent (tan), cotangent (cot), secant (sec), and cosecant (csc). Table 4.1 illustrates these functions and their calculations.

Values of Trigonometric Functions for Some Common Angles

Note that angles can be expressed in terms of degrees or radians. For example, a circle has a total of 360°, a straight line has 180°, and a right angle is 90°. The same circle has 2π radians, a straight line has π radians, and a right angle is 1/2 π radians. To convert an angle A expressed in degrees to radians, multiply by the factor $\frac{\pi}{180}$. To convert an angle in radians to degrees, multiply radians by the factor $\frac{180}{\pi}$.

Table 4.2 shows values of the trigonometric functions for common angles.

Trigonometric Identities

Some useful trigonometric identities include the following:

$$\begin{aligned} \sin^2(A) + \cos^2(A) &= 1 \\ 1 + \tan^2(A) &= \sec^2(A) \\ 1 + \cot^2(A) &= \csc^2(A) \end{aligned}$$

To find the cosine of an angle A when the sine of the angle is known, we can use the first identity as follows:

$$\begin{aligned} \sin^2(A) + \cos^2(A) &= 1 \\ \cos^2(A) &= 1 - \sin^2(A) \\ \sqrt{\cos^2(A)} &= \sqrt{1 - \sin^2(A)} \\ \cos(A) &= \sqrt{1 - \sin^2(A)} \end{aligned}$$

Table 4.1 Calculation of trigonometric functions.

Sine	$\sin(A) = \dfrac{\textit{opposite side}}{\textit{hypotenuse}} = \dfrac{a}{c}$	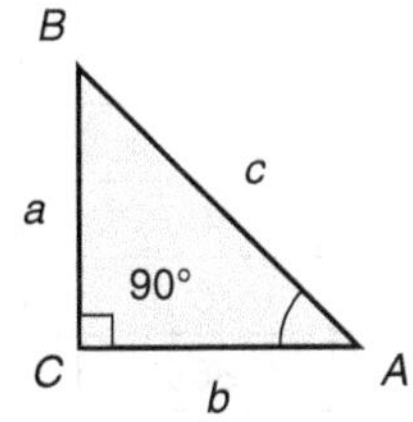
Cosine	$\cos(A) = \dfrac{\textit{adjacent side}}{\textit{hypotenuse}} = \dfrac{b}{c}$	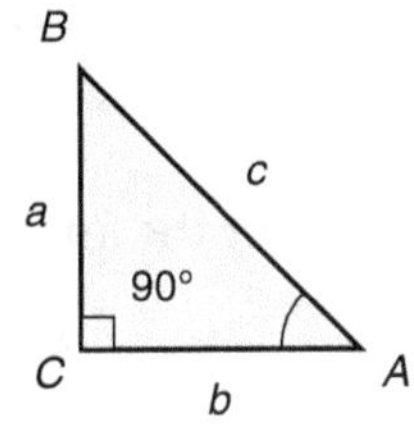
Tangent	$\tan(A) = \dfrac{\textit{opposite side}}{\textit{adjacent side}} = \dfrac{a}{b}$ $= \dfrac{\sin(A)}{\cos(A)}$	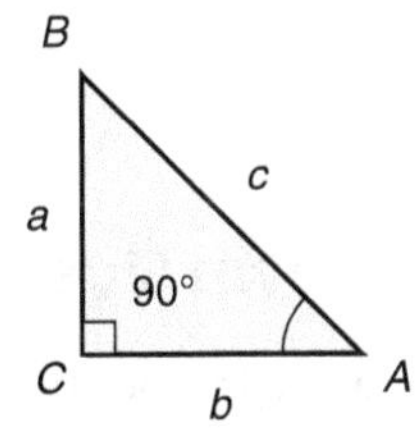
Cotangent	$\cot(A) = \dfrac{\textit{adjacent side}}{\textit{opposite side}} = \dfrac{b}{a}$ $= \dfrac{1}{\tan(A)} = \dfrac{\cos(A)}{\sin(A)}$	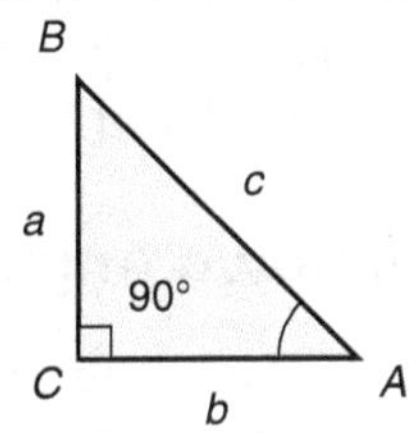
Secant	$\sec(A) = \dfrac{\textit{hypotenuse}}{\textit{adjacent side}} = \dfrac{c}{b}$ $= \dfrac{1}{\cos(A)}$	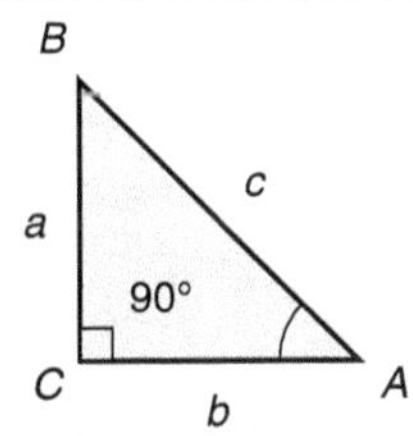
Cosecant	$\csc(A) = \dfrac{\textit{hypotenuse}}{\textit{opposite side}} = \dfrac{c}{a}$ $= \dfrac{1}{\sin(A)}$	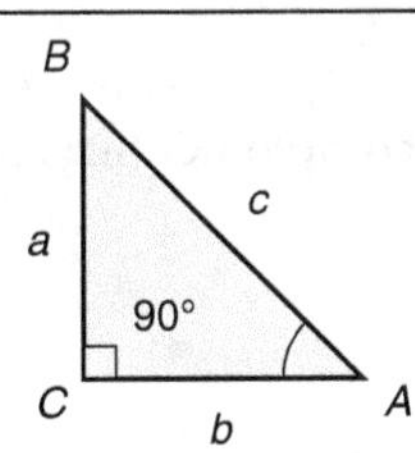

Table 4.2 Values of trigonometric functions for common angles.

Angle in degrees	Angles in radians	Sine	Cosine	Tangent	Cotangent	Secant	Cosecant
0	0	0	1.000	0	undefined	1.000	undefined
30	1/6 π	0.5000	0.8660	0.5774	1.7321	1.1547	2.000
45	1/4 π	0.7071	0.7071	1.000	1.000	1.4142	1.4142
60	1/3 π	0.8660	0.5000	1.7321	0.5774	2.000	1.1547
90	1/2 π	1.000	0	undefined	0	undefined	1.000

For example, for the angle A = 30°, and sin (30) known to be 0.5:

$$\sin^2(A)+\cos^2(A)=1$$
$$\cos^2(A)=1-0.5^2$$
$$\sqrt{\cos^2(A)}=\sqrt{1-0.5^2}$$
$$\cos(A)=0.866$$

Solving for Unknown Sides and Angles of a Right Triangle

Using the six basic trigonometric functions and the Pythagorean theorem, we can find the length of the unknown sides of a triangle, given the values of one side and one angle. For example, given angle A is 36°, and side *a* has length 6, angle B can be found, as well as the lengths of sides *b* and *c*, as shown in Figure 4.2.

Finding the Unknown Angle

Since the angles of a triangle sum to 180°, we can find the unknown angle via subtraction. For example, given the triangle in Figure 4.2, we can find the unknown angle B:

$$\angle A+\angle B+\angle C=180^\circ$$
$$49^\circ+\angle B+90^\circ=180^\circ$$
$$139^\circ+\angle B=180^\circ$$
$$\angle B+139^\circ-139^\circ=180^\circ-139^\circ$$
$$\angle B=41^\circ$$

Alternatively, we can also use the fact that the two acute angles of a right triangle are complementary, meaning that they sum to 90°:

$$\angle A+\angle B=90^\circ$$
$$49^\circ+\angle B=90^\circ$$
$$\angle B+49^\circ-49^\circ=90^\circ-49^\circ$$
$$\angle B=41^\circ$$

After finding all the angles, we can find the lengths of the unknown sides *b* and *c*.

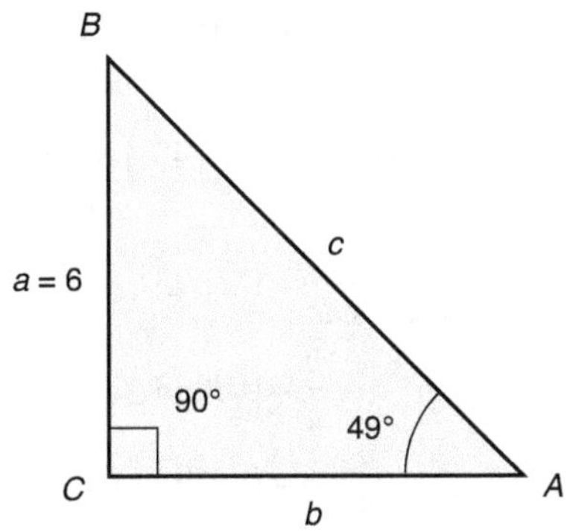

Figure 4.2 Finding angles in a right triangle.

FINDING THE UNKNOWN SIDES

We can find the lengths of the unknown sides using the trigonometric functions and a calculator:

$$\sin(A) = \sin(49^\circ) = 0.7547$$

$$\begin{aligned}
\frac{\textit{opposite side}}{\textit{hypotenuse}} &= 0.7547 \\
\frac{6}{c} &= 0.7547 \\
\frac{6}{c} \times c &= 0.7547 \times c \\
6 &= 0.7547c \\
6 \div 0.7547 &= 0.7547c \div 0.7547 \\
7.950 &= c
\end{aligned}$$

Now that we know the length of c, we can find the length of side b using the Pythagorean theorem:

$$\begin{aligned}
a^2 + b^2 &= c^2 \\
6^2 + b^2 &= (7.950)^2 \\
6^2 + b^2 - 6^2 &= (7.950)^2 - 6^2 \\
b^2 &= 27.2025 \\
\sqrt{b^2} &= \sqrt{27.2025} \\
b &= 5.216
\end{aligned}$$

Alternatively, we can use a trigonometric function to find the length of side *b*:

$$\cot(A) = \cot(49^{\circ}) = 0.8693$$

$$\frac{\textit{adjacent side}}{\textit{opposite side}} = 0.8693$$

$$\frac{b}{a} = 0.8693$$

$$\frac{b}{6} \times 6 = 0.8693 \times 6$$

$$b = 5.216$$

Inverse Trigonometric Functions

In some cases, we know the value of a trigonometric function for a particular angle, but do not know the degree measure of the angle. We can use an *inverse trigonometric function* to solve for the angle. These inverse functions are of the form:

$$\sin^{-1}(x), \cos^{-1}(x), \tan^{-1}(x), \cot^{-1}(x), \sec^{-1}(x), \csc^{-1}(x)$$

where x is the value of the trigonometric function. Each inverse function returns the angle measure associated with the value x. For example, given that sin(A) = 0.4580, we can find the angle A in degrees using a scientific calculator:

$$\angle A = \sin^{-1}(0.4580) = 27.26^{\circ}$$

OBLIQUE TRIANGLES

An *oblique triangle* has no 90° angles. Some special types of oblique triangles include the *equilateral triangle, isosceles triangle,* and the *scalene triangle,* as shown in Table 4.3.

Solving Oblique Triangles

Each triangle has three sides and three angles. In order to solve for an unknown side or angle in an oblique triangle, three pieces of information must be known. We will use the *laws of sines and cosines* to solve these problems.

Law of Sines

Given an oblique triangle such as the one in Figure 4.3, the law of sines states:

$$\frac{\sin(X)}{x} = \frac{\sin(Y)}{y} = \frac{\sin(Z)}{z}$$

This law can also be written in terms of its reciprocal:

$$\frac{x}{\sin(X)} = \frac{y}{\sin(Y)} = \frac{z}{\sin(Z)}$$

Table 4.3 Oblique triangles.

Equilateral triangle Three equal sides, three 60° angles	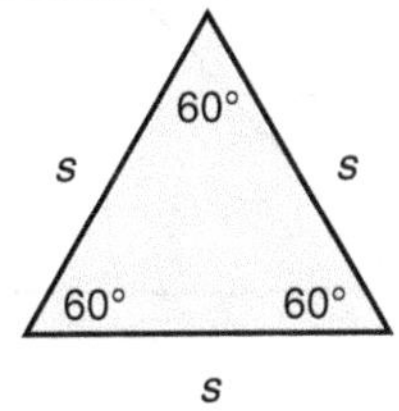
Isosceles triangle Two equal sides, two equal angles	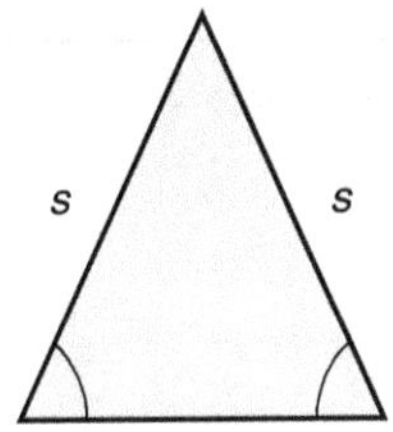
Scalene triangle All sides, all angles unique	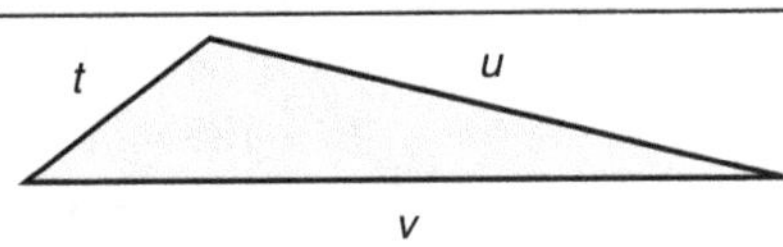

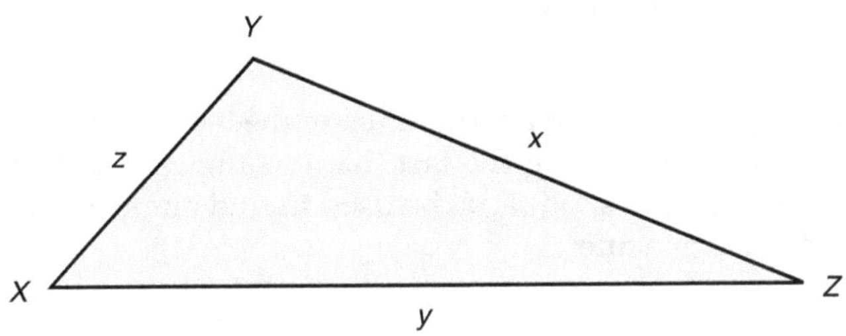

Figure 4.3 Oblique triangle.

Law of Cosines
Given an oblique triangle as in Figure 4.3, the law of cosines states:

$$x^2 = y^2 + z^2 - 2yz\cos(X)$$
$$y^2 = x^2 + z^2 - 2xz\cos(Y)$$
$$z^2 = x^2 + y^2 - 2xy\cos(Z)$$

Three Sides Known
Given three sides, we can find the values of the three angles using the law of cosines. For example, given the triangle in Figure 4.4, we can find the values of the three angles:

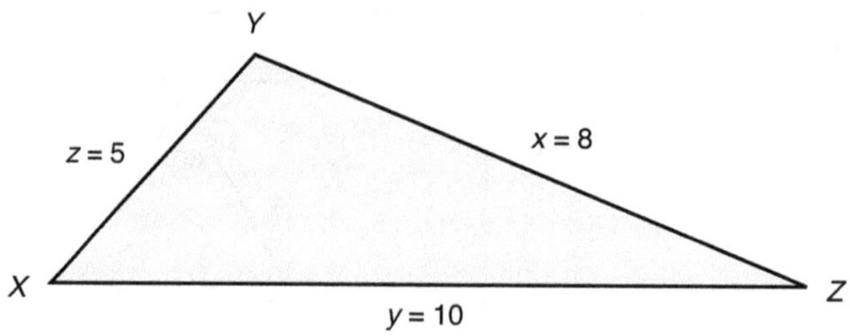

Figure 4.4 Oblique triangle with three sides known.

$$\begin{aligned}
x^2 &= y^2 + z^2 - 2yz\cos(X)\\
8^2 &= 10^2 + 5^2 - (2\times 10\times 5)\cos(X)\\
64 &= 125 - 100\cos(X)\\
64 - 125 &= 125 - 100\cos(X) - 125\\
-61 &= -100\cos(X)\\
-61 \div -100 &= -100\cos(X) \div -100\\
0.61 &= \cos(X)
\end{aligned}$$

$$\begin{aligned}
\cos^{-1}(0.61) &= \angle X\\
\angle X &= 52.41^\circ
\end{aligned}$$

The other angle Y can also be found using the law of cosines, and the remaining angle Z can be found using the fact that the angles of a triangle sum to 180°. Note that the law of sines should not be used to find angle Y since it is not known whether it is obtuse or acute:

$$\begin{aligned}
y^2 &= x^2 + z^2 - 2xz\cos(Y)\\
10^2 &= 8^2 + 5^2 - (2\times 8\times 5)\cos(Y)\\
100 &= 89 - 80\cos(Y)\\
100 - 89 &= 89 - 80\cos(Y) - 89\\
11 &= -80\cos(Y)\\
11 \div -80 &= -80\cos(Y) \div -80\\
-0.1375 &= \cos(Y)
\end{aligned}$$

$$\begin{aligned}
\cos^{-1}(-0.1375) &= \angle Y\\
\angle Y &= 97.90^\circ
\end{aligned}$$

Angle Z can now be found via subtraction:

$$\begin{aligned}
\angle Z &= 180^\circ - \angle X + \angle Y\\
\angle Z &= 180^\circ - 52.41^\circ + 97.90^\circ\\
\angle Z &= 29.69^\circ
\end{aligned}$$

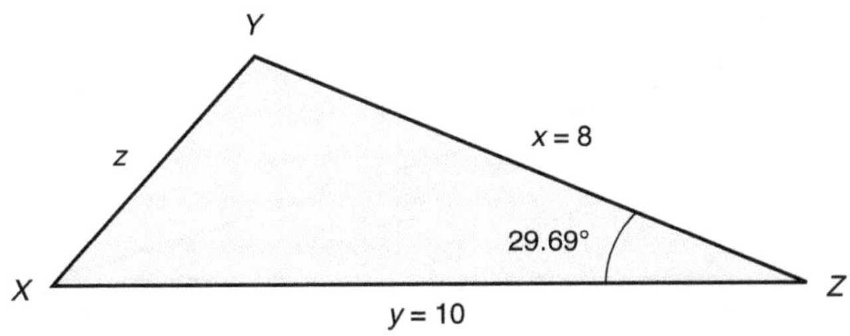

Figure 4.5 Oblique triangle with two sides and the angle between them known.

Two Sides and an Angle between Them Known

Given two sides and the angle between them, we can find the value of the remaining side and the other two angles. For example, given the triangle shown in Figure 4.5, we see that sides x and y are 8 and 10 units, respectively, and angle Z is 29.69°. We can find the length of side z using the law of cosines:

$$
\begin{aligned}
z^2 &= x^2 + y^2 - 2xy\cos(Z) \\
z^2 &= 8^2 + 10^2 - (2\times 8\times 10)\cos(29.69) \\
z^2 &= 164 - 160\cos(29.69) \\
z^2 &= 164 - 160(0.8687) \\
z^2 &= 25 \\
\sqrt{z^2} &= \sqrt{25} \\
z &= 5
\end{aligned}
$$

We can then apply the law of cosines to find angle X:

$$
\begin{aligned}
x^2 &= y^2 + z^2 - 2yz\cos(X) \\
8^2 &= 10^2 + 5^2 - (2\times 10\times 5)\cos(X) \\
64 &= 125 - 100\cos(X) \\
64 - 125 &= 125 - 100\cos(X) - 125 \\
-61 &= -100\cos(X) \\
-61 \div (-100) &= -100\cos(X) \div (-100) \\
0.61 &= \cos(X)
\end{aligned}
$$

$$
\begin{aligned}
\cos^{-1}(0.61) &= \angle X \\
\angle X &= 52.41
\end{aligned}
$$

Angle Y can be found through subtraction:

$$
\begin{aligned}
\angle Y &= 180 - \angle X - \angle Z \\
&= 180 - 29.69 - 52.41 \\
\angle Y &= 97.90
\end{aligned}
$$

Given two sides and the angle opposite one of the known sides, we can find the value of the remaining side and the other two angles.

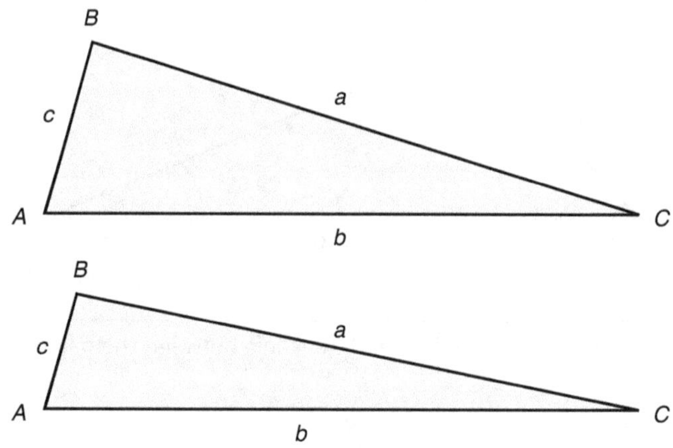

Figure 4.6 Oblique triangles with two sides and opposite angle known.

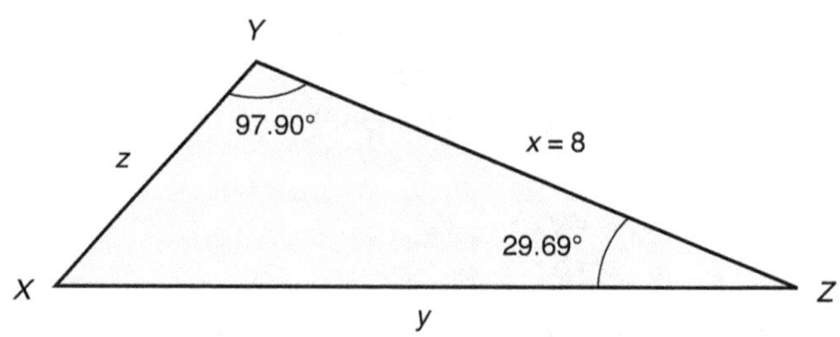

Figure 4.7 Oblique triangle with one side and two angles known.

Note that the case in which two sides and an opposite angle are known does not result in a unique solution. Refer to triangles ABC and ABD in Figure 4.6. Here, sides *a* and *b* as well as the opposite angle A are identical for both triangles. However, the length of the third side for each triangle is not the same (*c* and *d* are different lengths), demonstrating that the solution given two sides and an opposite angle is not unique.

One Side and Two Angles Known

Given one side and two angles, we can find the other two sides and the third angle. Here, side *x* is known, and angles Y and Z are known. Figure 4.7 shows an example.

Angle X can be calculated via subtraction:

$$\begin{aligned} \angle X &= 180^{*} - \angle Y - \angle Z \\ &= 180^{*} - 97.90^{*} - 29.69^{*} \\ \angle X &= 52.41^{*} \end{aligned}$$

With all angles now known, the lengths of the sides y and z can be found using the law of sines:

$$\frac{x}{\sin(X)} = \frac{y}{\sin(Y)} = \frac{z}{\sin(Z)}$$

$$\frac{8}{\sin(52.41)} = \frac{y}{\sin(97.90)}$$

$$\frac{8}{0.7924} = \frac{y}{0.9905}$$

$$10.10 = \frac{y}{0.9905}$$

$$10.10 \times 0.9905 = \frac{y}{0.9905} \times 0.9905$$

$$10 = y$$

$$\frac{8}{\sin(52.41)} = \frac{z}{\sin(29.69)}$$

$$\frac{8}{0.7924} = \frac{z}{0.4953}$$

$$10.10 = \frac{z}{0.4953}$$

$$10.10 \times 0.4953 = \frac{z}{0.4953} \times 0.4953$$

$$5 = z$$

Chapter 5

Measurement Systems

> Convert units within and between English and metric measurement systems (SI), such as inch to micro-inch and meter to millimeter (Apply).
>
> Body of Knowledge I.E

MEASUREMENT SYSTEMS

Three measurement systems are commonly used in industry: the International System of Units, abbreviated SI; the metric system; and the English system. The units of SI are internationally accepted standards and are generally the same as the units used in the metric system, with a few exceptions. Conversion of units within each of the SI and metric systems is relatively straightforward, since the systems are based on units that vary by factors of ten. For example:

$$1 \text{ liter} = 10 \text{ deciliters} = 100 \text{ centiliters} = 1000 \text{ milliliters}$$

In contrast, the English measurement system, still used by many companies in the United States, has many conversion factors. As a result, changing units within the English system can be quite cumbersome. For example:

$$1 \text{ quart} = 2 \text{ pints} = 4 \text{ cups} = 32 \text{ fluid ounces}$$

Quality inspectors should feel comfortable converting units within systems as well as between the English, SI, and metric systems. What follows is a listing of common measures and conversion factors for the three systems. A more comprehensive listing of measurement conversion factors can be found in Appendices.

ENGLISH SYSTEM MEASUREMENTS

To convert units within the English system, use the relationships found in Tables 5.1 through 5.6.

To convert length units within the English system, use the relationships given in Table 5.1. For example, to convert miles into feet, the table shows that we multiply the number of miles by a factor of 5280. Thus, for instance, to convert 2 miles into feet:

$$2 \textit{ miles} = 2 \times 5280 = 10{,}560 \textit{ feet}$$

Table 5.1 Converting measures of length.

From	To	Multiply by	Divide by
inch	microinch	1,000,000	
microinch	inch		1,000,000
feet	inches	12	
inches	feet		12
yards	feet	3	
feet	yards		3
miles	feet	5280	
feet	mile		5280
miles	yards	1760	
yards	miles		1760

Table 5.2 Converting measures of area.

From	To	Multiply by	Divide by
square feet	square inches	144	
square inches	square feet		144
square yards	square feet	9	
square feet	square yards		9
square yards	square inches	1296	
square inches	square yards		1296
acres	square yards	4840	
square yards	acres		4840
acres	square feet	43,560	
square feet	acres		43,560
square miles	acres	640	
acres	square miles		640

Similarly, to convert feet into miles we use the reverse operation; that is, we divide the feet by a factor of 5280. Thus:

$$10{,}560\ \textit{feet} = 10{,}560 \div 5280 = 2\ \textit{miles}$$

To convert area units within the English system, use the relationships given in Table 5.2. For example, to convert square feet into square inches, the table shows that we multiply the number of square feet by a factor of 144. Thus, for instance, to convert 2 square feet into square inches:

$$2\ \textit{square feet} = 2 \times 144 = 288\ \textit{square inches}$$

Table 5.3 Converting measures of volume.

From	To	Multiply by	Divide by
cubic feet	cubic inches	1728	
cubic inches	cubic feet		1728
cubic yards	cubic feet	27	
cubic feet	cubic yards		27
cubic yards	cubic inches	46,656	
cubic inches	cubic yard		46,656

Table 5.4 Converting measures of weight.

From	To	Multiply by	Divide by
pounds	ounces	16	
ounces	pounds		16
tons	pounds	2000	
pounds	ton		2000
long tons	pounds	2240	
pounds	long tons		2240

Similarly, to convert square inches into square feet, we use the reverse operation; that is, we divide the square inches by a factor of 144. Thus:

$$288 \text{ } \textit{square inches} = 288 \div 144 = 2 \text{ } \textit{square feet}$$

To convert volume units within the English system, use the relationships given in Table 5.3. For example, to convert cubic yards into cubic feet, the table shows that we multiply the number of cubic yards by a factor of 27. Thus, for instance, to convert 2 cubic yards into cubic feet:

$$2 \text{ } \textit{cubic yards} = 2 \times 27 = 54 \text{ } \textit{cubic feet}$$

Similarly, to convert cubic feet into cubic yards, we use the reverse operation; that is, we divide the cubic feet by a factor of 27. Thus:

$$54 \text{ } \textit{cubic feet} = 54 \div 27 = 2 \text{ } \textit{cubic yards}$$

To convert weight units within the English system, use the relationships given in Table 5.4. For example, to convert pounds into ounces, the table shows that we multiply the number of pounds by a factor of 16. Thus, for instance, to convert 2 pounds into ounces:

$$2 \text{ } \textit{pounds} = 2 \times 16 = 32 \text{ } \textit{ounces}$$

Similarly, to convert ounces into pounds, we use the reverse operation; that is, we divide the ounces by a factor of 16. Thus:

Table 5.5 Converting liquid measures.

From	To	Multiply by	Divide by
tablespoons	teaspoons	3	
teaspoons	tablespoons		3
fluid ounces	tablespoons	2	
tablespoons	fluid ounces		2
cups	tablespoons	16	
tablespoons	cups		16
cups	fluid ounces	8	
fluid ounces	cups		8
1 pint	cups	2	
cups	pints		2
pint	fluid ounces	16	
fluid ounces	pints		16
quarts	pints	2	
pints	quarts		2
quarts	cups	4	
cups	quarts		4
quarts	fluid ounces	32	
fluid ounces	quarts		32
gallons	quarts	4	
quarts	gallons		4
gallons	pints	8	
pints	gallons		8
gallons	cups	16	
cups	gallons		16
gallons	fluid ounces	128	
fluid ounces	gallons		128

$$32\ ounces = 32 \div 16 = 2\ pounds$$

To convert units of liquid measure within the English system, use the relationships given in Table 5.5. For example, to convert cups into tablespoons, the table shows that we multiply the number of pounds by a factor of 16. Thus, for instance, to convert 2 cups into tablespoons:

$$2\ cups = 2 \times 16 = 32\ tablespoons$$

Similarly, to convert tablespoons into cups, we use the reverse operation; that is, we divide the tablespoons by a factor of 16. Thus:

$$32\ tablespoons = 32 \div 16 = 2\ cups$$

To convert pressure units within the English system, use the relationships given in Table 5.6. For example, to convert pounds per square foot into pounds per square inch (psi), the table shows that we multiply the number of pound per square foot by a factor of 144. Thus, for instance, to convert 2 pounds per square foot into pounds per square inch:

$$2\ \textit{pounds per square foot} = 2 \times 144 = 288\ \textit{pounds per square inch}$$

Similarly, to convert pounds per square inch into pounds per square foot, we use the reverse operation; that is, we divide the pounds per square inch by a factor of 144. Thus:

$$288\ \textit{pounds per square inch} = 288 \div 144 = 2\ \textit{pounds per square foot}$$

SI/METRIC SYSTEM MEASUREMENTS

To convert length units within the metric or SI system, use the relationships given in Table 5.7. For example, to convert centimeters into millimeters, the table shows that we multiply the number of centimeters by a factor of 10. Thus, for instance, to convert 2 centimeters into millimeters:

$$2\ \textit{centimeters} = 2 \times 10 = 20\ \textit{millimeters}$$

Table 5.6 Converting measures of pressure.

From	To	Multiply by	Divide by
pounds per square inch (psi)	pounds per square foot		144
pounds per square foot	pounds per square inch	144	
pounds per square inch	atmospheres		14.7
atmospheres	pounds per square inch	14.7	
pounds per square foot	atmospheres		2116.3
atmospheres	pounds per square foot	2116.3	

Table 5.7 Converting measures of length—metric units.

From	To	Multiply by	Divide by
microns	millimeters		1000
millimeters	microns	1000	
millimeters	centimeters		10
centimeters	millimeters	10	
centimeters	meters		100
meters	centimeters	100	
millimeters	meters		1000
meters	millimeters	1000	
meters	kilometers		1000
kilometers	meters	1000	

Table 5.8 Converting measures of area—metric units.

From	To	Multiply by	Divide by
square millimeters	square centimeters		100
square centimeters	square millimeters	100	
square centimeters	square meters		10,000
square meters	square centimeters	10,000	
square meters	square kilometers		1,000,000
square kilometers	square meters	1,000,000	
square meters	ares		100
ares	square meters	100	

Table 5.9 Converting measures of volume—metric units.

From	To	Multiply by	Divide by
cubic millimeters	cubic centimeters		1000
cubic centimeter	cubic millimeters	1000	
cubic centimeters	cubic meters		1,000,000
cubic meters	cubic centimeters	1,000,000	

Similarly, to convert millimeters into centimeters, we use the reverse operation; that is, we divide the millimeters by a factor of 10. Thus:

$$20\ \textit{millimeters} = 20 \div 10 = 2\ \textit{centimeters}$$

To convert area units within the metric or SI system, use the relationships given in Table 5.8. For example, to convert square centimeters into square millimeters, the table shows that we multiply the number of square centimeters by a factor of 100. Thus, for instance, to convert 2 square centimeters into square millimeters:

$$2\ \textit{square centimeters} = 2 \times 100 = 200\ \textit{square millimeters}$$

Similarly, to convert square millimeters into square centimeters, we use the reverse operation; that is, we divide the square millimeters by a factor of 100. Thus

$$200\ \textit{square millimeters} = 200 \div 100 = 2\ \textit{square centimeters}$$

To convert volume units within the metric or SI system, use the relationships given in Table 5.9. For example, to convert cubic centimeters into cubic millimeters, the table shows that we multiply the number of cubic centimeters by a factor of 1000. Thus, for instance, to convert 2 cubic centimeters into cubic millimeters:

$$2\ \textit{cubic centimeters} = 2 \times 1000 = 2000\ \textit{cubic millimeters}$$

Similarly, to convert cubic millimeters into cubic centimeters, we use the reverse operation; that is, we divide the cubic millimeters by a factor of 1000. Thus:

$$2000\ \textit{cubic millimeters} = 2000 \div 1000 = 2\ \textit{cubic centimeters}$$

Table 5.10 Converting measures of mass.

From	To	Multiply by	Divide by
micrograms	milligrams		1000
milligrams	micrograms	1000	
milligrams	centigrams		10
centigrams	milligrams	10	
centigrams	grams		100
grams	centigrams	100	
grams	kilograms		1000
kilograms	grams	1000	

Table 5.11 Converting liquid measures—metric units.

From	To	Multiply by	Divide by
milliliters	centiliters		10
centiliters	milliliters	10	
centiliters	liters		100
liters	centiliters	100	
liters	kiloliters		1000
kiloliters	liters	1000	

To convert mass units within the metric or SI system, use the relationships given in Table 5.10. For example, to convert centigrams into milligrams, the table shows that we multiply the number of centigrams by a factor of 10. Thus, for instance, to convert 2 centigrams into milligrams:

$$2\ centigrams = 2 \times 10 = 20\ milligrams$$

Similarly, to convert milligrams into centigrams, we use the reverse operation; that is, we divide the milligrams by a factor of 10. Thus:

$$20\ milligrams = 20 \div 10 = 2\ centigrams$$

To convert units of liquid measure within the metric or SI system, use the relationships given in Table 5.11. For example, to convert liters into centiliters, the table shows that we multiply the number of liters by a factor of 100. Thus, for instance, to convert 2 liters into centiliters:

$$2\ liters = 2 \times 100 = 200\ centiliters$$

Similarly, to convert centiliters into liters, we use the reverse operation; that is, we divide the centiliters by a factor of 100. Thus

$$200\ centiliters = 200 \div 100 = 2\ liters$$

ENGLISH TO SI/METRIC CONVERSIONS

Use these relationships to convert from the English system to the SI/metric system, or from the SI/metric system to the English system.

To convert length units between the English and metric or SI systems, use the relationships given in Table 5.12. For example, to convert feet into centimeters, the table shows that we multiply the number of feet by a factor of 30.48. Thus, for instance, to convert 2 feet into centimeters:

$$2\,feet = 2 \times 30.48 = 60.96\ centimeters$$

Similarly, to convert centimeters into feet, we use the reverse operation; that is, we divide the centimeters by a factor of 30.48. Thus:

$$60.96\ centimeters = 60.96 \div 30.48 = 2\,feet$$

To convert area units between the English and metric or SI systems, use the relationships given in Table 5.13. For example, to convert square inches into square millimeters, the table shows that we multiply the number of square inches by a factor of 645.2. Thus, for instance, to convert 2 square inches into square millimeters:

$$2\ square\ inches = 2 \times 645.2 = 1290.4\ square\ millimeters$$

Similarly, to convert square millimeters into square inches, we use the reverse operation; that is, we divide the square millimeters by a factor of 645.2. Thus:

$$1290.4\ square\ millimeters = 1290.4 \div 645.2 = 2\ square\ inches$$

To convert volume units between the English and metric or SI systems, use the relationships given in Table 5.14. For example, to convert cubic inches into cubic centimeters, the table shows that we multiply the number of cubic inches by a factor of 16.38716. Thus, for instance, to convert 2 cubic inches into cubic centimeters:

Table 5.12 Converting measures of length—English and metric units.

From	To	Multiply by	Divide by
inches	centimeters	2.54	
centimeters	inches		2.54
feet	centimeters	30.48	
centimeters	feet		30.48
feet	millimeters	304.8	
millimeters	feet		304.8
yards	meters	0.9144	
meters	yards		0.9144
yards	centimeters	91.44	
centimeters	yards		91.44
miles	kilometers	1.609	
kilometers	miles		1.609
miles	meters	1609	
meters	miles		1609

Table 5.13 Converting measures of area—English and metric units.

From	To	Multiply by	Divide by
square inch	square centimeter	6.452	
square centimeter	square inch		6.452
square inch	square millimeter	645.2	
square millimeter	square inch		645.2
square foot	square meter	0.0929	
square meter	square foot		0.0929
square foot	square centimeter	929	
square centimeter	square foot		929
square yard	square meter	0.836	
square meter	square yard		0.836
acre	are	40.47	
are	acre		40.47
square mile	square kilometer	2.5899	
square kilometer	square mile		2.5899

Table 5.14 Converting measures of volume—English and metric units.

From	To	Multiply by	Divide by
cubic inch	cubic centimeter	16.38716	
cubic centimeter	cubic inch		16.38716
cubic foot	cubic meter	0.02832	
cubic meter	cubic foot		0.02832
cubic yard	cubic meter	0.7645	
cubic meter	cubic yard		0.7645

$$2\ \textit{cubic inches} = 2 \times 16.38716 = 32.77432\ \textit{cubic centimeters}$$

Similarly, to convert cubic centimeters into cubic inches, we use the reverse operation; that is, we divide the cubic inches by a factor of 16.38716. Thus:

$$32.77432\ \textit{cubic centimeters} = 32.77432 \div 16.38716 = 2\ \textit{cubic inches}$$

To convert weight and mass units between the English and metric or SI systems, use the relationships given in Table 5.15. For example, to convert pounds into grams, the table shows that we multiply the number of pounds by a factor of 453.6. Thus, for instance, to convert 2 pounds into grams:

$$2\ \textit{pounds} = 2 \times 453.6 = 907.2\ \textit{grams}$$

Similarly, to convert grams into pounds, we use the reverse operation; that is, we divide the grams by a factor of 453.6. Thus:

$$907.2\ \textit{grams} = 907.2 \div 453.6 = 2\ \textit{pounds}$$

Table 5.15 Converting measures of weight and mass.

From	To	Multiply by	Divide by
pound	gram	453.6	
gram	pound		453.6
pound	kilogram	0.4536	
kilogram	pound		0.4536
ton	kilogram	907.2	
kilogram	ton		907.2

Table 5.16 Converting liquid measures—English and metric units.

From	To	Multiply by	Divide by
quart	liter	0.946	
liter	quart		0.946
gallon	liter	3.785	
liter	gallon		3.785

Table 5.17 Converting temperatures—Celsius to Fahrenheit.

From	To	Multiply by	Add
°Celsius	°Fahrenheit	9/5	32

To convert units of liquid measure between the English and metric or SI systems, use the relationships given in Table 5.16. For example, to convert quarts into liters, the table shows that we multiply the number of quarts by a factor of 0.946. Thus, for instance, to convert 2 quarts into liters:

$$2\ quarts = 2 \times 0.946 = 1.892\ liters$$

Similarly, to convert liters into quarts, we use the reverse operation; that is, we divide the liters by a factor of 0.946. Thus:

$$1.892\ liters = 1.892 \div 0.946 = 2\ quarts$$

To convert temperature units from Celsius to Fahrenheit, use the relationship given in Table 5.17. To convert °Celsius to °Fahrenheit, the table shows that we multiply °Celsius by 9/5 and then add 32. Thus, for instance, to convert 20°Celsius to °Fahrenheit:

$$20^\circ\ Celsius = 20 \times \frac{9}{5} + 32 = 68^\circ\ Fahrenheit$$

Table 5.18 Converting temperatures—Fahrenheit to Celsius.

From	To	Subtract	Divide by
°Fahrenheit	°Celsius	32	9/5

Table 5.19 Converting temperatures—English and metric units.

From	To	Subtract	Add
°Rankine	°Fahrenheit	459.67	
°Fahrenheit	°Rankine		459.67
°Kelvin	°Celsius	273.15	
°Celsius	°Kelvin		273.15

Recall that 0°Celsius converts to 32°Fahrenheit, and that the boiling point of water is 100°Celsius or 212°Fahrenheit.

To convert temperature units from Fahrenheit to Celsius, use the relationship given in Table 5.18. To convert °Fahrenheit to °Celsius, the table shows that we subtract 32 from °Celsius and then divide by 9/5. Thus, for instance, to convert 68°Fahrenheit to °Celsius:

$$68^\circ\ Fahrenheit = (68 - 32) \div \frac{9}{5} = 20^\circ\ Celsius$$

Recall that 32 °Fahrenheit converts to 0°Celsius, and that the boiling point of water is 212°Fahrenheit or 100°Celsius.

To convert temperature units from Celsius to Kelvin, or Fahrenheit to Rankine, use the relationships given in Table 5.19. For example, to convert °Celsius to °Kelvin, the table shows that we add 273.15 to °Celsius. Thus, for instance, to convert 20°Celsius to °Kelvin:

$$20^\circ\ Celsius = 20 \times +273.15 = 293.15\ degrees\ Kelvin$$

Similarly, to convert °Kelvin to °Celsius, we use the reverse operation; that is, we subtract 273.15 from °Kelvin. Thus:

$$293.15^\circ\ Kelvin = 293.15 - 273.15 = 20\ degrees\ Celsius$$

More conversion factors are given in Appendix C.

Chapter 6

Numerical Conversions

> Use various numbering methods, such as scientific notation, decimals, and fractions, and convert values between these methods. (Apply)
>
> Body of Knowledge I.F

EXPONENTS

Before diving into scientific notation, we will cover both positive and negative exponents, as well as powers of 10.

Positive Exponents

Using positive exponents is a short-hand way to represent a number that is multiplied by itself. For example, given:

$$2 \times 2 \times 2 = 8$$

we can represent the multiplication statement in exponent form: 2^3. The first number, 2, represents the number being multiplied by itself. The superscript 3 is the *exponent*, and it tells us how many times the number 2 is multiplied by itself.

Conversely, an exponent can be converted to a multiplication statement. For example, 4^5 represents:

$$4 \times 4 \times 4 \times 4 \times 4 = 1024$$

A special case occurs when the exponent equals zero. Any non-zero number with exponent equal to 0 will equal 1. For example, $5^0 = 1$. This is read as "five to the zero power equals one."

For any number x with an exponent equal to 1, we say "x to the first power." For example, 5^1 is read "five to the first power."

For any number x with an exponent equal to 2, we say "x squared." For example, 5^2 is read as "five squared."

For any number x with an exponent equal to 3, we say "x cubed." For example, 5^3 is read as "five cubed."

If we have 5^4 we say "five to the fourth power." The number 5^5 is read "five to the fifth power," and so on.

Negative Exponents

We can also use exponents to represent multiplication statements such as:

$$\frac{1}{2} \times \frac{1}{2} \times \frac{1}{2} = \frac{1}{8}$$

The statement above is written as 2^{-3}. In this case we use a negative exponent to show that the number 2 is in the denominator. Note that 2^{-3} can also be written as $\frac{1}{2^3}$, meaning that 2^{-3} and 2^3 are reciprocals. We read 2^{-3} as "2 to the negative three power."

Powers of 10

Table 6.1 shows the powers of 10 and the numbers they represent. These powers of ten are used in scientific notation.

Scientific Notation

For very large or small values, it is often more convenient to represent a number in terms of scientific notation. Scientific notation uses powers of ten. For example, the number 100 can be written as 1×10^2. The decimal 0.01 can be shown as 1×10^{-2}.

Table 6.1 Powers of 10.

Exponent form	Number
10^{-6}	0.000001
10^{-5}	0.00001
10^{-4}	0.0001
10^{-3}	0.001
10^{-2}	0.01
10^{-1}	0.1
10^{0}	1
10^{1}	10
10^{2}	100
10^{3}	1000
10^{4}	10,000
10^{5}	100,000
10^{6}	1,000,000

Procedure for values greater than 1, or less than –1.
To display a number greater than 1 or less than –1 in. scientific notation, perform these three steps:

1. Move the decimal point to the left so that the original number is now displayed with a non-zero digit in the ones place.
   For example, the number 234 would be displayed as 2.34 when we move the decimal point to the left two times. The number –5567.8 would be displayed as –5.5678 when we move the decimal point three times to the left.

2. Next, display the number of places the decimal moved to the left as a positive power of ten.
   For example, the decimal was moved two times to the left to convert 234 to 2.34. We will denote this as 10^2.
   The decimal was moved three times to the left to convert –5567.8 to –5.5678. This is denoted as 10^3.

3. Put the two pieces together, linked by a multiplication sign.
   For example, the value 234 is displayed as 2.34×10^2 in scientific notation. The value –5567.8 is displayed as -5.5678×10^3 in scientific notation.

Procedure for values between –1 and +1
To display a number between zero and 1 or between -1 and 0 in. scientific notation, perform these three steps:

1. Move the decimal point to the right so that the original number is now displayed with a non-zero digit in the ones place.
   For example, the number 0.234 would be displayed as 2.34 when we move the decimal point to the right one time. The number –0.00556 would be displayed as –5.56 when we move the decimal point three times to the right.

2. Next, display the number of places the decimal moved to the right as a negative power of ten.
   For example, the decimal was moved one time to the right to convert 0.234 to 2.34. We will denote this as 10^{-1}.
   The decimal was moved three times to the right to convert –0.00556 to –5.56. This is denoted as 10^{-3}.

3. Put the two pieces together, linked by a multiplication sign.
   For example, the value 0.234 is displayed as 2.34×10^{-1} in scientific notation. The value –0.00556 is displayed as -5.56×10^{-3} in scientific notation.

Converting Scientific Notation Back to Original Form
To convert a number displayed in scientific notation back to its original form, take the first number in the scientific notation statement and move the decimal point to the right or left as indicated by the power of ten. For negative powers of ten, move the decimal to the left. For positive powers of ten, the decimal place is moved to the right.

For example, to convert 2.34×10^{-1} back to original form, we take 2.34 and move the decimal 1 place to the *left*—since the power of ten is negative—to get 0.234.

To convert 2.34×10^2 back to original form, we take 2.34 and move the decimal 2 places to the *right*—since the power of ten is positive—to get 234.

DECIMALS AND FRACTIONS

Converting Fractions to Decimals

Fractions can be converted to decimals by dividing denominator into the numerator. For example:

$$\frac{3}{8} = 8\sqrt{3} = 0.375$$

We read this decimal as "375 thousandths."

Some common fractions and their decimal equivalents are listed in Table 6.2.

Note that the decimal equivalents to the fractions 1/3, 1/6, 1/7, and 1/9 are infinitely repeating decimals, as shown below. A line drawn above a number or a sequence of numbers indicates that the number or sequence repeats infinitely.

$$\frac{1}{3} = 0.33\overline{3}$$
$$\frac{1}{6} = 0.166\overline{6}$$
$$\frac{1}{7} = 0.\overline{142857}$$
$$\frac{1}{9} = 0.11\overline{1}$$

Table 6.2 Common fractions and their decimal equivalents.

Fraction	Decimal
1/16	0.0625
1/10	0.1
1/8	0.125
1/5	0.2
1/4	0.25
3/8	0.375
2/5	0.4
1/2	0.5
3/5	0.6
5/8	0.625
3/4	0.75
4/5	0.8
7/8	0.875
15/16	0.9375

Converting Decimals to Fractions

To convert a decimal to a fraction, divide the decimal by one. Then, move the decimal point to the right in the numerator to make a whole number. Move the decimal place in the denominator the same number of times. For example, to convert the decimal 0.234 to a fraction, the decimal point in both the numerator and denominator is moved four places to the right:

$$\frac{0.0234}{1} = \frac{234}{10{,}000}$$

The resulting fraction can then be simplified if desired.

Chapter 7

Basic Statistics and Applications

1. **Measures of central tendency.** Calculate mean, median, and mode. (Apply)
2. **Measures of dispersion.** Calculate range, standard deviation, and variance. (Apply)
3. **Measures of proportion.** Calculate percentage and ratio measures for various data sets. (Apply)
4. **Graphical displays.** Define, interpret, and use scatter diagrams, tally sheets, and bar charts, to effectively display data in various situations. (Apply)
5. **Normal distribution.** Explain various characteristics of a normal distribution, such as symmetry, bell curve, and central tendency. (Apply)

Body of Knowledge I.G

The first three sections of this chapter present descriptive statistics: measures of central tendency, measures of dispersion, and measures of proportion. The fourth section covers graphical displays, while the final section presents a detailed discussion of the normal distribution.

Raw data itself provides little insight into underlying patterns and limits our ability to draw meaningful conclusions. This is where descriptive statistics becomes useful.

Descriptive statistics is the collection of tools and techniques for displaying and summarizing data. There are many ways in which to display and summarize data. Therefore, it is important to select the best way to facilitate an understanding of the data collected. In some cases, multiple statistics and presentations or views are required to gain the most insight from the data.

We can calculate metrics for a sample of data, as well as for data from an entire population. Measures computed using information from the entire population are referred to as *parameters*. Numerical measures computed from sample data are referred to as *statistics*. It is standard practice to denote parameters by the letters of the Greek alphabet and statistics by the letters of the English alphabet.

We divide these numerical measures into three categories: (1) measures of central tendency, (2) measures of dispersion, and (3) measures of proportion. *Measures of central tendency* give us information about the center of the data, *measures of dispersion* give us information about the variation within the data, and *measures of proportion* tell us the percentage of the data that falls into some specific interval or intervals.

MEASURES OF CENTRAL TENDENCY

Three common ways for quantifying the centrality of a population or sample include:

1. *Mean.* Although there are many types of statistical means, the most common is the arithmetic mean, or arithmetic average, of a data set. Typically, the mean will be reported to one more significant figure than the data. The calculation for the population mean, denoted by the Greek letter μ, is given in Equation 7.1.

   Population mean

$$\mu = \frac{\sum_{i=1}^{N} x_i}{N} \tag{7.1}$$

where N is the number of items in the population.

The sample mean is denoted by $\overline{X}$, read as "X bar." It is calculated using Equation 7.2.

Sample mean

$$\overline{X} = \frac{\sum_{i=1}^{n} x_i}{n} \tag{7.2}$$

where n is the sample size. It is usually recorded to one or two more decimal places than the original data.

2. *Median.* The *median* is the middle value of an ordered data set. If the data are composed of an odd number of data points, the median is the middle value of the ordered data set. If the data are composed of an even number of data points, the median is the average of the two middle values of the ordered data set.

3. *Mode.* The *mode* is the most frequently found value in a data set. Note: There may be more than one mode present, or the mode of a data set may not exist if all values have the same frequency.

Example 7.1

Consider the values given in Table 7.1 that represent a sample of diameters taken from a drilling operation. For convenience, this table has been sorted, but the raw data were collected in a random fashion. Compute all three measures of central tendency.

Table 7.1 Data for examples 7.1, 7.2, 7.4, and 7.7.

Number	Diameter	Number	Diameter
1	0.121	11	0.123
2	0.121	12	0.124
3	0.121	13	0.124
4	0.122	14	0.124
5	0.122	15	0.125
6	0.122	16	0.125
7	0.123	17	0.125
8	0.123	18	0.125
9	0.123	19	0.126
10	0.123	20	0.126
		TOTAL	**2.468**

From Table 7.1, we see that the total of the diameters is 2.468. Using Equation 7.2, we have:

$$\textit{Sample Mean} = \bar{X} = \frac{\sum_{i=1}^{n} x_i}{n} = \frac{2.468}{20} = 0.1234$$

There are an even number of data points in the ordered set, with n=20. Thus, the median is the average of the points 10 and 11 in. the ordered list:

$$\textit{Median} = \frac{0.123 + 0.123}{2} = 0.123$$

A tally sheet constructed from the data shows that the most frequent value in the data set is 0.123.

Value	Tally	Frequency
0.121	III	3
0.122	III	3
0.123	IIII	4
0.124	III	3
0.125	III	3
0.126	III	3

Thus, the mode = 0.123.

MEASURES OF DISPERSION

Although there are many ways to determine the dispersion, or spread, of data, the most common ways include the range, standard deviation, and variance.

Range. The *range* is simply the maximum value minus the minimum value, as shown in Equation 7.3.

Range

$$R = x_{Max} - X_{Min} \tag{7.3}$$

Note that the range typically increases as the sample size increases, unlike the standard deviation. For this reason, the standard deviation is the preferred measure of dispersion for large samples.

Standard deviation. The standard deviation is the square root of the variance and can be calculated for a population or a sample. The standard deviation is in the same units as the original data and is easier to interpret than the variance. This will be shown in the Empirical Rule section later in this chapter.

We can calculate the *population standard deviation* using Equation 7.4:

Population standard deviation

$$\sigma = \sqrt{\frac{\sum_{i=1}^{N}(x_i - \mu)^2}{N}} \tag{7.4}$$

where N is the population size and μ is the population mean. Note that this measure is not calculated often since data for the entire population is seldom available.

We can estimate the *sample standard deviation* using Equation 7.5:

Sample standard deviation

$$s = \sqrt{\frac{\sum_{i=1}^{n}(x_i - \overline{X})^2}{n-1}} \tag{7.5}$$

where n is the sample size and $\overline{X}$ is the sample mean.

We can also use the following form of the formula when performing hand calculations, shown in Equation 7.6:

Sample standard deviation, calculation formula

$$s = \sqrt{\frac{\sum_{i=1}^{n}x_i^2 - n\overline{X}^2}{(n-1)}} \tag{7.6}$$

The standard deviation is typically recorded to either one or two more significant digits than the raw data.

Variance. The variance of a data set gives the sum of squared deviations of the data from its mean. We can calculate the variance of either a population or a sample.

Population variance

$$\sigma^2 = \frac{\sum_{i=1}^{N}(x_i - \mu)^2}{N} \tag{7.7}$$

Sample variance

$$s^2 = \frac{\sum_{i=1}^{n}(x_i - \overline{X})^2}{n-1} \tag{7.8}$$

Sample variance, calculation formula

$$s=\sqrt{\frac{\sum_{i=1}^{n}x_i^2-n\bar{X}^2}{(n-1)}} \quad (7.9)$$

Example 7.2

Consider the sample of diameters taken from a drilling operation shown in Table 7.1. Compute the range, standard deviation, and variance of the data.

The range of the data is the maximum value minus the minimum value. From Table 7.1, we see the maximum and minimum values are 0.126 and 0.121. Applying Equation 7.3 yields:

$$R=x_{Max}-x_{Min}=0.126-0.121=0.005.$$

Since the data represent a sample, the standard deviation is calculated using either Equation 7.5 or 7.6:

$$s=\sqrt{\frac{\sum_{i=1}^{n}\left(x_i-\bar{X}\right)^2}{n-1}}=\sqrt{\frac{\sum_{i=1}^{27}\left(x_i-0.1234\right)^2}{(20-1)}}=0.00160$$

or, using the calculation formula,

$$s=\sqrt{\frac{\sum_{i=1}^{n}x_i^2-n\bar{X}^2}{(n-1)}}=\sqrt{\frac{0.3046-(20)0.1234^2}{(20-1)}}=0.00160$$

The sample variance can be calculated using Equation 7.8 or Equation 7.9:

$$s^2=\frac{\sum_{i=1}^{n}\left(x_i-\bar{X}\right)^2}{n-1}=\frac{\sum_{i=1}^{27}\left(x_i-0.1234\right)^2}{(20-1)}=0.00000256$$

$$s^2=\frac{\sum_{i=1}^{n}x_i^2-n\bar{X}^2}{(n-1)}=\frac{0.3046-(20)0.1234^2}{(20-1)}=0.00000256$$

Note: Readers using handheld calculators may obtain slightly different answers depending on the way calculations are performed, how the numbers are rounded, the number of significant digits used, and/or software used. To achieve the best results, keep all digits in the intermediate calculations, and then round at the end when reporting results.

MEASURES OF PROPORTION

Measures of proportion divide the data into various parts of certain percentages or help to locate the place of any data value in the whole data set. For example, more commonly used measures of proportion are percentiles and quartiles. Percentiles divide the data into 100 parts such that each part contains at the most 1% of the data, and quartiles divide the data into 4 parts such that each part contains at the most 25% of the data. From quartiles we can derive another measure, called the interquartile range, which gives the range of the middle 50% of the data values. The interquartile range may also be obtained by first writing the data in ascending order and then trimming 25% of the data values from both the lower end and the upper end.

Percentiles

Percentiles divide the data into 100 equal parts; each part contains at the most 1% of the data, and the percentiles are numbered from 1 to 99. For example, the median of a data set is the 50th percentile, which divides the data into two equal parts such that at the most 50% of the data fall below the median and at the most 50% of the data fall above it. There are several methods for computing percentiles. We present a straightforward, three-step method here.

Step 1: Sort the data from lowest to highest.

Step 2: Calculate L using

$$L = \left(\frac{p}{100}\right)n$$

where p is the percentile in question (e.g., 40, 50, 70) and n is the number of data values.

Step 3: *If L is a whole number*, the *kth* percentile is halfway between the *Lth* value in the sorted list and the next highest value. The *kth* percentile value is the average of these two values.

If L is not a whole number, change L by rounding up to the next larger whole number. The value of the *kth* percentile is then the *Lth* value, counting from the lowest in the sorted list.

Example 7.3

The following data give prices of fifteen three-bedroom homes in a popular suburb in thousands of dollars.

662, 248, 352, 463, 585, 651, 195, 276, 372, 451, 269, 273, 158, 555, 354

Find the 60th and 90th percentiles for this sample data.

Solution

Step 1: Write the data values in ascending order and note that $n = 15$.

Prices: 158, 195, 248, 269, 273, 276, 352, 354, 372, 451, 463, 555, 585, 651, 662

Step 2: Calculate L for the 60th and 90th percentiles:

$$L_{60} = \left(\frac{p}{100}\right)n = \left(\frac{60}{100}\right)15 = 0.6 \times 15 = 9$$

$$L_{90} = \left(\frac{p}{100}\right)n = \left(\frac{90}{100}\right)15 = 0.9 \times 15 = 13.5$$

Step 3: Since L_{60} is a whole number, the 60th percentile is halfway between the L = 9th and L + 1 = 10th values in the ordered list, starting from the lowest value.

Thus, the 60th percentile of the home price data is the average of \$372,000 and \$451,000.

$$\textit{60th percentile} = \frac{\$372{,}000 + \$451{,}000}{2} = \$411{,}500$$

That is, at most 60% of the home prices are less than \$411,500, and at most 40% of the home prices are more than \$411,500.

L_{90} is not a whole number, we round up to the next highest L value. Here, we round 13.5 up to 14. The 90th percentile is thus the 14th value in the ordered list.

$$90th\ percentile = \$651{,}000$$

At most 90% of the home prices are less than \$651,000, and at most 10% of the home prices are more than \$651,000.

We have shown how to determine the data value x for a given percentile p. Now we would like to find the percentile p corresponding to a given value x. This can be done by using Equation 7.10:

Percentile corresponding to a value *x*

$$p = \frac{(\#\ of\ data\ values \le x)}{(n+1)} \times 100 \tag{7.10}$$

Example 7.4

Using the home price data in Example 7.3, the percentile corresponding to the home price of \$300,000 is:

$$\begin{aligned} p &= \frac{(\#\ of\ data\ values \le x)}{(n+1)} \times 100 \\ &= \frac{(\#\ of\ data\ values \le 300)}{(15+1)} \times 100 \\ &= \frac{6}{16} \times 100 \\ &= 37.5 \end{aligned}$$

A home priced at \$300,000 is at the 37.5th percentile. In other words, at most 37.5% of the homes are priced at \$300,000, or at most 62.5% are priced higher than \$300,000.

Quartiles

In the previous section, we focused on percentiles, which divide the data into 100 equal parts. Some of the percentiles have special importance. These are the 25th, 50th, and 75th percentiles and are known as the first, second, and third quartiles (denoted by Q_1, Q_2, and Q_3), respectively. These quartiles are sometimes also known as lower, middle, and upper quartiles, respectively. Note that the second quartile is the same as the median. To determine the values of the different quartiles, one just has to find the 25th, 50th, and 75th percentiles.

Interquartile Range (IQR)

Often, we are more interested in finding information about the middle 50% of the data. A measure of dispersion relative to the middle 50% of the population or sample data is known as the *interquartile range* (IQR). This range is obtained by trimming 25% of the values from the bottom and 25% from the top. IQR is defined as:

Interquartile range

$$IQR = Q_3 - Q_1 \tag{7.16}$$

Example 7.5

Find the IQR for the housing data in Example 7.22, that is:

Home prices (in thousands of dollars): 662, 248, 352, 463, 585, 651, 195, 276, 372, 451, 269, 273, 158, 555, 354

Solution

In order to find the IQR, we first need to find the quartiles Q_1 and Q_3, or, equivalently, the 25th and 75th percentiles.

Step 1: Write the data values in ascending order and note that $n = 15$.

Prices: 158, 195, 248, 269, 273, 276, 352, 354, 372, 451, 463, 555, 585, 651, 662

Step 2: Calculate L for the 25th percentile:

$$L = \left(\frac{p}{100}\right)n = \left(\frac{25}{100}\right)15 = 0.25 \times 15 = 3.75$$

$$L = \left(\frac{p}{100}\right)n = \left(\frac{75}{100}\right)15 = 0.75 \times 15 = 11.25$$

Step 3: Since L for the 25th percentile is not a whole number, we round L up to the next largest whole number: 4. The 25th percentile is then the 4th value in the sorted list, starting from the lowest value. Likewise, the L for the 75th percentile is rounded up to 12.

The 25th percentile, or Q_1, is thus 269 and the 75th percentile, or Q_3, is 555.

The middle 50% of the home prices in the sample are between $269,000 and $555,000. The IQR in this example is:

$$IQR = \$555{,}000 - \$269{,}000 = \$286{,}000$$

The IQR gives the range of variation among the middle 50% of the population. It is potentially a more meaningful measure of dispersion than the range, since it is not affected by extreme values that may be present in the data. By trimming 25% of the data from the bottom and 25% from the top, we eliminate the extreme values that may be present in the data set. Very often, the IQR is used as a measure of comparison for comparing two or more data sets arising from similar studies.

Box-and-Whisker Plot

The *box-and-whisker plot* (also called a *box plot*) is a graphical display of the "five number summary" of the data. The five numbers that are displayed are:

1. Minimum value
2. First quartile
3. Second quartile (also known as the median)
4. Third quartile
5. Maximum value

Since the quartiles are displayed, the IQR is automatically displayed. Commonly, the original box-and-whisker plot is modified to identify outliers. Depending on the software being used, other descriptive statistics may be available for display as well. Figure 7.1 shows an example of a box plot, which was constructed using the data from Table 7.1. However, an additional data point was added to depict the outlier at 0.117.

With a brief visual review of Figure 7.1, we quickly determine many of the descriptive statistics calculated previously.

We see that 50% of the data is between $Q_1=0.122$ and $Q_3=0.125$, and that the IQR is 0.003.

We can also make a judgment about the symmetry of the data from inspecting the plot:

- If the second quartile (median) is close to the center of the box and each of the whiskers is approximately of equal length, the distribution is symmetric.
- If the upper box is substantially larger than the left box and/or the upper whisker is much longer than the lower whisker, the distribution is right, or positively, skewed.
- If the lower box is substantially larger than the upper box and/or the lower whisker is much longer than the upper whisker, the distribution is left, or negatively, skewed.

Upon inspecting Figure 7.1, we notice that the data are compact, but positively skewed, and that an outlier is present.

Box-and-whisker plots are quite appealing and provide a lot of information about a data set in just a single diagram. They are also extremely useful as a way of comparing different samples or groups of data.

Empirical Rule

If the data have a distribution that is approximately bell-shaped, the following rule, known as the *empirical rule*, can be used to compute the percentage of data that will fall within k standard deviations from the mean ($k=1, 2, 3$).

- Approximately 68% of the data will fall within one standard deviation of the mean, that is, between $(\mu-1\sigma)$ and $(\mu+1\sigma)$.
- Approximately 95% of the data will fall within two standard deviations of the mean, that is, between $(\mu-2\sigma)$ and $(\mu+2\sigma)$.

- Approximately 99.7% of the data will fall within three standard deviations of the mean, that is, between $(\mu - 3\sigma)$ and $(\mu + 3\sigma)$.

Figure 7.2 illustrates these features of the empirical rule.

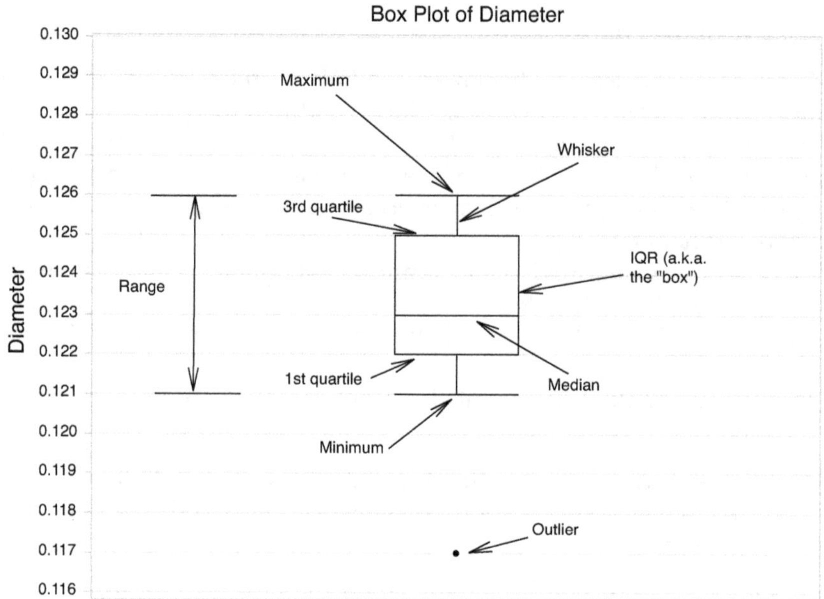

Figure 7.1 Basic components of a box-and-whisker plot (box plot).

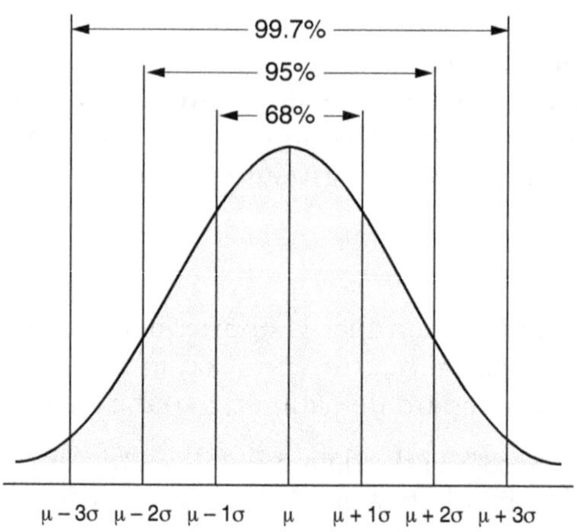

Figure 7.2 Application of the empirical rule.

Example 7.6

A soft drink filling machine is used to fill 16 oz soft drink bottles. Since the amount of beverage slightly varies from bottle to bottle, it is believed that the actual amount of beverage in the bottle forms a bell-shaped distribution with a mean of 15.8 oz and a standard deviation of 0.15 oz. Use the empirical rule to find the percentage of bottles that contain between 15.5 oz and 16.1 oz of beverage.

Solution

We are given $\mu = 15.8$ oz and $\sigma = 0.15$ oz. We are interested in finding the percentage of bottles that contain between 15.5 oz and 16.1 oz of beverage. We note that both 15.5 and 16.1 are 2 standard deviations away from the mean:

$$15.5 = \mu - 2\sigma = 15.8 - (2 \times 0.15)$$
$$16.1 = \mu + 2\sigma = 15.8 + (2 \times 0.15)$$

Comparing Figure 7.2 with Figure 7.3, we can conclude that approximately 95% of the bottles contain between 15.5 oz and 16.1 oz of beverage, since 15.5 and 16.1 are two standard deviations away from the mean.

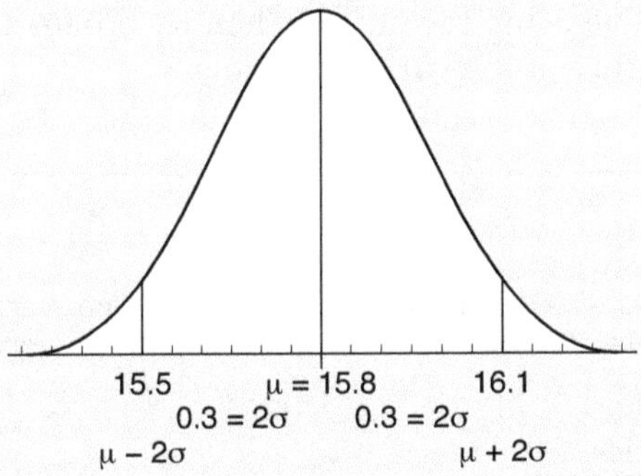

Figure 7.3 Amount of soft drink contained in a bottle.

GRAPHICAL DISPLAYS

In this section, we will discuss a variety of graphical methods for depicting data, including scatter plots, tally sheets, bar charts, Pareto charts, histograms, stem-and-leaf plots, and dot plots. Graphical methods are quite appealing because they provide a visual representation of the data that provides valuable insights.

Scatter Plots

A *scatter plot* is a plot of two variables, one on the *y*-axis and the other on the *x*-axis. The resulting graph allows visual examination for patterns to determine whether the variables show any relationship or if there is just random "scatter." A pattern, or lack thereof, aids in choosing the appropriate type of model for estimation. It is important to note that evidence of a pattern does not imply that a causal relationship exists between the variables. Note that the scatter plot is one of the seven basic tools of quality.

Patterns on a scatter plot may suggest straight lines (linear) or curves (nonlinear). If the plotted points suggest a straight-line relationship between the two variables, then we can measure the strength of the linear relationship by using the *sample correlation coefficient*, denoted by the letter r. A positive correlation means that the line moves from the bottom left to the upper right. A negative correlation means that the line moves from the top left to the bottom right. If all the points fall exactly on a straight line that tips up on the right end, then $r = 1$. If all the points fall on a straight line that tips down on the right end, then $r = -1$. The value of r is always between –1 and +1, inclusive. This may be stated as $-1 \leq r \leq 1$. Therefore, the first step in analyzing a set of data for correlation is to construct a scatter diagram. If the points on this diagram tend to form a straight line, it makes sense to calculate the value of r.

The fact that a correlation exists does not imply that a cause-and-effect relationship exists. To conclude that there is a cause-and-effect relationship, we look to logic, scientific theory, or a designed experiment for the cause-and-effect relationships. When we obtain a high correlation coefficient, it means that we can predict one variable using the other. When we obtain a low correlation coefficient, the predictive ability of the variables is low.

Example 7.7

Suppose that it is discovered that an injection molding machine is producing parts with pitted surfaces. The following potential causes have been suggested:

- Mold pressure
- Coolant temperature
- Mold cooldown time
- Mold squeeze time

Values of each of these variables, as well as the quality of the surface finish, were collected across ten batches. Table 7.2 shows these data.

Table 7.2 Surface finish data for Example 7.7.

Batch number	Mold pressure, PSI	Coolant temperature, °F	Cooldown time, minutes	Squeeze time, minutes	Surface finish, microinches
1	220	102.5	14.5	0.72	37
2	200	100.8	16.0	0.91	30
3	410	102.6	15.0	0.90	40
4	350	101.5	16.2	0.68	32
5	490	100.8	16.8	0.85	27
6	360	101.4	14.8	0.76	35
7	370	102.5	14.3	0.94	43
8	330	99.8	16.5	0.71	23
9	280	100.8	15.0	0.65	32
10	400	101.2	16.6	0.96	30

Four graphs have been plotted, as shown in Figure 7.4. In each graph, *surface finish* is along the vertical, or *y*, axis and one variable (that is, mold pressure, coolant temperature, mold cooldown time, or mold squeeze time) is along the horizontal, or *x*, axis. Each data point on a graph represents one batch.

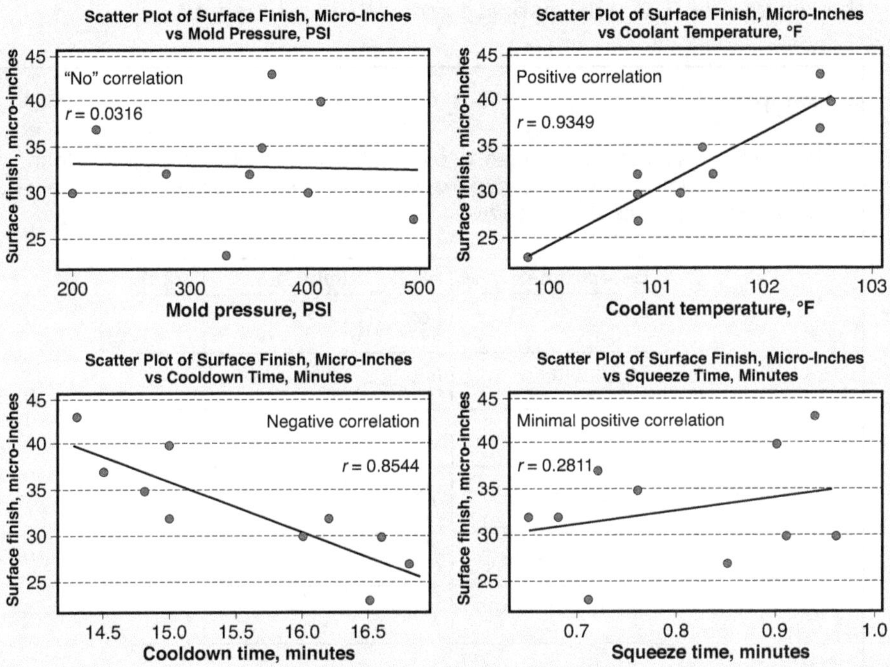

Figure 7.4 Scatter diagrams for Example 7.7.

Let's briefly examine each of the graphs:

- *Scatter plot of surface finish versus mold pressure.* Notice the overall dispersion of the dot points. They are well dispersed and do not appear to present any noticeable pattern. The regression line for the variables is the solid black line in the middle of the graph. Notice that it is almost flat. This indicates that the y-value variation is almost completely unrelated to any change in the x-value. Consequently, there is little to no correlation.
- *Scatter plot of surface finish versus coolant temperature.* The dot points are more tightly clustered together and clearly present a noticeable linear pattern. The addition of the regression line confirms a visually strong correlation coefficient.
- *Scatter plot of surface finish versus cooldown time.* The dot points in this graph are somewhat polarized in the lower and upper ends of the cooldown times, appearing to contribute to the correlation, which in this case is negative in nature.
- *Scatter plot of surface finish versus squeeze time.* Like the mold pressure, there is a lot of dispersion in the dot points, but a minor appearance of a linear pattern is present. The dot points are polarized and loosely clustered at the ends. An experienced eye might be able to determine the correlation to be minimally positive, but the aid of the regression line is certainly helpful.

Tally Sheets

A *tally sheet*, also known as a check sheet, is a data collection and analysis tool that counts the number of occurrences of various events or problems. For example, an inspector can use a tally sheet to track the occurrences of different types of defects they encounter. The tally mark are generally hand drawn.

Example 7.8

Figure 7.5 shows a tally sheet used to collect data on the number of errors found during inspection of clothing orders ready for shipment. We see that the most frequent packing errors are wrong items or wrong sizes.

<table>
<tr><th>Defect type</th><th>Tally</th><th>Total</th></tr>
<tr><td>Wrong item</td><td>||||| ||||</td><td>9</td></tr>
<tr><td>Missing item</td><td>||</td><td>2</td></tr>
<tr><td>Extra item</td><td>|</td><td>1</td></tr>
<tr><td>Wrong size</td><td>||||| |</td><td>6</td></tr>
<tr><td>Wrong color</td><td>|</td><td>1</td></tr>
<tr><td>Defective product</td><td>|</td><td>1</td></tr>
</table>

Figure 7.5 Tally sheet of packing errors found during inspection.

Example 7.9

Tague (2024) presents an example of a tally sheet used to track reasons for a clerk's telephone interruptions by day of the week. The data were collected over several weeks, with the results shown in Figure 7.6.

<table>
<tr><th></th><th colspan="6">Day</th></tr>
<tr><th>Reason</th><th>Mon</th><th>Tues</th><th>Wed</th><th>Thurs</th><th>Fri</th><th>Total</th></tr>
<tr><td>Wrong number</td><td>|||||</td><td>||</td><td>|</td><td>|||||</td><td>|||||||</td><td>20</td></tr>
<tr><td>Info request</td><td>||</td><td>||</td><td>||</td><td>||</td><td>||</td><td>10</td></tr>
<tr><td>Boss</td><td>|||||</td><td>||</td><td>|||||||</td><td>|</td><td>||||</td><td>19</td></tr>
<tr><td>Total</td><td>12</td><td>6</td><td>10</td><td>8</td><td>13</td><td>49</td></tr>
</table>

Figure 7.6 Cause of telephone interruptions by day of week.

Here, we can see that Mondays and Fridays have the most telephone interruptions, with wrong numbers and calls from the boss being the most frequent types of phone interruptions.

Bar Charts

Bar charts are commonly used to study one or more populations when they are classified into various categories, for example, by sector, by region, or over different time periods. We may want or need to know more about the sales of a company by sector, by region, or over different time periods. A bar chart is constructed by creating categories that are represented by intervals of equal length on a horizontal axis, or x-axis. Within each category, we indicate the number of observations as a frequency of the corresponding category, which is then represented by a rectangular bar of length proportional to the frequency. Example 7.10 illustrates the construction of a bar chart.

Example 7.10

The following table contains a company's sales revenue (in thousands of dollars) in each of its five sales regions for May:

Region	Sales $
Northeast	32.4
Southeast	120.2
Midwest	97.3
Mountain	55.7
West	50.2

Figure 7.7 shows a bar chart of these data.

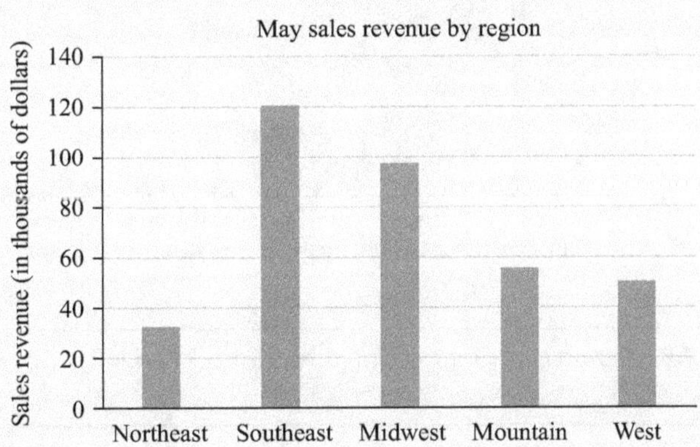

Figure 7.7 Bar chart for annual sales revenue by region.

Pareto Charts

We can augment a tally sheet by creating a *Pareto chart,* a special type of bar chart in which the categories on the x-axis are arranged by decreasing order of frequency.

Vilfredo Pareto, an Italian economist, developed his eponymous chart to show the distribution of wealth in nineteenth century Europe. He found that 80% of the wealth was controlled by just 20% of the population.

In the quality field, the Pareto chart allows the user to quickly identify which defects occur more frequently. This allows the user to prioritize resources to eliminate the defects with the greatest effect on the quality of the product. In this sense, the Pareto chart helps separate the *vital few* categories from the *trivial many.* Specifically, the *Pareto Principle,* or the 80/20 rule, states that 80% of a problem is caused by roughly 20% of the causes.

Example 7.11

Figure 7.8 shows a Pareto chart created for the packing errors tallied in Example 7.8. The chart shows that the top two error types are wrong items and extra items. In fact, these two errors comprise 15/20, or 75%, of all the packing errors. To significantly decrease errors, a team should focus on eliminating wrong item and extra item errors.

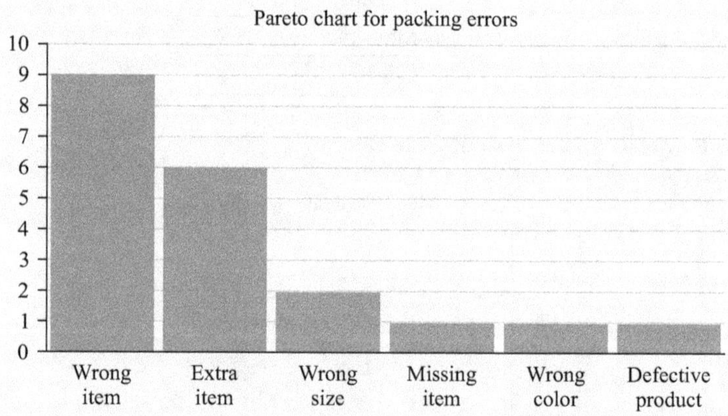

Figure 7.8 Pareto chart for packing errors introduced in Example 7.8.

Example 7.12

We can present the sales revenue data from Example 7.10 using a Pareto chart, as shown in Figure 7.9. Using this chart, we can clearly see that the Southeast and Midwest regions contributed the most revenue.

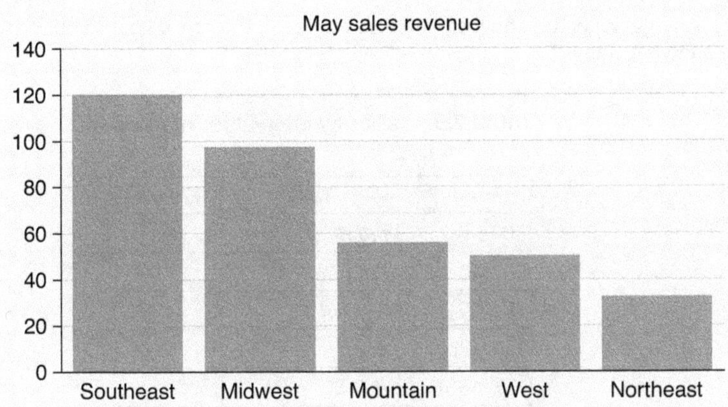

Figure 7.9 Pareto chart for May sales revenue by region.

Histograms

Histograms are used to show the shape, central tendency, and spread of data graphically. Such information is not apparent from the raw data.

Histograms are constructed by dividing the raw data into nonoverlapping intervals or "buckets" and tallying the number of data points that fall in each interval. Histograms are most often created using spreadsheet or statistical software; manual construction can be tedious.

The intervals or buckets are displayed on the x-axis, and the number of data points that falls in each class (the frequency) is plotted on the y-axis.

Example 7.13

The survival time (in hours) of 50 parts involved in a field test under extreme operating conditions was collected.

34	61	87	120	134
46	75	98	123	145
79	46	80	83	108
86	47	86	90	118
79	50	79	75	160
40	54	56	76	165
44	72	115	138	69
32	64	117	141	75
75	64	87	60	67
64	89	116	77	66

(continued)

An interval width equal to 14 is used to break the raw data into buckets, and the tallies are shown in Table 7.3.

Table 7.3 Tallies by interval for the survival time of parts.

Interval	Tally	Frequency
[30–54)	///// ///	8
[54–78)	///// ///// ///// //	17
[78–102)	///// ///// //	12
[102–126)	///// //	7
[126–150)	//	4
[150–174)	/	2
Total		50

Figure 7.10 shows the resulting histogram.

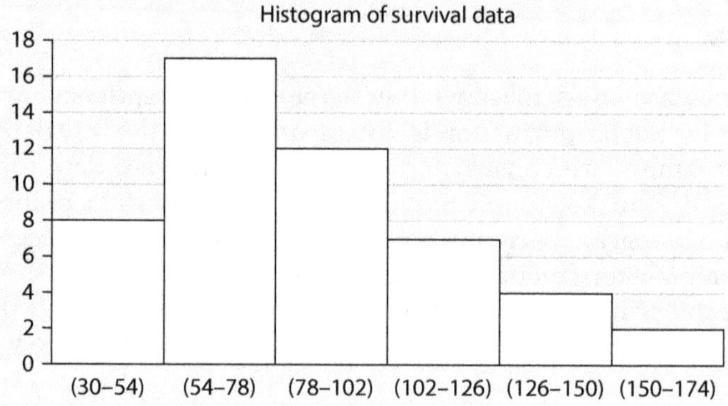

Figure 7.10 Frequency histogram for survival time of parts under extreme operating conditions.

We can see that the most frequent survival times are between 54 and 78 hours. The shape of the distribution is not symmetric. Using the distributions in Figure 7.11 as a guide, we can conclude that the survival times are right-skewed, a feature often attributed to survival data.

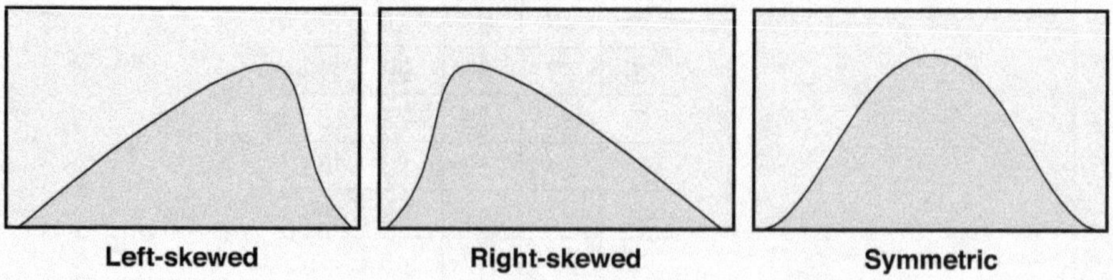

Figure 7.11 Common shapes of distributions curves.

Stem-and-Leaf Plots

A *stem-and-leaf plot* is a graphical method for displaying quantitative data in a format like a histogram while preserving the identity of the raw data.

The stem-and-leaf plot is useful for small sets of data that are easy to sort manually, but it quickly becomes unwieldy for large data sets. Pyzdek (2003) recommends data sets smaller than 200. However, the size of the data set is irrelevant when computer software is involved.

The process for developing a stem-and-leaf plot is quite straightforward, as described by Tague (2023) pp 302–303:

1. Gather and sort the data in ascending order.
2. Determine the values for the "stem" and the "leaves."

   Decide which digits in the data are changing. Out of this group, choose the two or three digits on the left that are the most significant. Of these, the digit on the right will be the leaf, and the one or two on the left will be the stem.

   Note: It is possible that some stems will not have any leaves, depending on the nature of the data set.
3. Draw a vertical line on a page. On the left side of the line, write the stems in ascending order in a column.
4. To the right of the line, write opposite each stem the leaf of each data value that has a stem. Do not use any digits to the right of the leaf.
5. Add a key or legend to indicate how to read the stem-and-leaf plot. A common example of the key or legend is "6|3=63."

Example 7.14

Using the data in Table 7.4, construct a stem-and-leaf plot.

Table 7.4 Data for Example 7.11.

Number	Sorted data	Number	Sorted data	Number	Sorted data
1	0.11	10	0.24	19	0.39
2	0.17	11	0.24	20	0.42
3	0.18	12	0.25	21	0.42
4	0.18	13	0.26	22	0.44
5	0.19	14	0.33	23	0.47
6	0.19	15	0.34	24	0.48
7	0.21	16	0.36	25	0.54
8	0.22	17	0.37	26	0.55
9	0.24	18	0.38	27	0.56

(continued)

We will select the leaf to be the digit in the second decimal place and the stem will be everything in the first decimal place. Figure 7.12 shows the output. The first column contains the stems in increments of 0.10; the following values are the leaves in increments of 0.01. For example, the first row of the plot is made up of the data values 0.11, 0.17, 0.18, 0.18, 0.19, and 0.19.

Stem-and-Leaf Display: Data

```
Stem-and-leaf of Data N = 27
Leaf Unit = 0.010

        1   178899
        2   1244456
        3   346789
        4   22478
        5   456
```

Figure 7.12 Stem-and-leaf plot for Example 7.14.

A stem-and-leaf plot can easily be constructed using paper and pencil and is an effective method for showing the distribution of the data while still preserving individual data values.

Dot Plots

A *dot plot* is one of the easiest graphs to construct and this plot does not make use of the frequency distribution table. In a dot plot, each observation is plotted on a real line. For illustration, we consider Example 7.15.

Example 7.15

The number of defective motors received in 20 different shipments is recorded, as shown.

8, 12, 10, 16, 6, 25, 21, 15, 17, 5, 26, 21, 29, 8, 10, 21, 10, 17, 15, 13

To construct a dot plot, a horizontal line is first drawn, the scale of which begins at the smallest observation (5 in. this case) or smaller and ends with the largest observation (29 in. this case) or larger. Then each value is plotted on the scale, as shown in Figure 7.13.

Dot plots are more useful when the sample size is small. A dot plot gives us, for example, information about how far the data are scattered and where most of the observations are concentrated. For instance, in Figure 7.13, we see that the minimum number of defective motors is 5 and the maximum number received in any shipment is 29. Also, we can see that there were 3 shipments with 10 defectives and 3 with 21 defectives.

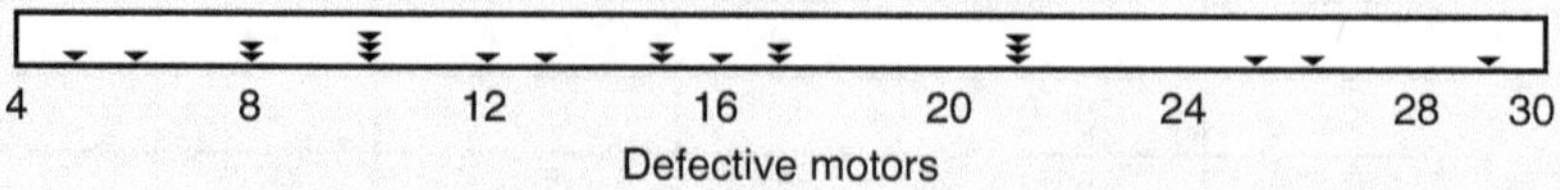

Figure 7.13 Dot plot for the data on defective motors received in 20 different shipments.

NORMAL DISTRIBUTION

The following section was adapted from McShane-Vaughn, *The ASQ Certified Six Sigma Black Belt Handbook* (2022). It is hard to underestimate the importance of the normal distribution in statistical analysis. It has vast quality applications. Consider the following scenarios:

- A design engineer uses the normal distribution to develop tolerances for a new metal part.
- A quality inspector estimates the proportion of units that will fall out of specification if the current process mean shifts by 1.5 standard deviations.
- A quality engineer calculates the capability of a process to report to the customer.
- An R&D engineer tests whether a new design has significantly reduced the average weight of an aerospace component.
- A machine operator tracks the average and range of strength readings of a welded assembly using an X bar and R chart.

The Normal Distribution Function

The normal distribution has two parameters: μ, a location parameter, and σ, a scale parameter. Notation for a normally distributed random variable X is X ~ N(μ |,σ). The parameter μ can take on any value, but σ must be positive. Equation 7.11 shows the normal probability density function.

Equation 7.11 Normal probability density function

$$f(x) = \frac{1}{\sqrt{2\pi\sigma^2}} e^{-\frac{(x-\mu)^2}{2\sigma^2}}, -\infty < x < \infty$$

This function produces the characteristic bell-shaped curve that we have all seen, an example of which is shown in Figure 7.14.

Figure 7.15 shows that changing μ moves the center of the distribution along the *x*-axis, while increasing σ makes the distribution more spread out. Note that the normal distribution is always symmetric.

The Standard Normal Distribution

If X is normally distributed with mean μ and variance σ^2, we can standardize X to have a mean equal to 0 and a variance equal to 1 by computing what is known as a Z score, as shown in Equation 7.12. By using the Z transformation, we are given a relative location of a data point x from a normal distribution: The point is $|z|$ standard deviations away from the mean. A negative Z indicates that the point is below the mean, while a positive Z indicates that the point is above the mean. A Z equal to 0 indicates that the data point is equal to the mean.

Z score

$$Z = \frac{(X-\mu)}{\sigma} \quad (7.12)$$

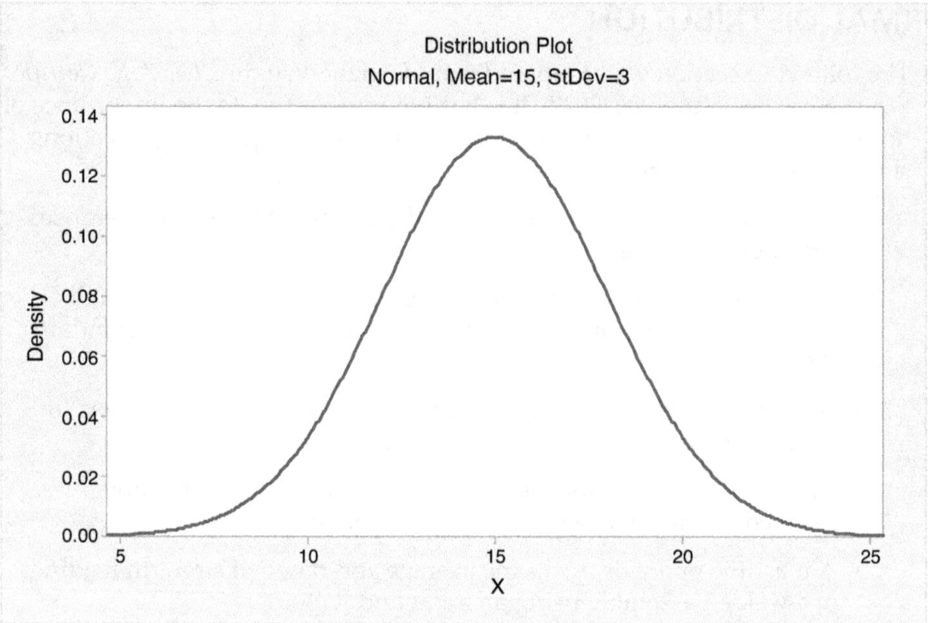

Figure 7.14 Normal distribution curve with $\mu = 15$, $\sigma = 3$.

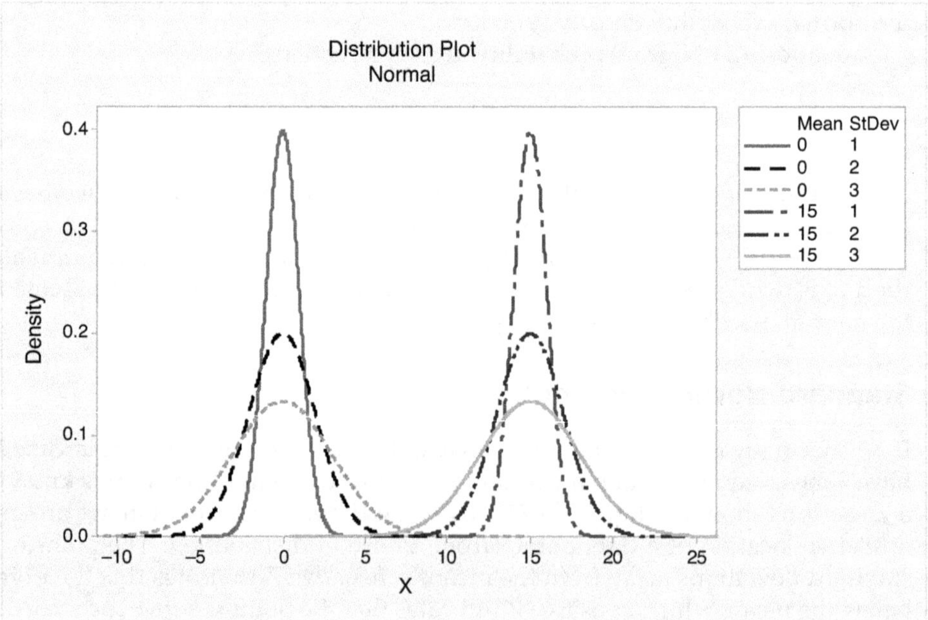

Figure 7.15 Normal distribution with various μ and σ.

The Standard Normal Probability Density Function. The random variable Z follows a standard normal distribution with E(Z) = 0 and Var(Z) = 1, written as $Z \sim N(0, 1)$. The probability density function is thus simplified, as shown in Equation 7.13.

Probability density of the standard normal distribution

$$f(z)=\frac{1}{\sqrt{2\pi}}e^{-\frac{z^2}{2}},-\infty<z<\infty \tag{7.13}$$

Areas Under the Curve

Figure 7.16 shows the graph of the standard normal probability density. It is centered at the mean, which is equal to 0, and has a standard deviation equal to 1.

For continuous distributions, probabilities are calculated as areas under the curve. The normal and standard normal distributions do not have a closed form solution for the area under the curve, so we rely on tabulated values that have been integrated numerically to find probabilities. To determine a probability for a normal distribution with mean μ and variance σ^2, we first convert X to a standard normal distribution, $Z \sim N(0,1)$. Then we can use a cumulative standard normal table based on the Z scale to find probabilities. Why standardize X to Z? If we didn't, there would have to be a separate table of probabilities for every possible combination of μ and σ^2 (an infinite number of tables).

For all normal distributions, the probability of the random variable X lying with one standard deviation of the mean is 0.6827. The probability of X lying within two standard deviations of the mean is 0.9545, and that of X lying within three standard deviations of the mean is 0.9973. These probabilities are shown as areas under the curve in Figure 7.17, Figure 7.18, and Figure 7.19, respectively.

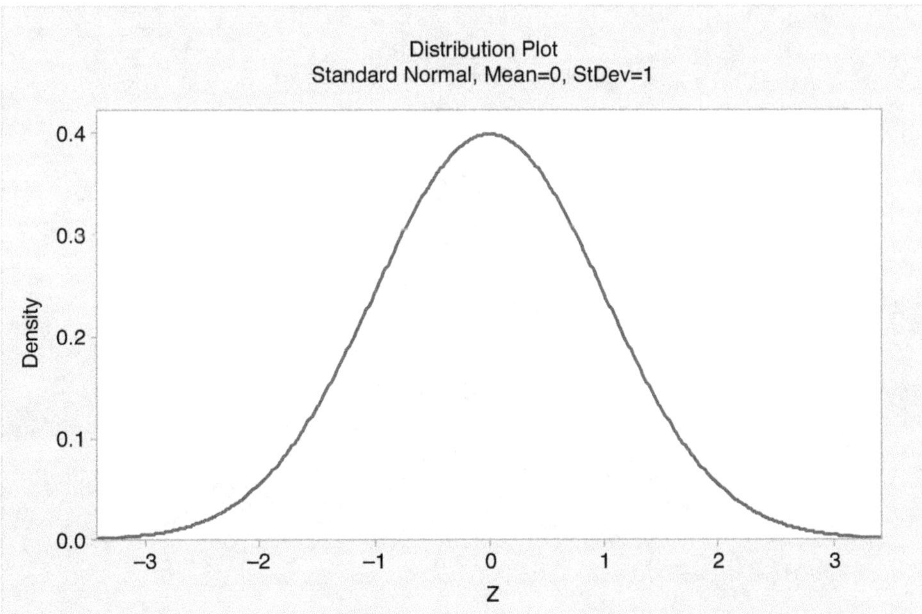

Figure 7.16 Standard normal probability density.

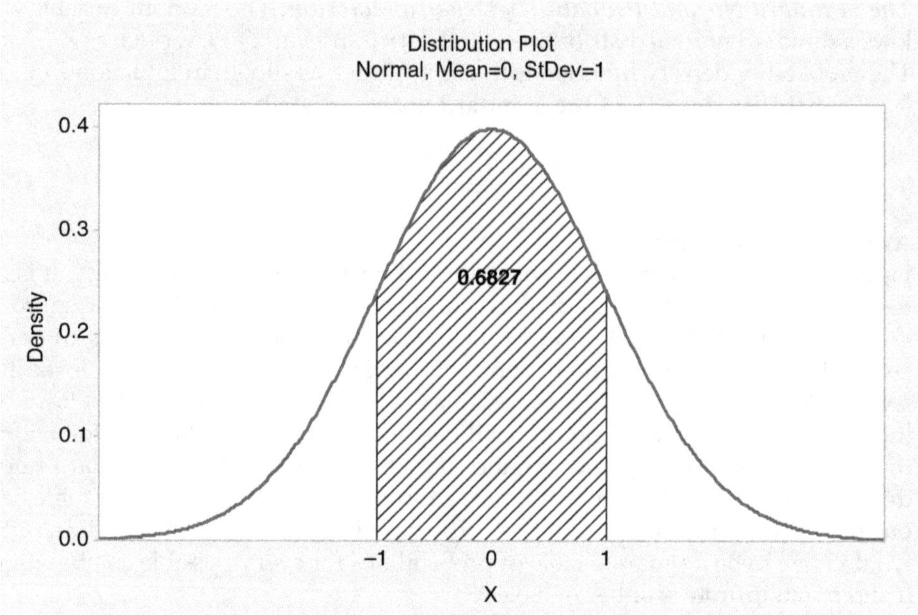

Figure 7.17 Area under the standard normal curve, ±1 standard deviation from the mean.

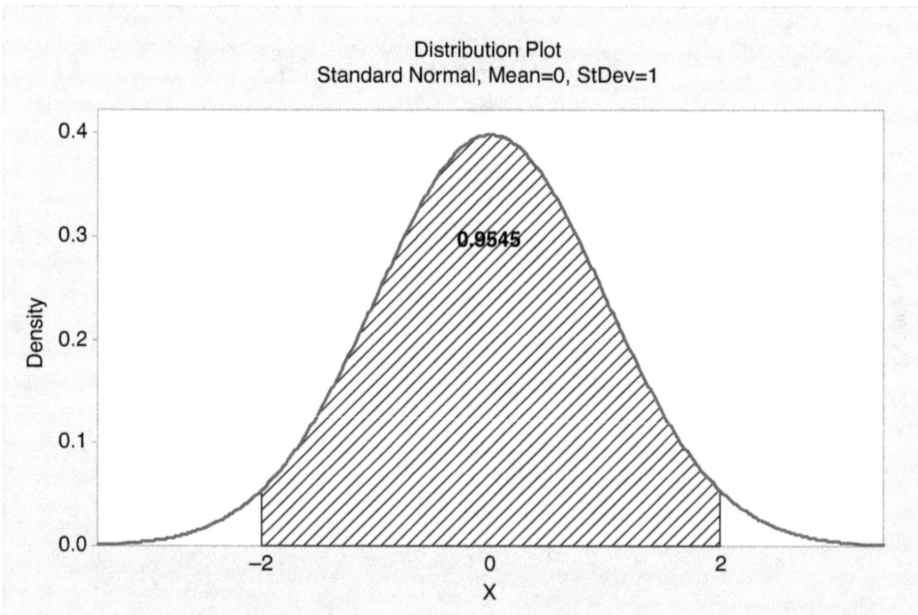

Figure 7.18 Area under the standard normal curve, ±2 standard deviations from the mean.

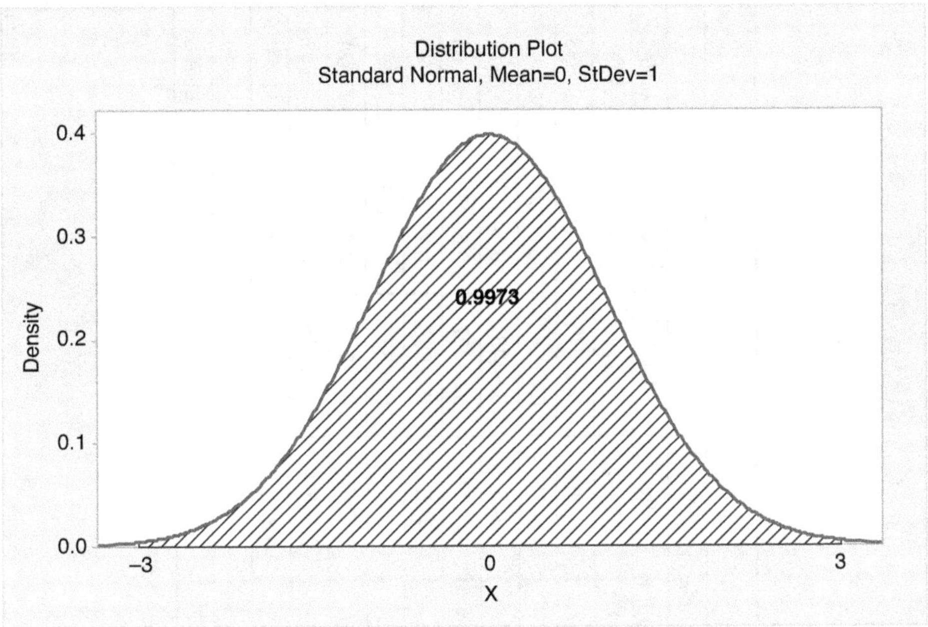

Figure 7.19 Area under the standard normal curve, ±3 standard deviations from the mean.

Using the Cumulative Standard Normal Table

The cumulative standard normal table in Appendix C gives the probability that the random variable Z is equal to or less than some value *a*. Note that for continuous distributions, there is no probability associated with a single point, that is, $\Pr(Z=a)=0$. Therefore, the statements $\Pr(Z \le a)$ and $\Pr(Z < a)$ are equivalent.

Example 7.16

Let $X \sim N(10, 4)$. If we want to know the probability that X is less than the value 12.1, we first convert to the Z scale:

$$Z=\frac{(X-\mu)}{\sigma}=\frac{(12.1-10)}{2}=1.05$$

1. Table 7.5 shows a portion of the cumulative standard normal table. To find $\Pr(Z \le 1.05)$, we look up the value 1.05 and read off the corresponding probability from inside the table. The value $Z=1.05$ is found by using the first column to find 1.0 and then by looking across the top of the table to find the hundredths place 0.05. The corresponding value 0.85314 is the probability that Z is less than or equal to 1.05.

(continued)

Table 7.5 Cumulative probabilities of the standard normal distribution, partial table.

Standard normal Z table values in chart show P(Z<z)										
	0.00	**0.01**	**0.02**	**0.03**	**0.04**	**0.05**	**0.06**	**0.07**	**0.08**	**0.09**
0.00	0.50000	0.50399	0.50798	0.51197	0.51595	0.51994	0.52392	0.52790	0.53188	0.53586
0.10	0.53983	0.54380	0.54776	0.55172	0.55567	0.55962	0.56356	0.56749	0.57142	0.57535
0.20	0.57926	0.58317	0.58706	0.59095	0.59483	0.59871	0.60257	0.60642	0.61026	0.61409
0.30	0.61791	0.62172	0.62552	0.62930	0.63307	0.63683	0.64058	0.64431	0.64803	0.65173
0.40	0.65542	0.65910	0.66276	0.66640	0.67003	0.67364	0.67724	0.68082	0.68439	0.68793
0.50	0.69146	0.69497	0.69847	0.70194	0.70540	0.70884	0.71226	0.71566	0.71904	0.72240
0.60	0.72575	0.72907	0.73237	0.73565	0.73891	0.74215	0.74537	0.74857	0.75175	0.75490
0.70	0.75804	0.76115	0.76424	0.76730	0.77035	0.77337	0.77637	0.77935	0.78230	0.78524
0.80	0.78814	0.79103	0.79389	0.79673	0.79955	0.80234	0.80511	0.80785	0.81057	0.81327
0.90	0.81594	0.81859	0.82121	0.82381	0.82639	0.82894	0.83147	0.83398	0.83646	0.83891
1.00	0.84134	0.84375	0.84614	0.84849	0.85083	0.85314	0.85543	0.85769	0.85993	0.86214
1.10	0.86433	0.86650	0.86864	0.87076	0.87286	0.87493	0.87698	0.87900	0.88100	0.88298

Example 7.17

Finding the Probability That Z Is Greater Than the Value a

If we want the probability that Z is *greater than* some value a, we can use the complement rule. Using the table, we can find the probability that Z is *less than* a, and then subtract the result from one. For example, to find $Pr(Z > 1.05)$, we use the table and the complement to find $1 - Pr(Z \leq 1.05) = (1-0.8531) = 0.1469$, as shown in Figure 7.20.

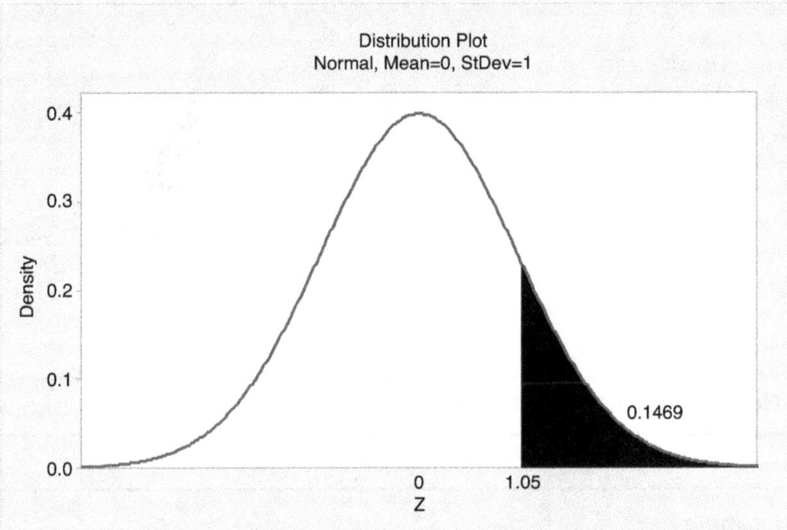

Figure 7.20 $Pr(Z > 1.05)$.

Example 7.18

Finding the Probability That Z Is between Two Values

The probability that Z is within an interval can be solved by first sketching the normal curve and shading the area of interest. The sketch makes it clear what values need to be looked up and what needs to be done with them.

For example, to find $Pr(1.20 < Z < 2.08)$, we can create the sketch found in Figure 7.21.

We see that we will need to find the area between the values 1.20 and 2.08. If we look up 2.08 in. the standard normal table, that will give us the area from 2.08 all the way down to $-\infty$ That area is more than we want. If we subtract off the area that is less than 1.20, we will be left with the shaded-in area between these two numbers, which is what we need to solve the problem.

$$\begin{aligned} Pr(1.20 < Z < 2.08) &= Pr(Z < 2.08) - Pr(Z < 1.20) \\ &= 0.9812 - 0.8849 \\ &= 0.0963 \end{aligned}$$

(continued)

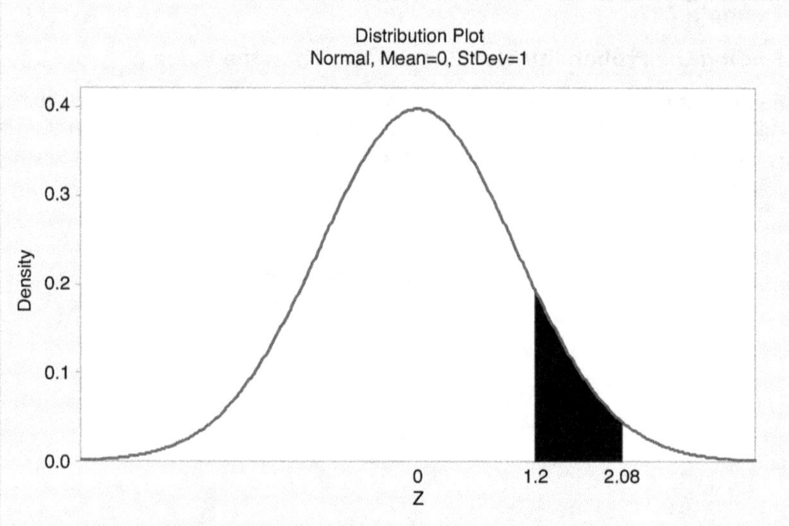

Figure 7.21 Pr(1.20<Z<2.08).

Example 7.19

Finding the Probability That Z Is Less Than a Negative Value

The standard normal distribution is centered on 0, which means half of its values are negative. Most cumulative standard normal tables do not include entries for negative Z values, but we have provided one in this handbook. We can find the probability of Z less than any negative value directly using the table in Appendix C. Once again, sketching is a vital step in understanding what area we are seeking. For example, if we need to find Pr(Z<–1.40), we can create a sketch, as shown in Figure 7.22.

We find Pr(Z<–1.40)=0.0808 from Appendix C.

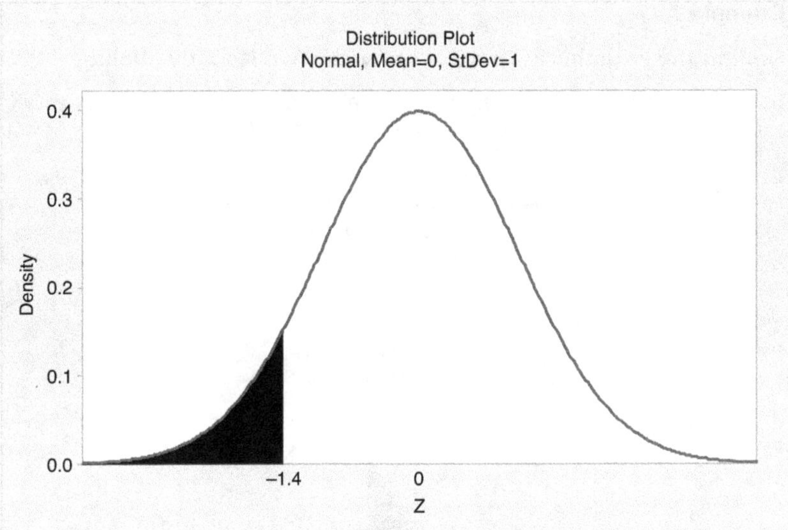

Figure 7.22 Pr(Z<–1.40).

Example 7.20

Finding the Probability That Z Is Greater Than a Negative Value

To find the probability that Z is greater than a negative value, we sketch and use the complement rule. For example, to find Pr(Z>–1.88), we first sketch as in Figure 7.23. Using Appendix C and the complement rule, we have Pr(Z>–1.88) = 1 – Pr(Z<–1.88) = 1–0.0301 = 0.9699.

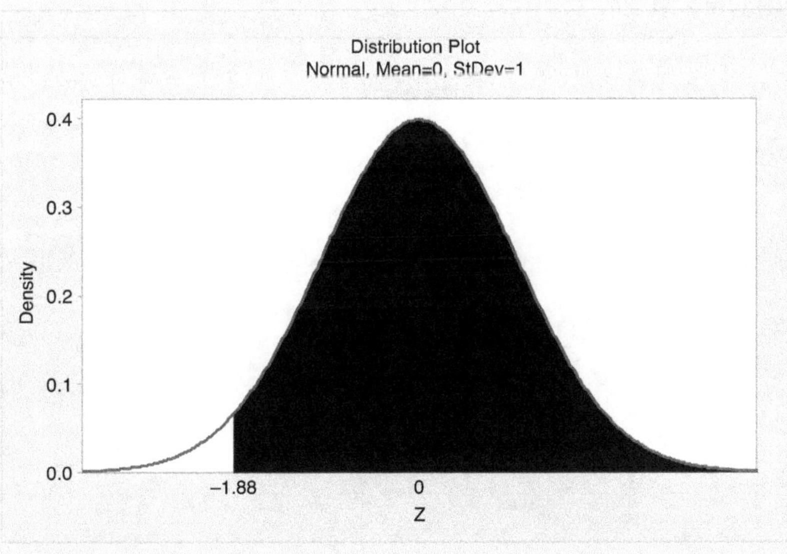

Figure 7.23 Pr(Z>–1.88).

Example 7.21

Finding the Probability That Z Is between Two Negative Values

To find the probability that Z lies in an interval bound by two negative values, we can sketch and use the table in Appendix C. For example, to find Pr(–2.34<Z<–0.42), we use the sketch in Figure 7.24 and write:

$$\begin{aligned} Pr(-2.34 < Z < -0.42) &= Pr(Z < -0.42) - Pr(Z < -2.34) \\ &= 0.4904 - 0.1628 \\ &= 0.3276 \end{aligned}$$

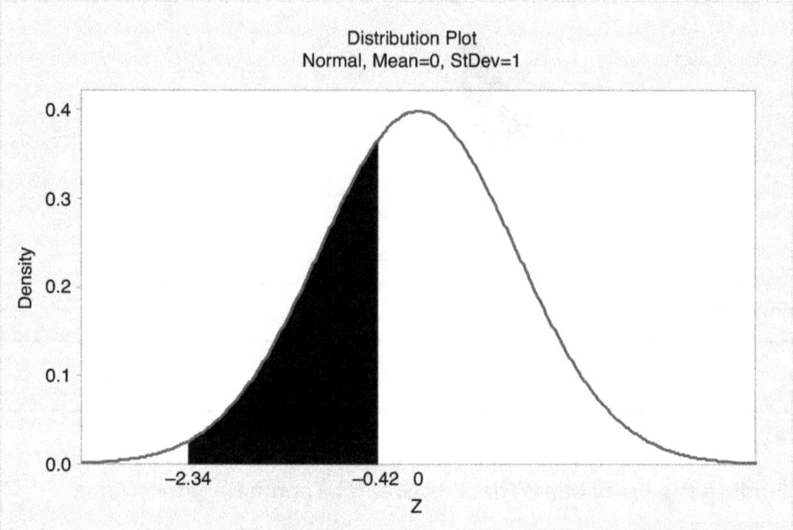

Figure 7.24 Pr(–2.34<Z<–0.42).

Example 7.22

Finding the Probability That Z Is between a Negative and a Positive Value

To find Pr(−1.00 < Z < 1.55), we have the sketch in Figure 7.25:

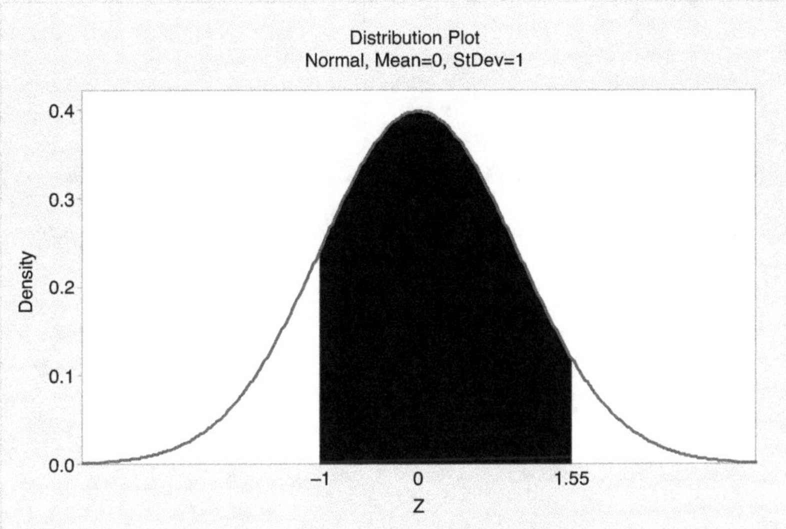

Figure 7.25 Pr(−1.00 < Z < 1.55).

Using Appendix C, we can solve:

$$Pr(Z < 1.55) - Pr(Z < -1.00) = 0.9394 - (0.1587)$$
$$= 0.7807$$

Section II

Metrology

Metrology is defined as the science of precision measurements. It has many applications in science and industry and employs a wide range of measuring instruments. In the competitive manufacture of precision-engineered products, where a high degree of quality is required, it is particularly important that proper measuring instruments be employed to provide accurate, reliable, and cost-effective inspection results. Thus, the selection of appropriate measuring and gauging instruments is becoming more critical in the ever-increasing requirement to stay competitive and meet customer demand for higher-quality products.

Chapter 8

Common Gauges and Measurement Instruments

1. ***Variable gauges.*** Identify and use variable gauges, including micrometers, calipers, dial indicators, and Coordinate Measuring Machines (CMM). Understand linear scales, such as steel rule and gauge blocks. Use borescopes, thermometers, and temperature probes. (Apply)
2. ***Attribute gauges.*** Identify and use attribute gauges, such as thread plugs, progressive gauges, ring gauges, flush pins, pin gauges, and radius gauges. (Apply)
3. ***Transfer gauges.*** Identify and use transfer gauges, including small-hole gauges, telescoping gauges, and spring calipers. (Apply)
4. ***Measurement scales.*** Identify and use measurement scales, such as dial, digital, and vernier scales. (Apply)

Body of Knowledge II.A

INTRODUCTION

Dimensional metrology is concerned with the measurement or gauging of a variety of workpiece characteristics, including length, diameter, thickness, angle, taper, roughness, concentricity, and profile. Different sensing technologies may be employed to measure or gauge those characteristics, depending on the requirements for accuracy and other considerations. There are basically five different technologies that may be used individually or in combination to perform these inspection functions:

1. *Mechanical.* Small displacements are amplified by a mechanical system.
2. *Electronic.* Utilize an electric or electronic phenomenon such as electrical resistance.
3. *Air or pneumatic.* Small variations made in the dimension are measured with respect to a reference dimension and are shown by a variation in air pressure or the velocity of airflow.

4. *Light waves.* Utilizing the phenomenon of the interference of light waves to provide a standard. Such a standard is the wavelength of a monochromatic light, expressed in terms of the meter.
5. *Electron beam.* Stabilized lasers are used as working standards for dimensional measurements, providing a precise and stable frequency for the standard.

In general, the mechanical and electronic types of measuring and gauging instruments have sensing devices or probes that come into contact with the workpiece, and are referred to as *contact instruments.* Air instruments, while employing contacting elements, rely on air pressure difference to effect measurement. Thus, they are basically *noncontact instruments.* Although different technologies are involved in the light-wave and electron-beam instruments, they both utilize a variety of optical systems. Thus, they are often grouped together as *optical noncontact instruments.*

VARIABLE GAUGES

Most of the basic or general-purpose linear measuring instruments are typified by the steel rule, the vernier caliper, or the micrometer caliper.

Steel Rules

Steel rules are used effectively as line measuring devices, which means that the ends of a dimension being measured are aligned with the graduations of the scale, from which the length is read directly. A depth rule for measuring the depth of slots, holes, and so on, is a type of steel rule. Steel rules are also incorporated into vernier calipers, where they are adapted to end-measuring operations. These are often more accurate and easier to apply than in-line measuring devices.

Verniers

The *vernier caliper* shown in Figure 8.1 typifies the type of instrument using the vernier principle of measurement. The main or beam scale on a typical metric vernier caliper is numbered in increments of 10 mm, with the smallest scale division equivalent to 1 mm. The vernier scale slides along the edge of the main scale and is separated into 50 divisions, and these 50 divisions are the same in total length as the 49 divisions on the main scale. Each division on the vernier scale is equal to 1/50 of (49×1), or 0.98 mm, which is 0.02 mm less than each division on the main scale.

Aligning the zero lines on both scales would cause the first lines on each scale to be 0.02 mm apart, the second lines 0.04 mm apart, and so on. A measurement on a vernier is designated by the positions of its zero line and the line that coincides with a line on the main scale. For example, the metric scale in Figure 8.1a shows a reading of 12.42 mm. The zero index of the vernier is located just beyond the line at 12 mm on the main scale, and line 21 (after 0) coincides with a line on the main scale, indicating the zero index is 0.42 mm beyond the line at 12 mm. Thus, $12.00 + 0.42 = 12.42$ mm.

The vernier caliper illustrated in Figure 8.1 also has an inch scale so that it can be used interchangeably for either inch or millimeter measurements. The smallest division on the main scale represents 0.25 in. and the vernier is divided into

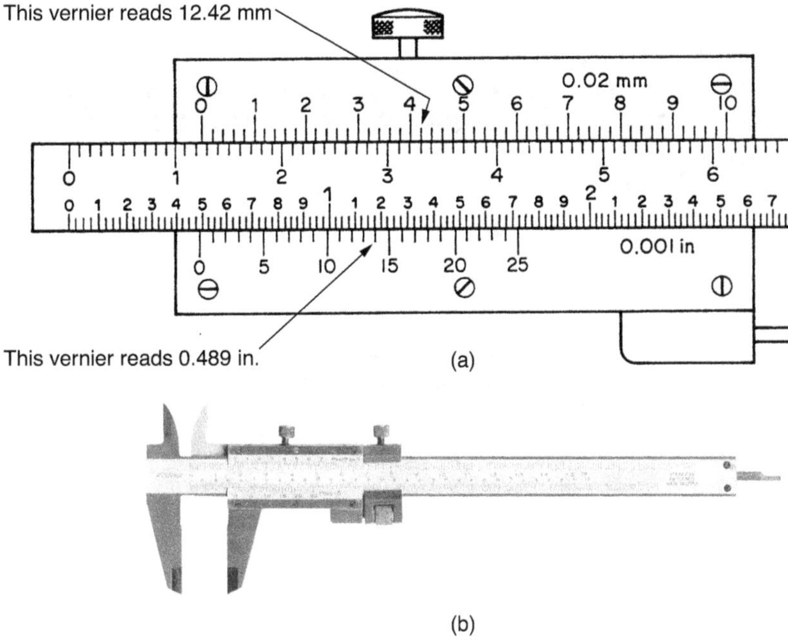

Figure 8.1 Fine-adjustment style vernier caliper.
Source: Fred V. Fowler Company.

0.001 in. increments. Thus, the measurement illustrated is 0.475 in. away from the main scale plus 0.014 in. from the vernier scale, for a total of 0.489 in.

The vernier caliper shown in Figure 8.1b consists of a steel rule with a pair of fixed jaws at one end and a pair of sliding jaws affixed to a vernier. Outside dimensions are measured between the lower jaws, and inside dimensions are measured over the tips of the upper jaws.

The *digital reading caliper* shown in Figure 8.2 provides LCD readouts in either millimeters or inches and operates on a microprocessor-based system. The caliper has a measuring range of 0–152 mm (0–6 in.) with readings in increments of 0.013 mm (0.0005 in.). The unit is capable of retaining a reading in the display when the tool is used in an area where visibility is restricted.

The *vernier height gauge* is similar to a vernier caliper except the fixed jaw has been replaced by a fixed base, and the sliding jaw may have a scriber attached to it for layout work, or a dial indicator for measuring or comparing operations. A more sophisticated version of the vernier height gauge is shown in Figure 8.3. This instrument can easily measure in two dimensions, in either angular or polar coordinates. It can measure external, internal, and distance dimensions, as well as perpendicularity, flatness, straightness, centers, and diameters.

Vertical measurements are made on the gauge shown in Figure 8.3 in. either metric or inch units to a resolution of 0.0005 mm (0.00002 in.) by an optoelectronic sensor moving over a high-accuracy glass scale. The gauge head moves on a cushion of air generated by a completely self-contained pneumatic system. Dedicated function

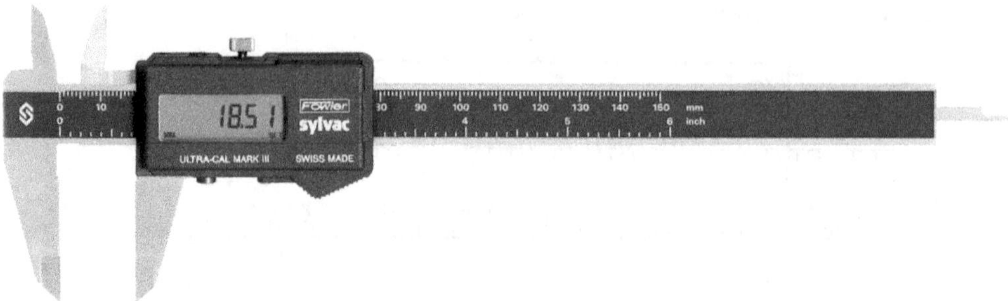

Figure 8.2 LCD digital-reading caliper with 0–152 mm (0–6 in) range.
Source: Fred V. Fowler Company.

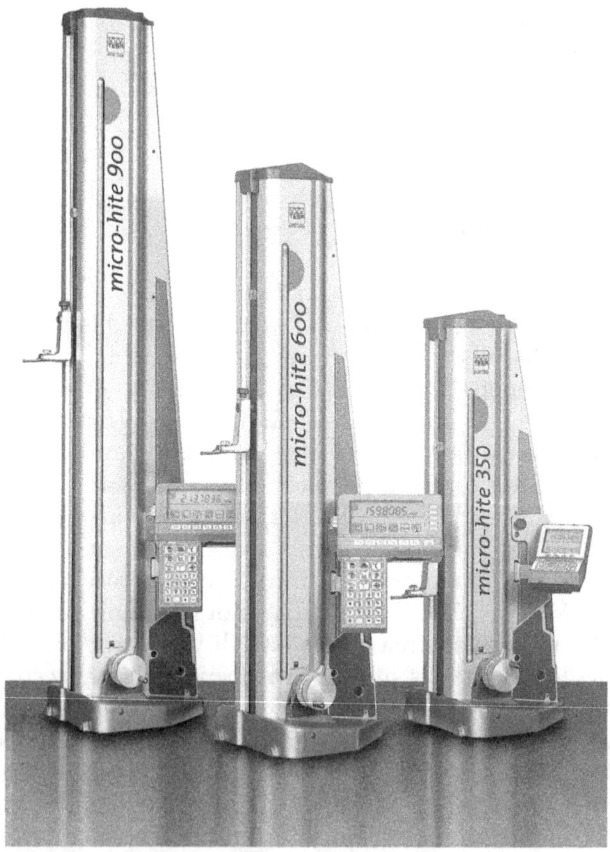

Figure 8.3 Digital-reading, single-axis height gauge for two-dimensional measurements.
Source: Brown and Sharpe Manufacturing Company.

programs, along with a keypad and interactive LCD display, are designed to guide operators smoothly and efficiently through a variety of measurement operations. The gauge can be used for direct data transfer to other data collection devices.

Micrometers

The *micrometer caliper* shown in Figure 8.4 is representative of the type of instrument using a precision screw as a basis for measuring. The measuring elements consist of a fixed anvil and a spindle that moves lengthwise as it is turned.

The thread on the spindle of a typical metric micrometer has a lead of 1/2 or 0.5 mm, so one complete revolution of the thimble produces a spindle movement of this amount. The graduated scale on the sleeve of the instrument has major divisions of 1.0 mm and minor divisions of 0.5 mm. Thus, one revolution of the spindle causes the beveled edge of the thimble to move through one small division on the sleeve scale. The periphery of the beveled edge of the thimble is graduated into 50 equal divisions, each space representing 1/50 of a complete rotation of the thimble, or a 0.01 mm movement of the spindle. Micrometers with scales in inch units operate in a similar fashion. Typically, the spindle thread has a lead of 0.025 in., and the smallest division on the sleeve represents 0.025 in. The periphery of the beveled edge of the thimble is graduated into 25 equal divisions, each space representing 1/25 of a complete rotation of the thimble or a spindle movement of 0.001 in.

A reading on a micrometer is made by adding the thimble division that is aligned with the longitudinal sleeve line to the largest reading exposed on the sleeve scale. For example, in Figure 8.5 the thimble has exposed the number 10, representing 10.00 mm, and one small division worth 0.50 mm. The thimble division 16 is aligned with the longitudinal sleeve line, indicating that the thimble has moved 0.16 mm beyond the last small division on the sleeve. Thus, the final reading is obtained by summing the three components: 10.00 + 0.50 + 0.16 = 10.66 mm.

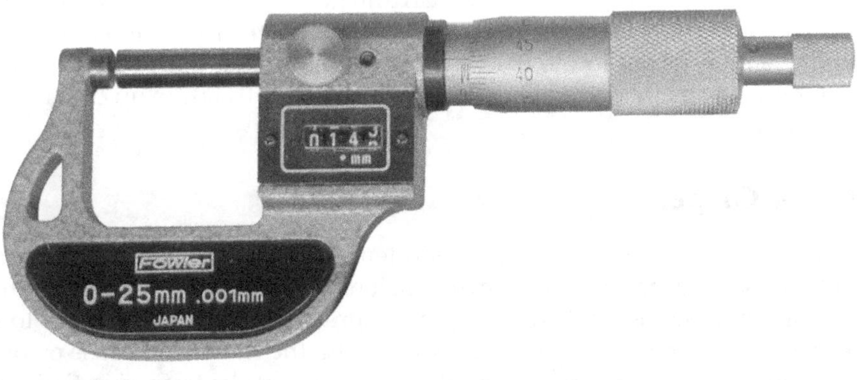

Figure 8.4 A 0–25 mm micrometer caliper.
Source: Fred V. Fowler Company.

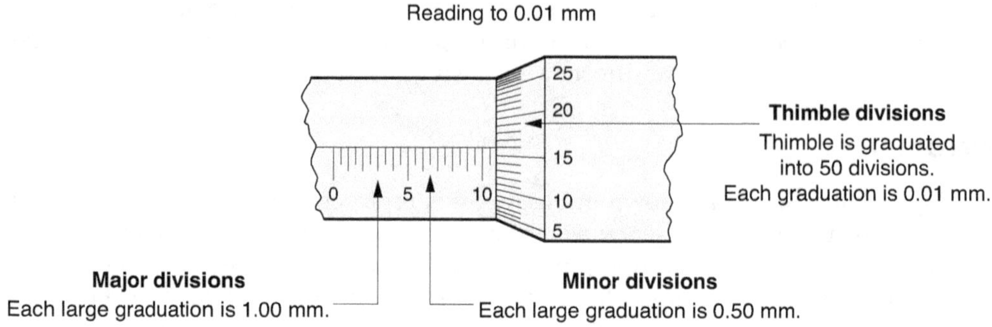

Figure 8.5 Micrometer reading of 10.66 mm.
Source: Reprinted by permission from Society of Manufacturing Engineers, *Manufacturing Processes and Materials.* 4th ed. 2000.

Vernier Scales

A *vernier micrometer caliper,* such as that represented by the scales shown in Figure 8.6, has a vernier scale on the sleeve, permitting measurement to 0.001 mm. The vernier scale shown has 10 divisions over a length equivalent to 19 divisions around the periphery of the thimble. Thus, the difference in length of a division on the vernier scale and two divisions on the thimble is $0.02-(1/10)(19\times 0.01)=0.001$ mm. Thus, the reading illustrated in Figure 8.6 is $10.00+0.50+0.16+0.006=10.666$ mm.

Digital Micrometers

Micrometers with digital readouts make readings faster and easier for inspection personnel regardless of their degree of experience. The digital micrometer shown in Figure 8.4 represents one instrument of this type for use in measuring to a resolution of 0.001 mm. The instrument shown in Figure 8.7 has a digital readout with a resolution to 0.0001 in. When equipped with vernier scales, the resolution may be increased to 0.001 mm (commonly 0.0001 in. in case of an inch-reading device).

Micrometer Calipers

The *micrometer caliper,* or *mike* as it is often called, is an end-measuring instrument for use in measuring outside dimensions. Although the mike is fairly easy to apply, its accuracy depends on the application of the proper amount of torque to the thimble. Too much torque is likely to spring the frame and cause error. Thus, it is important that personnel using these instruments be trained in their use and be required to periodically check their measurements against a standard to minimize errors. The indicating micrometer shown in Figure 8.8 has a built-in dial indicator to provide a positive indication of the measuring pressure applied. The instrument can be used like an indicating snap gauge.

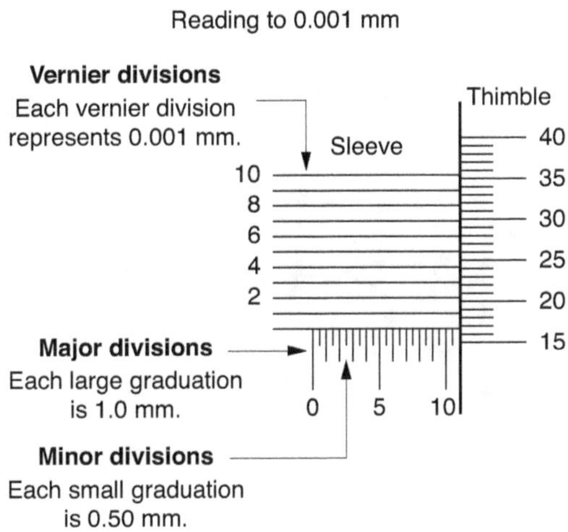

Figure 8.6 Scales of a vernier micrometer showing a reading of 10.666 mm.
Source: Reprinted by permission from Society of Manufacturing Engineers, *Manufacturing Processes and Materials.* 4th ed. 2000.

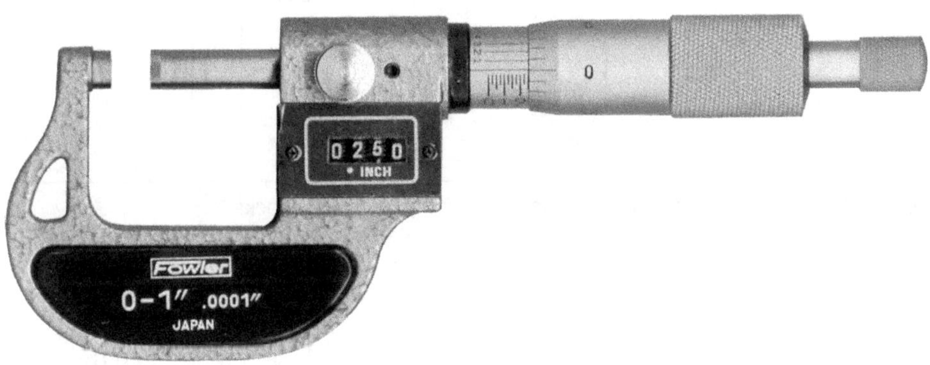

Figure 8.7 A digital micrometer.
Source: Fred V. Fowler Company.

Gauge Blocks

Gauge blocks are dimensional measurement standards used to calibrate other measuring devices. They come in sets of different grades depending on the desired measurement accuracy required. Each set has many blocks of incremental lengths. These blocks are stacked together (wringed) to build a desired length.

The process of building a desired length of a stack of gauge blocks is *wringing*. After the reference block is built, the blocks are said to be "wrung" together. This process, as shown in the schematic in Figure 8.9, is as follows:

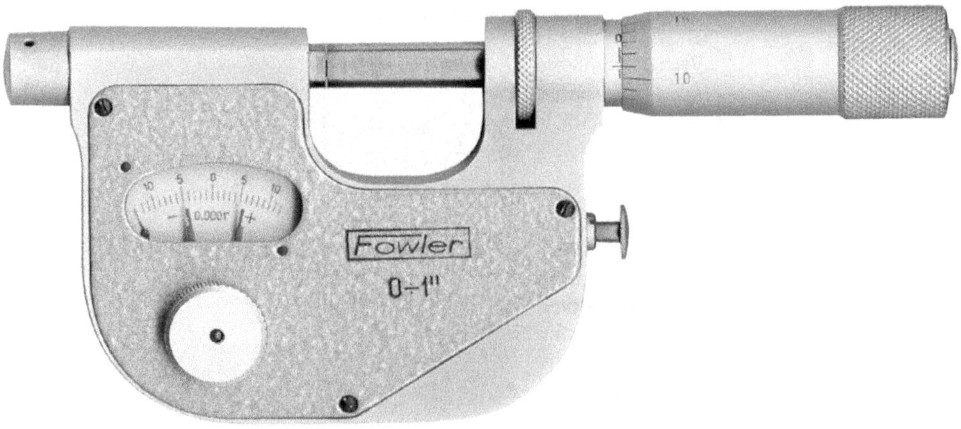

Figure 8.8 An indicating micrometer.
Source: Fred V. Fowler Company.

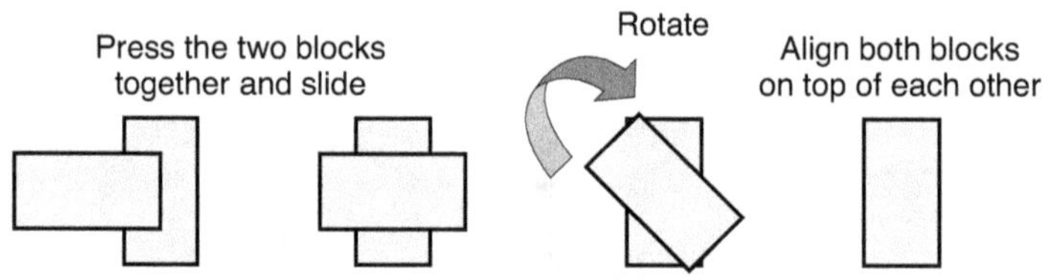

Figure 8.9 A schematic showing the process of wringing gauge blocks.

- The blocks should be clean.
- Lubricate the surfaces of the two blocks lightly and then clean them, creating a thin film of oil that allows one block to slide over the other and allows the blocks to be separated after the measurement.
- Slide the surfaces of the blocks together as shown. Apply pressure while sliding the blocks. The blocks should strongly adhere to each other after being rotated into place.

Dial Indicators

Dial indicators are used for many kinds of measuring and gauging operations. They also serve to check machines and tools, alignments, and cutter runout. Dial indicators are often incorporated into special gauges in measuring instruments.

Borescopes

A *borescope* is an optical device (such as a prism or optical flat) used to inspect an inaccessible space (such as an engine cylinder) (*Merriam-Webster's Dictionary*). Borescopes are used for inspecting the locations and dimensions of the parts that are difficult and/or costly to access.

Thermometers and Temperature Probes

A *thermometer* is an instrument for measuring temperature, either digital or analog, produced in many shapes to accommodate different industrial, metrological, medical, or other applications. *Temperature probes* are metallic or plastic sensors conventionally used to measure temperatures when running experiments in metals, meat, and other substances. Applications vary in biology, chemistry, material science, and other medical fields.

Coordinate Measuring Machines

The *coordinate measuring machine* (CMM) is a flexible measuring device capable of providing highly accurate dimensional position information along three mutually perpendicular axes. This instrument is widely used in manufacturing industries for the post-process inspection of a large variety of products and their components. It is also very effectively used to check dimensions on a variety of process tooling, including mold cavities, die assemblies, assembly fixtures, and other work-holding or tool-positioning devices.

Over the last decade, CMMs have become a primary means of dimensional quality control for manufactured parts of complex form where the volume of production does not warrant the development of functional gauging. The advent of increasingly inexpensive computing power and more fully integrated manufacturing systems will continue to expand the use of these machines into an even larger role in the overall quality assurance of manufactured parts.

CMMs can most easily be defined as physical representations of a three-dimensional rectilinear coordinate system. They now represent a significant fraction of the measuring equipment used for defining the geometry of different-shaped workpieces. Most dimensional characteristics of many parts can be measured within minutes with these machines. Similar measurements would take hours using older measuring equipment and procedures. CMMs have several advantages besides flexibility and speed:

1. Different features of a part can be measured in one setup. This eliminates errors introduced due to setup changes.
2. All CMM measurements are taken from one geometrically fixed measuring system, eliminating the accumulation of errors resulting from using functional gauging and transfer techniques.
3. The use of digital readouts eliminates the necessity for the interpretation of readings, such as with the dial or vernier-type measuring scales.

4. Most CMMs have automatic data recording, which minimizes operator influence.
5. Part alignment and setup procedures are greatly simplified by using software supplied with computer-assisted CMMs. This minimizes the setup time for measurement.
6. Data can be automatically saved for further analysis.

Coordinate Measuring Machine Classification. Although CMMs can be thought of as representations of a simple rectilinear coordinate system for measuring the dimensions of different-shaped workpieces, they naturally are constructed in many different configurations, all of which offer different advantages. CMMs provide a means for locating and recording the coordinate location of points in their measuring volumes. Traditional CMMs are classified according to their configurations, as follows and shown in Figure 8.10 (ASME 1985):

1. *Cantilever configuration,* in which the probe is attached to a vertical machine ram (z-axis) moving on a mutually perpendicular overhang beam (y-axis) that moves along a mutually perpendicular rail (x-axis). Cantilever configuration is limited to small- and medium-sized machines. It provides for easy operator access and the possibility of measuring parts longer than the machine table.
2. *Bridge-type configuration,* in which a horizontal beam moves along the x-axis, carrying the carriage that provides the y motion. In other configurations, the horizontal beam (bridge structure) is rigidly attached to the machine base, and the machine table moves along the x-axis. This is called *fixed bridge* configuration. A bridge-type coordinate measuring machine provides more rigid construction, which in turn provides better accuracy. The presence of the bridge on the machine table makes it more difficult to load large parts.
3. *Column-type configuration,* in which a moving table and saddle arrangement provides the x and y motions and the machine ram (z-axis) moves vertically relative to the machine table.
4. *Horizontal-arm configuration* features a horizontal probe ram (z-axis) moving horizontally relative to a column (y-axis), which moves in a mutually perpendicular motion (x-axis) along the machine base. This configuration provides the possibility for measuring large parts. Other arrangements of the horizontal-arm configuration feature a fixed horizontal-arm configuration in which the probe is attached and moving vertically (y-axis) relative to a column that slides along the machine base in the x direction. The machine table moves in a mutually perpendicular motion (z-axis) relative to the column.
5. *Gantry-type configuration* comprises a vertical ram (z-axis) moving vertically relative to a horizontal beam (x-axis), which in turn moves along two rails (y-axis) mounted on the floor. This configuration provides easy access and allows the measurement of large components.

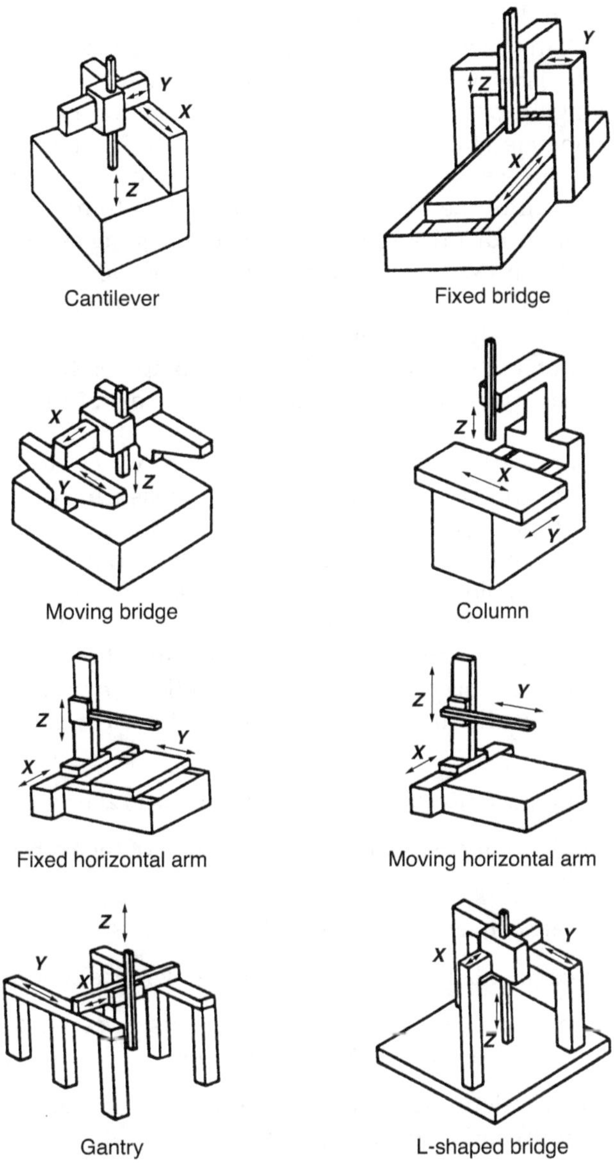

Figure 8.10 CMM classifications.

6. *L-shaped bridge configuration* comprises a ram (z-axis) moving vertically relative to a carriage (x-axis), which moves horizontally relative to an L-shaped bridge moving in the y direction.

In addition physical configuration, CMMs can also be classified according to their mode of operation: manually oriented, computer-assisted, or direct computer-controlled. In manual machines, the operator moves the probe along the machine's axes to establish and manually record the measurement values provided by digital readouts (DROs). In some machines, digital printout devices are used.

Computer-assisted CMMs can be either manually positioned (free-floating mode) by moving the probe to measurement locations, or manually driven by providing power-operated motions under the control of the operator. In either case, a computer accomplishes the data processing. Some computer-assisted CMMs can perform some or all of the following functions: inch to metric conversion, automatic compensation for misalignment, storing of premeasured parameters and measurement sequences, data recording, means for disengagement of the power drive to allow manual adjustments and manipulations of the machine motions, and geometric and analytical evaluations.

Direct computer-controlled CMMs use a computer to control all machine motions and measuring routines and to perform most of the routinely required data processing.

These machines are operated in much the same way as Computer Numerical Control (CNC) machine tools. Both control and measuring cycles are under program control. Offline programming capability is also available.

The effective use of computers for CMM applications is a principal feature differentiating available CMM systems. The value of a measurement system depends a great deal on the sophistication and ease of use of the associated software and its functional capabilities. The functional capabilities of a CMM software package depend on the number and types of application programs available. The following is a list of the many different types of system software available for CMMs:

1. Printout instructions, measurement sequence, zero reference, and so on
2. Automatic compensation for misalignment of the workpiece with the machine axes
3. Coordinate conversion between Cartesian and polar coordinates
4. Tolerance calculations providing out-of-tolerance conditions
5. Defining geometric elements such as points, lines, circles, planes, cylinders, spheres, cones, and their intersections
6. Automatic redefinition of coordinate systems or machine axes, and printout of origin and inspection planes
7. Inspection of special shapes or contours, such as gears and cams
8. Multiple-point hole checking using least squares techniques for determining best fit center, mean diameter, roundness, and concentricity
9. Evaluating geometric tolerance conditions by defining type of form and positional relationship, such as roundness, flatness, straightness, parallelism, or squareness.
10. Hold diameter and location checking considering maximum and minimum material conditions as defined in ANSI Y14.5.
11. Friendly operator interfaces for self-teaching or part programs

12. Other software for statistical analysis include graphic data display, histograms, integration of areas under a curve, contour plotting, automatic part or lot acceptance or rejection based on statistical evaluation, and so on

ATTRIBUTE GAUGES

In some operations, it is uneconomical to attempt to obtain absolute sizes during each inspection operation. In many cases, it is only necessary to determine whether one or more dimensions of a mass-produced part are within specified limits. For this purpose, a variety of inspection instruments referred to as *gauges* are employed. However, the distinction between gauging devices and measuring devices is not always clear, as there are some instruments referred to as gauges that do not give definite measurements.

To promote consistency in manufacturing and inspection, gauges may be classified as working, inspection, or reference or master gauges.

Working gauges are used by the machine operator or shop inspector to check the dimensions of parts as they are being produced. They usually have limits based on the piece being inspected.

Inspection gauges are used by personnel to inspect purchased parts when received or manufactured parts when finished. These gauges are designed and made so as not to reject any product previously accepted by a properly designed and functioning working gauge.

Reference, or *master, gauges* are used only for checking the size or condition of other gauges, and they represent as exactly as possible the physical dimensions of the product.

A gauge may have a single size and be referred to as a *nonlimit gauge*, or it may have two sizes and be referred to as a *limit gauge*. A limit gauge, often called a *go/no-go gauge*, establishes the high and low limits prescribed by the tolerance on a dimension. A limit gauge may be either double-end or progressive. A *double-end gauge* has the "go" member at one end and the "no-go" member at the other. Each end of the gauge is applied to the workpiece to determine its acceptability. The go member must pass into or over an acceptable piece, but the no-go member should not. A *progressive gauge* has both the go and the no-go members at the same end so that a part may be gauged with one movement.

Some gauges are fixed in size, while others are adjustable over certain size ranges. *Fixed gauges* are usually less expensive initially, but they have the disadvantage of not permitting adjustment to compensate for wear.

Most gauges are subjected to considerable abrasion during their application and therefore must be made of materials resistant to wear. High-carbon and alloy steels have been used as gauge materials for many years because of their relatively high hardenability and abrasion resistance. Further increased surface hardness and abrasion resistance may be obtained through the use of chrome plating or cemented carbides as surface material on gauges. Some gauges are made entirely of cemented carbides, or they have cemented carbide inserts at certain wear points. Chrome plating is also used as a means of rebuilding and salvaging worn gauges.

Common Gauges

Typical common functional gauges can be classified on the basis of whether they are used to check outside dimensions, inside dimensions, or special features. Some examples of typical gauges are shown in Figure 8.11. They include:

- Ring and snap gauges for checking outside dimensions
- Plug gauges for checking inside dimensions
- Other gauges for checking other geometrical shapes, such as tapers, threads, and splines

Typical *plug gauges*, such as the ones shown in the top view of Figure 8.11a, consist of a hardened and accurately ground steel pin with two gauge members: the go gauge member and the no-go gauge member. *Progressive plug gauges* (bottom view of Figure 8.11a) combine both go and no-go members into one. The design of the gauge member and the method used to attach it to the handle depend on its size, as shown in Figure 8.11b. The gauge members are usually held in the handle by a threaded collet and bushing (view 1); a taper lock, where gauge members have a taper shank on one end that fits into the end of the handle (view 2); or a trilock, where the gauge members have a hole drilled through the center and are counterbored on both ends to receive a standard socket-head screw (view 3). One way of checking a hole for out-of-roundness is to have flats ground on the side of the gauge member, as shown in Figure 8.11c.

Ring Gauges. *Ring gauges*, such as those shown in Figure 8.11d, are used for checking the limit sizes of a round shaft. They are generally used in pairs: the go gauge for checking the upper limit of the part tolerance and the no-go gauge for checking the lower limit. The no-go ring has a groove in the outside diameter of the gauge to distinguish it from the go ring. It is possible that a shaft's ends are larger than the middle, or it could suffer an out-of-roundness condition. This situation cannot be detected with a standard cylindrical ring gauge. Such an out-of-roundness condition can be checked by a ring gauge that has the inside diameter relaxed, such as the one shown in Figure 8.11e.

Snap Gauges. A *snap gauge* is another fixed gauge with the gauging members specially arranged for measuring diameters, thickness, and lengths. A typical (may also be called *adjustable*) *external measuring snap gauge* is shown in Figure 8.11f. It consists of a C-frame with gauging members in the jaw of the frame. Figure 8.11g shows other types of snap gauges. Threads can be checked with thread plug gauges, thread ring gauges, thread snap gauges, or a screw thread micrometer. *Thread snap gauges* have two pairs of gauging elements combined in one gauge. With appropriate gauging elements, these gauges may be used to check the maximum and minimum material limits of external screw threads in one pass.

An example of a thread snap gauge is shown in Figure 8.11h. In some cases, special snap gauges may be desired. The example in Figure 8.11i illustrates the use of a special double-end snap gauge for inspecting the outside diameter of a narrow groove.

Spline Gauges. A *spline gauge* is often used to inspect splined workpieces prior to assembly. External splines are checked with internal-toothed rings, whereas internal splines are checked with external-toothed plugs. Figure 8.11j shows the

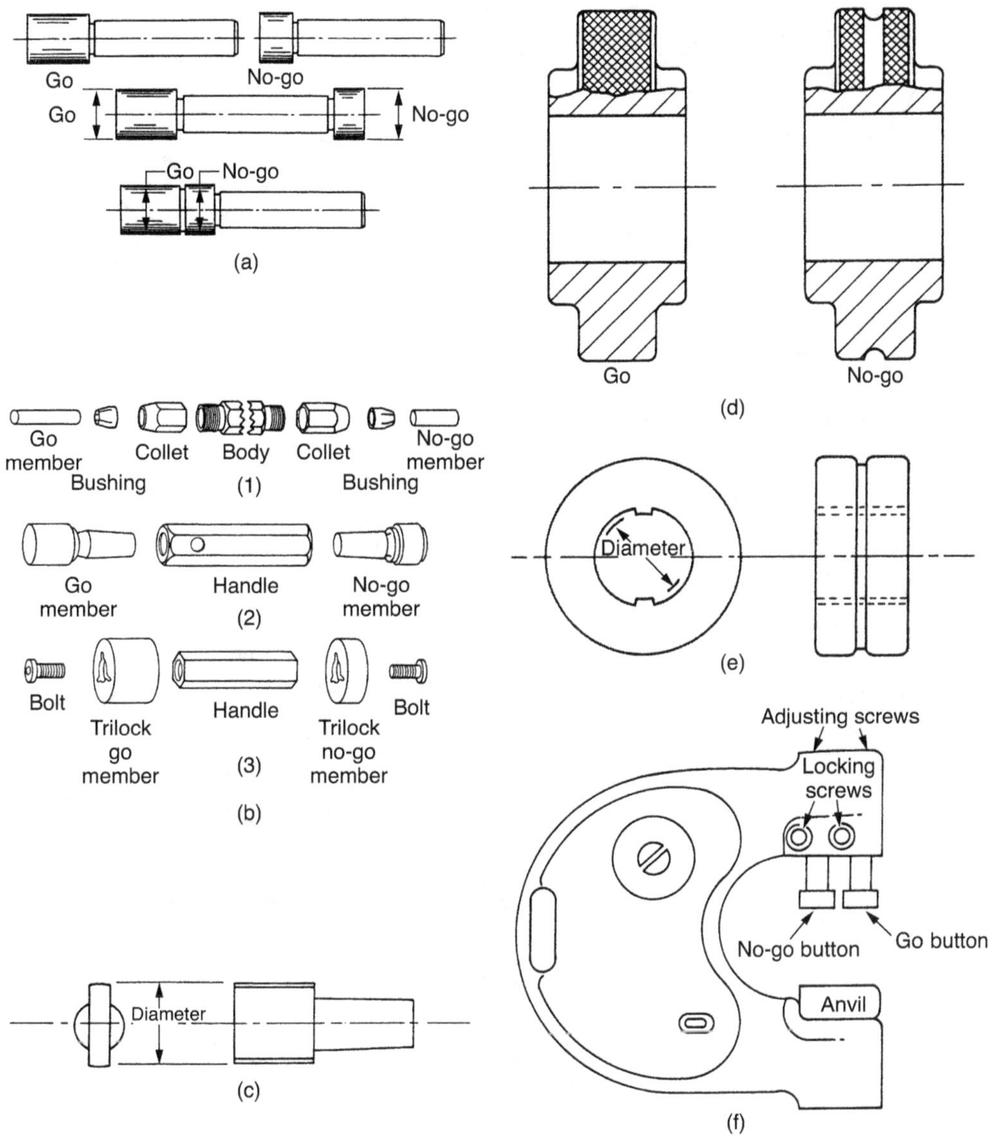

Figure 8.11 Examples of typical gauges.
Source: Reprinted by permission from Society of Manufacturing Engineers, *Manufacturing Processes and Materials.* 4th ed. 2000.

two basic types of fixed-limit spline gauges: composite and sector gauges. *Composite gauges* have the same number of teeth as that of the part. *Sector gauges* have only two sectors of teeth 180° apart. These gauges are further subdivided into go and no-go gauges. View 1 of Figure 8.11j shows a go composite ring gauge; view 2 shows a no-go sector ring gauge.

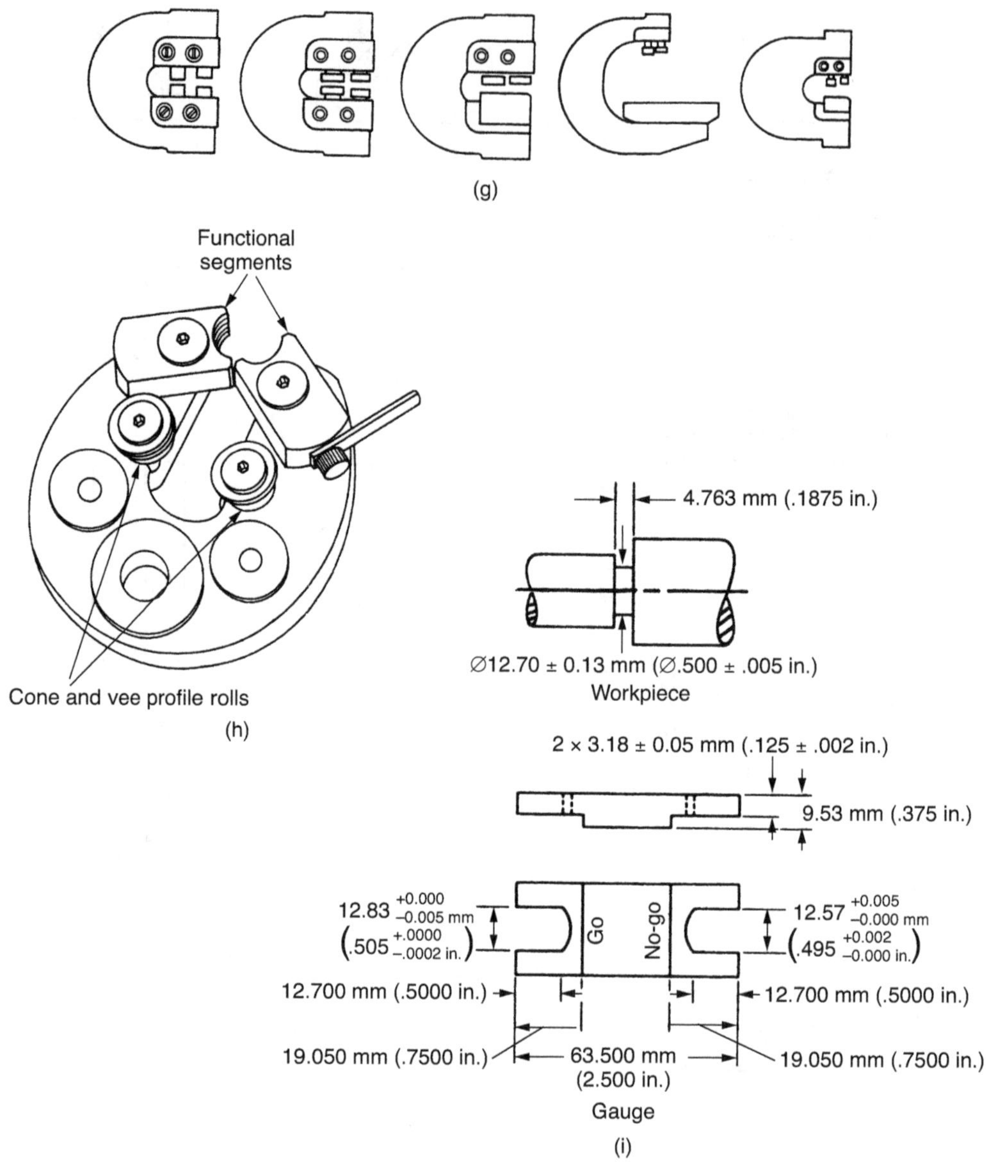

Figure 8.11 *(continued)*

Screw Thread Micrometer. A *screw thread micrometer,* such as the one shown in Figure 8.11k, has a specially designed spindle and anvil so that externally threaded parts can be measured. Screw thread micrometers are generally designed to measure threads within a narrow range of pitches. *Thread plug gauges* are similar in design to cylindrical plug gauges except that they are threaded. They are designed to check internal threads. Typical thread plug gauges, such as those shown in Figure 8.11l, consist of a handle and one or two thread gauge

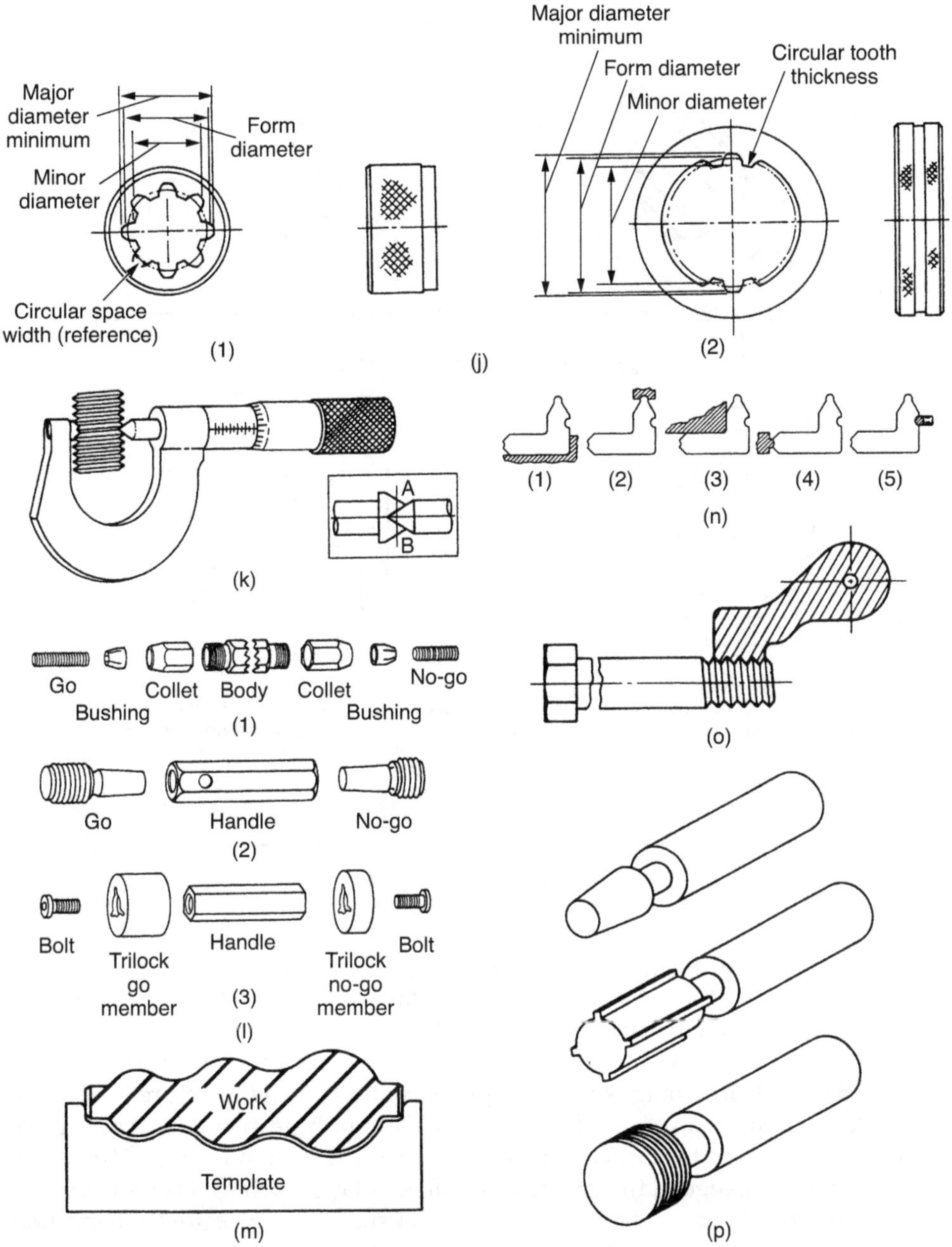

Figure 8.11 *(continued)*

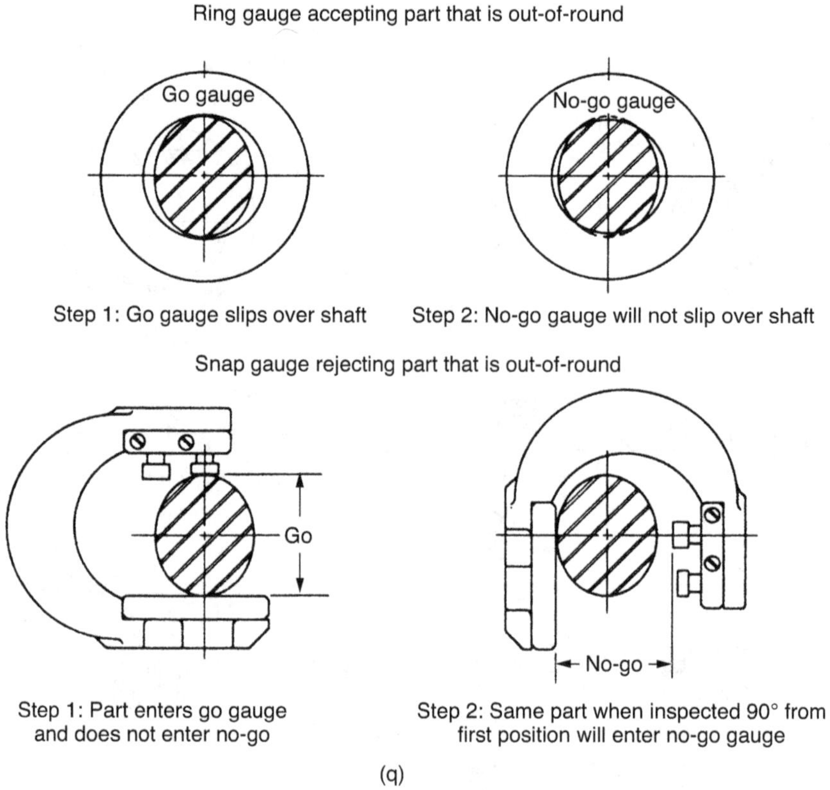

Figure 8.11 *(continued)*

members. Depending on the size of the gauging member, the member can be held in the handle using a threaded collet and bushing design (view 1), a taper lock design (view 2), or a trilock design (view 3).

Templates. To check a specified profile, *templates* may be used. They may also be used to control or gauge special shapes or contours in manufactured parts. These templates are normally made from thin, easy-to-machine materials. An example of a contour template for inspecting a turned part is shown in Figure 8.11m. To visually inspect or gauge radii or fillets, special templates, such as those shown in Figure 8.11n, may be used. The five basic uses of such templates are (1) inspection of an inside radius tangent of two perpendicular planes, (2) inspection of a groove, (3) inspection of an outside radius tangent to two perpendicular planes, (4) inspection of a ridge segment, and (5) inspection of roundness and diameter of a shaft.

Screw Pitch Gauges. The pitch of a screw may be checked with a *screw pitch gauge*. To determine the pitch, the gauge is placed on the threaded part, as shown in Figure 8.11o. A drawback of using screw pitch gauges is their inability to give an adequate check on thread form for precision parts.

Special Gauges. It is sometimes necessary to design special gauges for checking special part features such as square, hexagonal, or octagonal holes. Figure 8.11p shows some special plug gauges for checking the profile or taper of holes. As an inspection tool, a snap gauge is sometimes a better choice than a ring gauge. Figure 8.11q illustrates how a ring gauge may accept an out-of-roundness condition that would otherwise be rejected by a snap gauge. *Flush pin gauges* check the limits of dimensions between surfaces in a specified position.

TRANSFER GAUGES

Transfer-type linear measuring devices are typified by the spring caliper, spring divider, firm joint caliper, telescoping gauge, and small-hole gauge.

The outside caliper is used as an end measure to measure or compare outside dimensions, while the inside caliper is used for inside diameters, slot and groove widths, and other internal dimensions. They are quite versatile but, due to their construction and method of application, their accuracy is somewhat limited.

Telescope gauges are T-shaped instruments. They consist of a pair of telescoping tubes connected to a handle. The tubes are called *legs* or *plungers*. They are closed at their outer ends and spring-loaded to push them apart. The knob on the end of the handle provides careful adjustment of a locking action. The spacing of the outer end duplicates the distance between the reference and measured points of the part features. Telescope gauges are made in sizes to measure holes from 1/2 to 6 inches.

MEASUREMENT SCALES

Measuring instruments may use analog or digital displays. Some instruments use both. The micrometer shown in Figure 8.4 has an analog display, where measurement can be defined on a continuous display of values. The micrometer shown in Figure 8.8, however, uses a digital display, which makes the reading much easier. Depending on the application, measurement scales can be either graduated or nongraduated. Examples of graduated scales are those used on rules, vernier calipers, micrometers, dial indicators, and so on. Some tools, such as surface plates, the sine bar, calipers, straightedges, and so on, use nongraduated scales.

Chapter 9

Special Gauges and Applications

> Identify and describe the following tools and components. (Remember)
>
> 1. ***Electronic gauging tools:*** oscilloscopes and multimeters
> 2. ***Automatic gauging components:*** machine vision, ultrasonic, X-ray, and laser
> 3. ***Pneumatic gauging components:*** probes and rings
> 4. ***Force gauging:*** torque wrenches and load cells
> 5. ***Environment instrumentation:*** chart recorders and data loggers
>
> Body of Knowledge II.B

ELECTRONIC GAUGING TOOLS

Certain gauges are called *electric limit gauges* because they have the added feature of a rack stem that actuates precision switches. The switches connect lights or buzzers to show limits and also may energize sorting and corrective devices.

An *electronic gauge* gives a reading in proportion to the amount a stylus is displaced. It may also actuate switches electronically to control various functions. Figure 9.1 shows an example of an electronic gauge and diagrams of the most common kinds of gauge heads. The *variable inductance* or *inductance-bridge transducer* has an alternating current fed into two coils connected to a bridge circuit. The reactance of each coil changes as the position of the magnetic core is changed. This changes the output of the bridge circuit. The *variable transformer* or *linear variable displacement transformer* (LVDT) transducer has two opposed coils into which currents are induced from a primary coil.

The net output depends on the displacement of the magnetic core. The deflection of a strain gauge transducer is sensed by the changes in length and the resistance of strain gauges on its surface. This is also a means for measuring forces. Displacement of a variable capacitance head changes the air gap between plates of a condenser connected in a bridge circuit. In every case, an alternating current is fed into the gauge as depicted in Figure 9.1e. The output of the gauge head circuit is

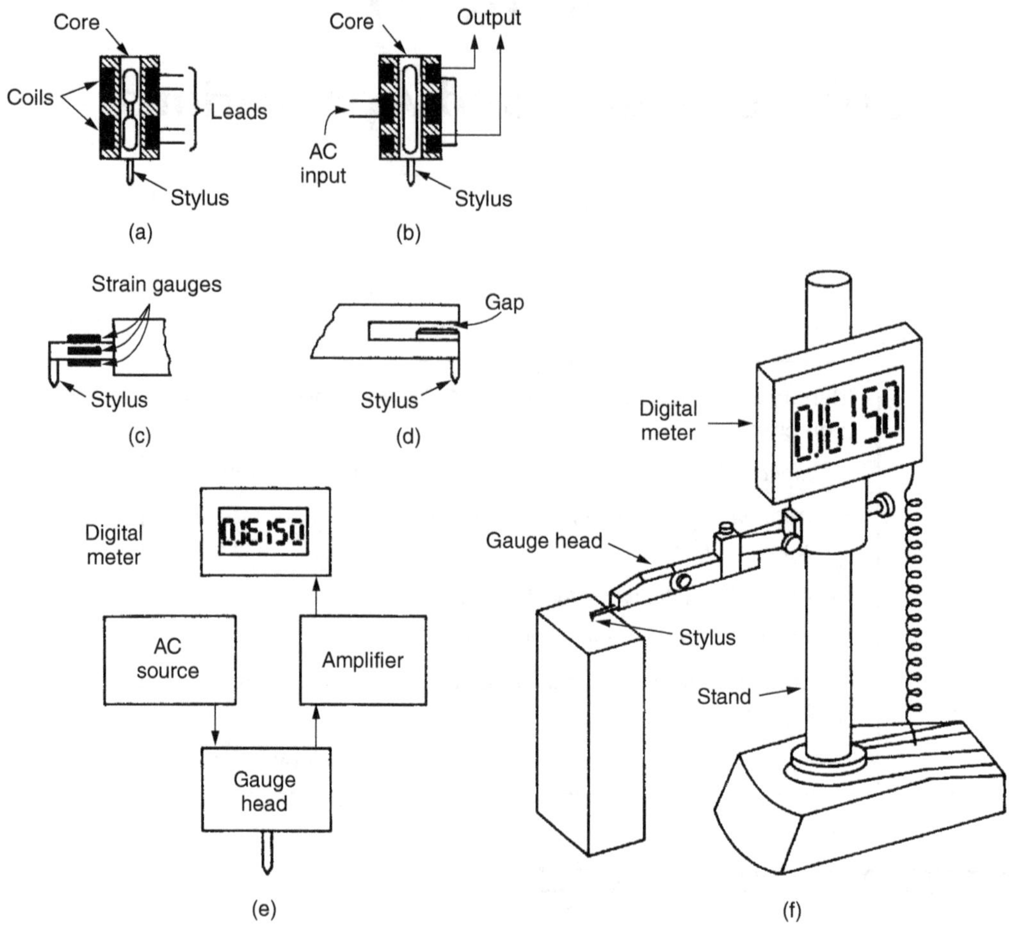

Figure 9.1 Elements of electronic gauges.
Source: Reprinted by permission from Society of Manufacturing Engineers, *Manufacturing Processes and Materials.* 4th ed. 2000.

amplified electronically and displayed on a dial or digital readout. In some cases, the information from the gauge may be recorded on tape or stored in a computer.

Advantages

Electronic gauges have several advantages:

1. They are very sensitive (they commonly read to a few micrometers).
2. Output can be amplified as much as desired; a high-quality gauge is quite stable.
3. They can be used as an absolute measuring device for thin pieces up to the range of the instrument.

The amount of amplification can easily be switched, and three or four ranges are common for one instrument. Two or more heads may be connected to one amplifier to obtain sums or differences of dimensions, as for checking thickness, parallelism, and so on.

Other types of electronic test equipment commonly used in practice are oscilloscopes, multimeters, and pyrometers. Brief descriptions of such instruments follow. For additional information, see Hickman (2001), Mazur (2009), and Burgess (2006).

Oscilloscopes

The basic principle of oscillography is the representation, by graphical means, of a voltage that is varying (Hickman 2001).

A typical oscilloscope is box shaped with a visual screen, attached connectors, and control knobs and buttons on the front panel. A grid is normally drawn on the screen. Each square in the grid is known as a *division*. One of the most common uses of the oscilloscope is to troubleshoot the failure of electronic equipment by graphically showing signals that indicate failure or malfunctioning.

Multimeters

A *multimeter* is an electronic gauge that combines more than one function in a single unit. Multimeters use either analog or digital displays. A multimeter is commonly used for fault discovery, in fieldwork of electronic or telecommunications technicians, or as a basic workshop instrument. Standard measurements of a multimeter include voltage, current, and resistance (see Mazur 2009).

AUTOMATIC GAUGING COMPONENTS

As industrial processes become automated, gauging must keep pace. Automated gauging is performed in two general ways. The first is in-process or on-the-machine control by continuous gauging of the work. The second is post-process or after-the-machine gauging control. Here, the parts coming off the machine are passed through an automatic gauge. A control unit responds to the gauge to sort pieces by size, and adjusts or stops the machine if parts are found out of limits.

Interferometry

Light waves of any kind are of invariable length and are the standard for ultimate measures of distance. Basically, all interferometers divide a light beam and send it along two or more paths. Then the beams are recombined and always show interference in some proportion to the differences between the lengths of the paths. One of the simplest illustrations of this phenomenon is the optical flat and a monochromatic light source of known wavelength.

The optical flat is a plane lens, usually a clear fused quartz disk, that is about 51–254 mm (2–10 in.) in diameter and 13–25 mm (0.5–1 in.) thick. The faces of a flat are accurately polished to nearly true planes; some have surfaces within 25 nm (0.000001 in.) of true flatness.

Helium is commonly used in industry as a source of monochromatic or single wavelength light because of its convenience. Although helium radiates a number of wavelengths of light, the portion that is emitted with a wavelength of 587 nm (0.00002313 in.) is so much stronger than the rest that the other wavelengths are practically unnoticeable.

Principle of Operation. The principle of light wave interference and the operation of the optical flat are illustrated in Figure 9.2a, wherein an optical flat is shown resting at a slight angle on a workpiece surface. Energy, in the form of light waves, is transmitted from a monochromatic light source to the optical flat. When a ray of light reaches the bottom surface of the flat, it is divided into two rays. One ray is reflected from the bottom of the flat toward the eye of the observer, while the other continues downward and is reflected, and loses one-half wavelength upon striking the top of the workpiece. If the rays are in phase when they re-form, their energies reinforce each other and they appear bright. If they are out of phase, their energies cancel each other and they are dark. This phenomenon produces a series of light and dark fringes or bands along the workpiece surface and the bottom of the flat, as illustrated in Figure 9.2b.

The distance between the workpiece and the bottom surface of the optical flat at any point determines which effect takes place. If the distance is equivalent to some whole number of half wavelengths of the same monochromatic light, the reflected rays will be out of phase, thus producing dark bands. This condition exists at positions x and z of Figure 9.2a. If the distance is equivalent to some odd number of quarter wavelengths of the light, the reflected rays will be in phase with each other and produce light bands. The light bands would be centered between the dark bands. Thus, a light band would appear at position y in Figure 9.2a.

Since each dark band indicates a change of one-half wavelength in distance separating the work surface and flat, measurements are made by simply counting the number of these bands and multiplying that number by one-half the wavelength of the light source.

This procedure may be illustrated by the use of Figure 9.2b as follows:

- The diameter of a steel ball is compared with a gauge block of known height
- Assume a monochromatic light source with a wavelength of 0.5875 μm (23.13 μin.)
- From the block, it is obvious that the difference in elevation of positions A and B on the flat is $(4 \times 0.5875)/2 = 1.175$ μm ($[4 \times 23.13]/2$ or 46.26 μin.)
- By simple proportion, the difference in elevation between points A and C is equal to $(1.175 \times 63.5)/12.7 = 5.875$ μm ($[46.26 \times 2.5]/0.5 = 231.3$ μin.)
- Thus, the diameter of the ball is $19.05 + 0.005875 = 19.055875$ mm ($0.750 + 0.0002313 = 0.7502313$ in.)

Optical Flats. Optical flats are often used to test the flatness of surfaces. The presence of interference bands between the flat and the surface being tested is an indication that the surface is not parallel with the surface of the flat.

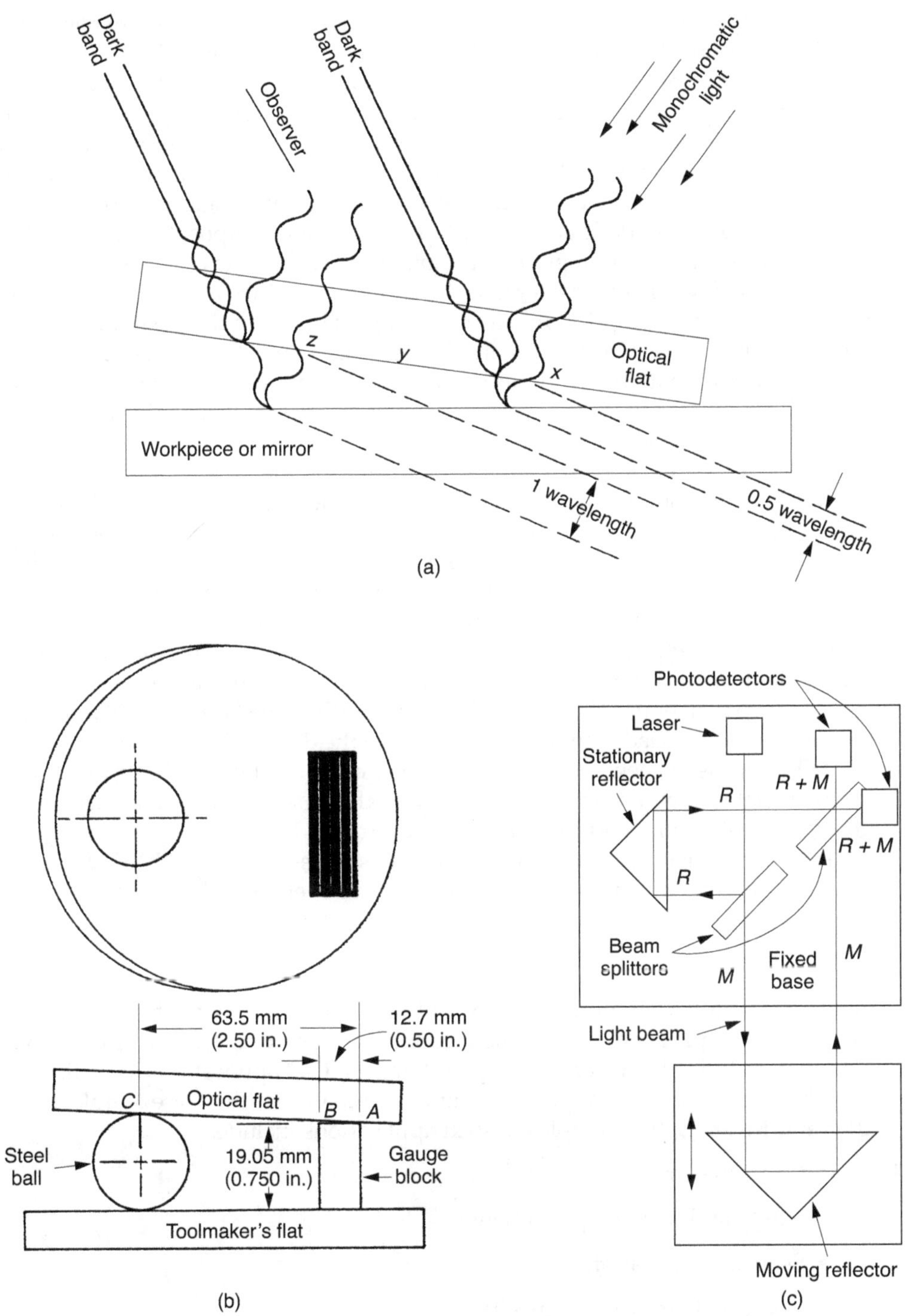

Figure 9.2 (a) Light wave interference with an optical flat, (b) application of an optical flat, (c) diagram of an interferometer.
Source: Reprinted by permission from Society of Manufacturing Engineers, *Manufacturing Processes and Materials*. 4th ed. 2000.

The way dimensions are measured by interferometry can be explained by moving the optical flat of Figure 9.2a in a direction perpendicular to the face of the workpiece or mirror. It is assumed that the mirror is rigidly attached to a base and that the optical flat is firmly held on a true slide. As the optical flat moves, the distance between the flat and the mirror changes along the line of traverse, and the fringes appear to glide across the face of the flat or mirror. The amount of movement is measured by counting the number of fringes and fraction of a fringe that pass a mark. It is difficult to precisely superimpose a real optical flat on a mirror or the end of a piece to establish the end points of a dimension to be measured. This difficulty is overcome in sophisticated instruments by placing the flat elsewhere and by optical means reflecting its image in the position relative to the mirror in Figure 9.2a. This creates interference bands that appear to lie on the face of and move with the workpiece or mirror. The image of the optical flat can be merged into the planes of the workpiece surfaces to establish beginning and end points of dimensions.

Machine Tool Application. A simple interferometer for measuring movements of a machine tool slide to nanometers (millionths of an inch) is depicted in Figure 8.2c. A strong light beam from a laser is split by a half mirror. One component becomes the reference R and is reflected solely over the fixed machine base. The other part M travels to a reflector on the machine side and is directed back to merge with ray R at the second beam splitter. Their resultant is split and directed to two photodetectors. The rays pass in and out of phase as the slide moves. The undulations are converted to pulses by an electronic circuit; each pulse stands for a slide movement equal to one-half the wavelength of the laser light. The signal at one photodetector leads the other according to the direction of movement.

When measurements are made to nanometers (millionths of an inch) by an interferometer, they are meaningful only if all causes of error, such as temperature, humidity, air pressure, oil films, impurities, and gravity, are closely controlled. Consequently, a real interferometer is necessarily a highly refined and complex instrument; only its basic elements have been described here.

Machine Vision

One of the most common applications of machine vision is in the inspection of manufactured products such as semiconductor components. Machine vision is also used in the automotive industry and has medical applications as well.

There are numerous applications of machine vision (see, for example, Davies 2005 and Myler 1999). Examples of such applications include:

- Coordinate measurements of large components
- Safety and hazard applications
- Security applications
- Coordinate measuring machines
- Automated guided vehicles
- Automated manufacturing

- Medical imaging processing
- Complex surgical procedures
- Robotics
- Product failure identification

X-Ray Inspection

The quality control role played by X-ray systems in the nondestructive testing and inspection of electronic assemblies—especially today's printed circuit boards, densely packed with area array components—is well understood. X-ray imaging is especially valuable in rapid prototyping and reverse-engineering applications in electronics, avionics, advanced materials research, and casting and other manufacturing industries. Unlike machine vision and optical inspection equipment, which can image only the surface of samples, X-ray systems penetrate substrate materials to expose hidden features. They are ideal for detecting and evaluating features in complex micro-electro-mechanical systems (MEMS) and micro-optoelectro-mechanical systems (MOEMS) devices, and for revealing cracks, voids, delamination, and other crucial component anomalies in prototype electronic assemblies, medical devices, and castings during preproduction (Rademaker 2005).

Once a need for X-ray inspection has been determined, the user must decide whether two-dimensional (2-D) or three-dimensional (3-D) imaging is best for the particular task at hand. While 2-D X-ray imaging is sufficient in many cases, 3-D X-ray images can reveal complex inner structures in an intuitive format so that anomalies can be viewed in their 3-D positions (Rademaker 2005).

PNEUMATIC GAUGING COMPONENTS

A *pneumatic gauge* is a means of measuring, comparing, or checking dimensions by sensing the flow of air through the space between a gauge head and workpiece surface. The gauge head is applied to each workpiece in the same way, and the clearance between the two varies with the size of the piece. The amount the airflow is restricted depends on the clearance. Four basic types of air gauge sensors are shown in Figure 9.3. All have a controlled constant-pressure air supply.

The *back-pressure gauge* responds to the increase in pressure when the airflow is reduced. It can magnify from 1000:1 to over 5000:1, depending on the range, but is somewhat slow because of the reaction of air to changing pressure.

The *differential gauge* is more sensitive. Air passes through this gauge in one line to the gauge head and in a parallel line to the atmosphere through a setting valve. The pressure between the two lines is measured.

There is no time lag in the *flow gauge*, where the rate of airflow raises an indicator in a tapered tube. The dimension is read from the position of the indicating float. This gauge is simple, does not have a mechanism to wear, is free from hysteresis, and can amplify to over 500,000:1 without accessories.

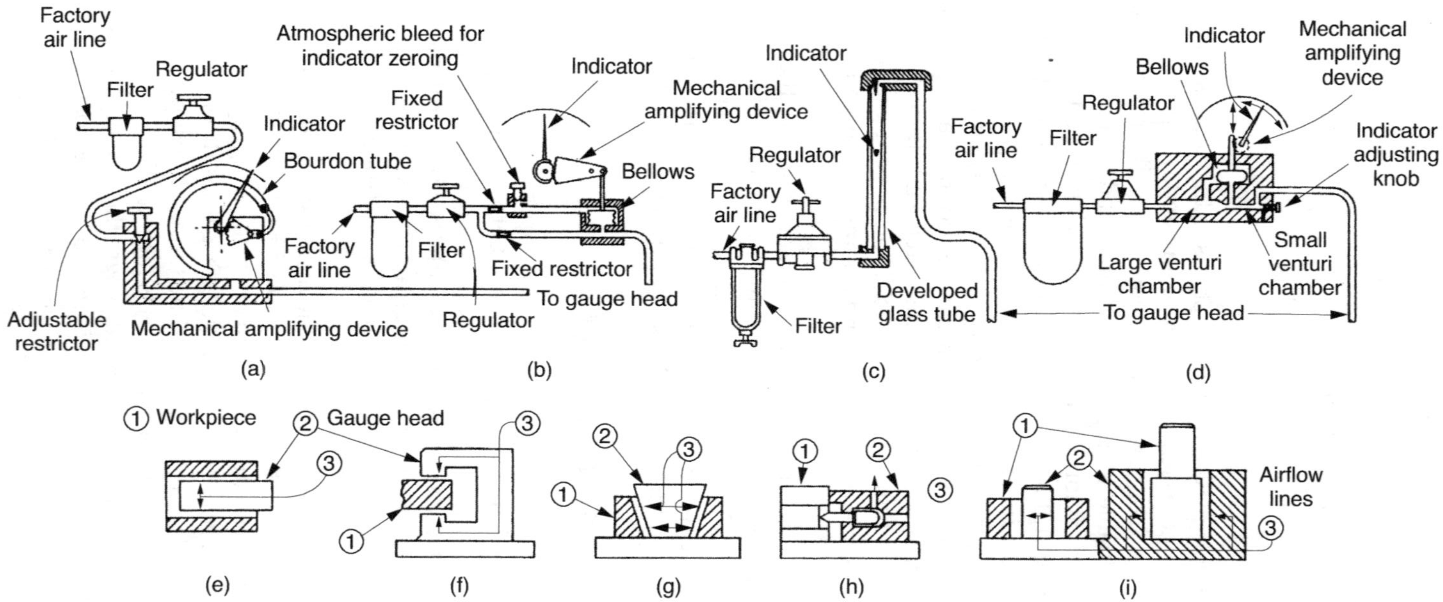

Figure 9.3 Diagram of air gauge principles.
Source: Reprinted by permission from Society of Manufacturing Engineers, *Manufacturing Processes and Materials*. 4th ed. 2000.

The *venturi gauge* measures the drop in pressure of the air flowing through a venturi tube. It combines the elements of the back-pressure and flow gauges and is fast, but sacrifices simplicity.

A few of the many kinds of gauge heads and applications are also shown in Figure 9.3. An air gauge is basically a comparator and must be an asset to a master for dimension or to two masters for limits. The common single gauge head is the plug. Practically all inside and outside linear and geometric dimensions can be checked by air gauging.

Air match gauging, as depicted in Figure 9.3i, measures the clearance between two mating parts. This provides a means of controlling an operation to machine one part to a specified fit with the other. A *multidimension gauge* has a set of cartridge or contact gauge heads (Figure 9.3h) to check several dimensions on a part at the same time. The basic gauge sensor can be used for a large variety of jobs, but a different gauge head and setting master are needed for almost every job and size.

A major advantage of a pneumatic gauge is that the gauge head does not have to tightly fit the part. A clearance of up to 0.08 mm (0.003 in.), and even more in some cases, between the gauge head and workpiece is permissible. Thus, no pressure is needed between the two, which causes wear, and the gauge head may have a large allowance for any wear that does occur. The flowing air helps keep surfaces clean.

The lack of contact makes air gauging particularly suitable for checking against highly finished and soft surfaces. Because of its loose fit, a pneumatic gauge is easy and quick to use. An inexperienced inspector can measure the diameter of a hole to 25 nm (0.000001 in.) in a few seconds with a pneumatic gauge. The same measurement (to 25 μm [0.001 in.]) with a vernier caliper by a skilled inspector may take up to one minute. The faster types of pneumatic gauges are adequate for high-rate automatic gauging in production.

FORCE GAUGING

Torque is the force applied on an object from a distance. A torque wrench is a simple tool that is used in applications such as a nut or a bolt needing tightening, or adjustment to a desired value. The application of force in this case may be manual or motorized depending on the type of application and nut or bolt size. To produce high quality forged parts, *load cells* are used. These are devices that enable accurate and consistent force application in forging, and aid in providing consistent application and process monitoring and control.

ENVIRONMENTAL INSTRUMENTATION

Some quality assurance applications for monitoring environmental conditions include data loggers and chart recorders. These are devices to record and display data, either physically on paper (chart recorders) or digitally (data loggers).

SUMMARY OF GAUGE USES AND APPLICATIONS

Table 9.1 provides a useful summary of the uses and applications of the gauges presented in Chapter 8.

Table 9.1 Summary of commonly used gauges and their applications.

Gauge	Uses and applications
Steel rule	Used effectively as a line measuring device, which means that the ends of a dimension being measured are aligned with the graduations of the scale, from which the length is read directly.
Depth rule	Used for measuring the depth of slots, holes, and so on.
Vernier caliper	Typifies instruments using the vernier principle of measurement. Outside dimensions are measured between the lower jaws, inside dimensions over the tips of the upper jaws.
Digital-reading caliper	Provides LCD readouts in either millimeters or inches and operates by a microprocessor-based system. Capable of retaining a reading in the display when the tool is used in an area where visibility is restricted.
Vernier height gauge	Similar to a vernier caliper except that the fixed jaw has been replaced by a fixed base, and the sliding jaw may have a scriber attached to it for layout work, or a dial indicator for measuring or comparing operations.
Digital micrometer	Used in measuring diameters, thicknesses, inside dimensions, heights, and outside dimensions.
Dial indicator	Used for applications such as measuring heights, flatness, diameters, and so on.
Optical comparator	Used for measuring complicated or difficult shapes and other configurations.
Gauge block	Dimensional measurement standards that are used to calibrate other measuring devices. They come in sets of different grades depending on the desired measurement accuracy required. Each set has many blocks of incremental lengths. These blocks are stacked together (wringed) to build a desired length.
Working gauge	Used by the machine operator or shop inspector to check the dimensions of parts as they are being produced. They usually have limits based on the piece being inspected.
Inspection gauge	Used by personnel to inspect purchased parts when received, or manufactured parts when finished. These gauges are designed and manufactured so as not to reject any product previously accepted by a properly designed and functioning working gauge.
Reference or master gauge	Used for checking the size or condition of other gauges, and representing as exactly as possible the physical dimensions of the product.
Limit gauge	Often called a go/no-go gauge, establishes the high and low limits prescribed by the tolerance on a dimension. A limit gauge may be either double-end or progressive.
Ring gauge	Used for checking outside dimensions, such as the limit sizes of a round shaft.
Plug gauge	Used for checking inside dimensions.
Snap gauge	A fixed gauge with the gauging members specially arranged for measuring diameters, thickness, and lengths.
Spline gauge	An instrument for inspecting splined workpieces prior to assembly.
Screw-thread micrometers	Designed to measure threads within a narrow range of pitches.
Template	Used to control or gauge special shapes or contours in manufactured parts, or to check a specified profile.
Screw-pitch gauge	Used to check the pitch of a screw.

(continued)

Table 9.1 Summary of commonly used gauges and their applications. (continued)

Gauge	Uses and applications
Oscilloscope	Used for troubleshooting electronic equipment failure by graphically showing signals that indicate failure or malfunction.
Multimeter	An electronic gauge that combines more than one function in a single unit. Multimeters use either analog or digital displays. Common uses include fault discovery, fieldwork of electronic or telecommunications technicians, or as a basic workshop instrument. Standard measurements taken using a multimeter include voltage, current, and resistance.
Pyrometer	A noncontact instrument for measuring an object's temperature by measuring its self-emission and emissivity.
Pneumatic gauge	Used for measuring, comparing, or checking dimensions by sensing the flow of air through the space between a gauge head and workpiece surface.
Optical flats	Often used to test the flatness of surfaces. The presence of interference bands between the flat and the surface being tested is an indication that the surface is not parallel with the surface of the flat.
Machine vision	Used for inspection of manufactured products such as semiconductor components and automotive parts, and also used in medical applications.
X-ray systems	Used in nondestructive testing and inspection of electronic assemblies, especially printed circuit boards.

Chapter 10

Gauge Selection, Handling, and Use

1. ***10:1 rule.*** Apply the 10:1 rule: inspection measurements require better than the tolerance of a dimension by a factor of 10, and calibration standards require better than the inspection measurements by a factor of 10. (Apply)
2. ***Gauge selection.*** Select gauges according to the feature or characteristic to be measured, the applicable tolerance, and the accuracy, environment, resolution, and capability of the test instrument. Determine whether the type of measurement should be direct, differential, or transfer. (Apply)
3. ***Gauge handling, preservation, and storage.*** Identify and apply various methods of cleaning, handling, and storing gauges. (Apply)
4. ***Gauge correlation.*** Identify and apply methods for establishing the correlation between measurement instruments, such as gauge-to-gauge or manual-to-automated process. (Apply)

Body of Knowledge II.C

THE RULE OF TEN (10:1 RULE)

The *rule of ten*—sometimes referred to as the *10:1 rule* or the *gauge maker's rule*—serves as a baseline or beginning for the gauge selection process. It states that inspection measurements should be better than the tolerance of a dimension by a factor of 10, and that calibration standards should be better than inspection measurements by a factor of 10.

As an example, if the tolerance on a shaft diameter is ±0.025 mm (±0.0010 in.), then the increment of measurement on the inspection instrument should be as small as 0.025/10 = 0.0025 mm (0.00010 in.).

Similarly, the increment of measurement of the calibration standard for that inspection instrument should be as small as 0.025/10 = 0.0025 mm (0.00010 in.).

GAUGE SELECTION

Once the smallest increment of measurement for an instrument has been determined, candidate instruments need to be evaluated in terms of the degree of satisfaction they offer relative to the following performance criteria:

1. *Accuracy.* The ability to measure the true magnitude of a dimension
2. *Linearity.* The accuracy of the measurements of an instrument throughout its operating range
3. *Magnification.* The amplification of the output reading on an instrument over the actual input dimension
4. *Repeatability.* The ability of the instrument to achieve the same degree of accuracy on repeated applications (often referred to as *precision*)
5. *Resolution.* The smallest increment of measurement that can be read on an instrument
6. *Sensitivity.* The smallest increment of difference in dimension that can be detected by an instrument
7. *Stability or drift.* The ability of an instrument to maintain its calibration over a period of time

In addition to the above, other selection criteria may include factors such as the shape and size of the measured part or workpiece, workpiece material, capabilities of the metrology laboratory, and environment (such as the temperature, atmospheric pressure, humidity, and other conditions of the metrology laboratory).

Consideration of these factors, along with cost and operation convenience, should help in selecting an appropriate measuring or gauging device for a particular inspection operation. For operating convenience, most instruments are or can be equipped with discrete digital readout devices. Most of these can be connected to microprocessors or computers for data recording and analysis.

GAUGE HANDLING, PRESERVATION, AND STORAGE

The following guidelines for the control of measuring and monitoring devices are taken from section 7.6 of ISO 9000:2000:

- Determine the monitoring and measurement to be undertaken and the monitoring and measuring devices needed to provide evidence of conformity of product to determined requirements (7.2.1).
- Establish processes to ensure that monitoring and measurement can and will be carried out in a manner that is consistent with the monitoring and measurement requirements.
- Where necessary to ensure valid results, the measuring equipment shall:
  - Be calibrated or verified at specified intervals or prior to use against traceable standards.

- Where no standards exist, the basis for calibration or verification shall be recorded
  - Be adjusted or readjusted as necessary
  - Be identified to enable calibration status to be determined
  - Be safeguarded from adjustments
  - Be protected from damage and deterioration during handling, maintenance, and storage
- Assess and record the validity of previous measuring results when equipment is found to be nonconforming to requirements.
- Take appropriate action on any product affected.
- Record results of calibration and verification.
- Software—confirm ability prior to initial use and reconfirm as necessary, when used in monitoring and measurement of specified requirements.
- Define and implement effective and efficient measuring and monitoring processes, including verification of products and processes to ensure satisfaction of interested parties. This includes:
  - Surveys
  - Simulations
  - Other measurement and monitoring activities
- Measuring and monitoring processes should include confirmation that the devices are fit for use and are maintained to suitable accuracy and accepted standards, and be able to identify the status of the devices.
- Consider means to eliminate potential errors from processes.

In addition, section 8.2.4 discusses the monitoring and measurement of products. The following are points that management should consider:

- Monitor and measure the characteristics of the product to verify that requirements have been met.
- Perform monitoring and measurement at appropriate stages in the realization process in accordance with planned arrangements (7.1).
- Maintain evidence of conformity to acceptance criteria.
- Records shall indicate the person(s) authorizing release of product (4.2.4).
- One can't release and deliver product until all planned arrangements (7.1) have been completed (relevant authority and/or customer waiver).
- Establish and specify measurement requirements for products (including acceptance criteria). Plan and perform in order to verify that the requirements of interested parties have been achieved and used to improve the realization process.

- When selecting measurement methods, consider:
  - Types of product characteristics
  - Equipment, software, and tools required
  - Location of suitable measurement points in the realization process sequence
  - Qualification of people, materials, products, processes, and the quality management system (QMS)
  - Final inspection
  - Recording the results

GAUGE CORRELATION

Gauge correlation is about making sure the data collected from different gauges are credible. With so much focus these days on collecting and analyzing data, and on implementing Six Sigma and other initiatives to ensure that design goals are met, it's possible that critical details such as the level of accuracy of the data have gone unnoticed (Jenkins 2003).

ISO/IEC 17025 sets the guidelines of gauge correlation and measurement techniques. Several factors contribute to gauge correlation, such as measurement uncertainty, validation of data collection systems, interlaboratory assessment, and part features and characteristics, as well as equipment and operator variability. A gauge repeatability and reproducibility (gauge R&R) study is an effective way to isolate the different types of variation in a measurement. The gauge R&R study is presented in detail in Chapter 13.

Although ISO/IEC 17025 doesn't—and indeed can't—mandate perfect measurements, it does provide some assurance that laboratory measurement quality is evaluated, defined, and controlled. The following concepts address several issues to consider when evaluating a testing or calibration supplier, given the current state of accreditation (Jenkins 2003):

- *Estimated measurement uncertainty.* These values are needed when determining calibration or test adequacy and subsequent conclusions. They are also needed when performing actual measurements on certain parts or assemblies. If a part or an assembly is close to being out of specification, the estimated uncertainty may be the deciding factor in the acceptance/rejection decision.
- *Calibration and/or test data.* Prior to submitting a device or object for testing or calibration, a list of the measurements that will be taken and a sample data report should be available.
- *Ensuring quality of calibration and testing results.* ISO/IEC 17025 lists many methods that labs can use to ensure the quality of their calibration and testing results. One of these, proficiency testing, has since become a requirement by many accreditation bodies.

Chapter 11

Surface Plate Tools and Techniques

1. ***Surface plate equipment.*** Select and use height gauges, V-blocks, and other indicators to measure various types of features. Understand the care, cleaning, calibration, and lapping of a surface plate. (Apply)
2. ***Angle measurement instruments.*** Identify and use protractors, sine bars, and angle blocks. (Apply)

Body of Knowledge II.D

SURFACE PLATE EQUIPMENT

Surface Plate

A *surface plate* provides a true reference plane from which measurement can be made. A cast-iron surface plate is a heavy-ribbed, boxlike casting that stands on three points (establishing a plane) and has a thick and well-supported flat top plate. New plates generally have an average of 18 bearing spots on an area of 6.5 cm^2 ($\approx$ 1 $in.^2$) that do not vary from a true plane by more than 0.005 mm (0.0002 in.). The use of natural stones for surface plates is becoming increasingly popular because of their hardness, resistance to corrosion, minimum response to temperature change, and nonmagnetic qualities. Figure 11.1 shows a granite surface plate used in inspection work.

Care and Maintenance of Granite Surface Plates. *Quality* magazine published useful guidelines for the care and maintenance of granite surface plates in its May 2003 issue. The following is a summary of these guidelines.

1. Be sure the granite surface plate is supported on the support system that comes furnished with the plate. The support system is normally a hard rubber pad attached to the bottom of the plate that forms a non-distortable three-point support system. The pads are installed during manufacturing, and the plate rests on them throughout lapping, inspection, and shipping. The pads are a critical factor in surface plate accuracy and must not be removed.

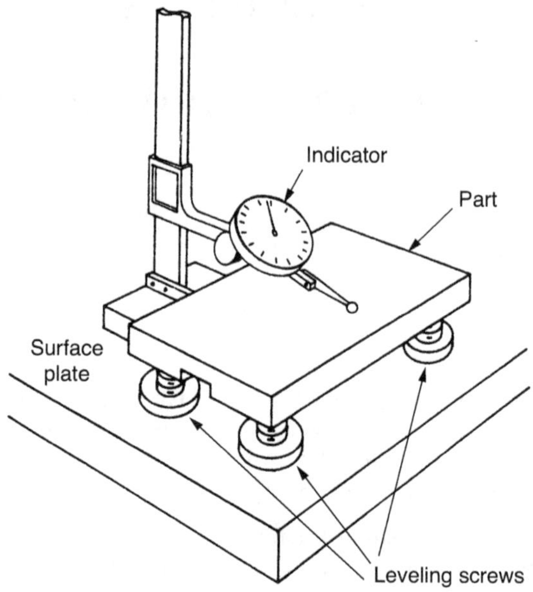

Figure 11.1 Granite surface plate for checking the flatness of a part, with dial indicator and leveling screws.
Reprinted by permission from Society of Manufacturing Engineers, *Manufacturing Processes and Materials*. 4th ed. 2000.

2. Do not move or lift a granite surface plate with a forklift by placing the metal forks against the granite. Always have protective padding between the metal forks and the granite. Preferably, use nylon slings to move it.
3. After setup, granite surface plates require little care and maintenance. Keep the surface clean and free from excess buildup of dust, dirt, grease, grime, or other foreign particles. Using a dirty surface plate will cause inaccuracies in tool readings and also cause the plate to wear faster. Environment and usage have a lot to do with the frequency of cleaning; however, as a rule, if a plate is used daily, it should be cleaned daily.
4. When using a granite surface plate, carefully set the workpiece to be measured on the plate. A sudden jolt or blow to the plate with a heavy metal object can damage the surface. Also, when inspecting workpieces, especially small parts, try to use different areas of the surface plate. Inspecting in the same spot on the surface plate month after month, year after year, will wear the plate in that spot.
5. Surface plates should be checked on a regular basis for wear. The plates can be checked with a reading gauge that is designed for this purpose.

Other important issues include:

- Calibration of overall flatness traceable to NIST can be performed periodically using an autocollimator.

- When inspection indicates that the overall accuracy of a surface plate is out of tolerance, the plate should be relapped.
- When using threaded inserts in a granite surface plate, use the minimum torque required on hold-down bolts, but do not exceed limits. A caution label is attached to all granite surface plates that are furnished with threaded inserts.

Dial Indicators

Indicating gauges and comparators magnify the amount a dimension deviates above or below a standard to which the gauge is set. Most indicate in terms of actual units of measurement, but some show only whether a tolerance is within a given range. The ability to measure to 25 nm (0.00001 in.) depends on the magnification, resolution, and accuracy of the setting gauges and staging of the workpiece and instrument.

Dial indicators are simple mechanical indicating gauges. Figure 11.2 depicts an example of a dial indicator. Movement of stem *A* is transmitted from the rack to compound gear trains *B* and *C* to pointer *D*, which moves around a dial face. Springs exert a constant force on the mechanism and return the pointer to its original position after the object being measured is removed.

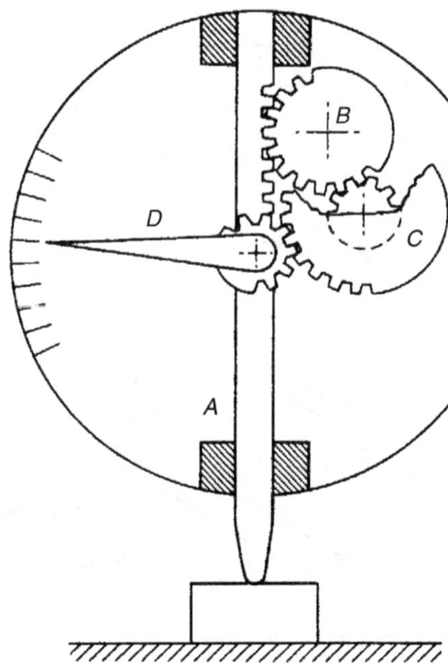

Figure 11.2 Simple dial indicator mechanism.
Reprinted by permission from Society of Manufacturing Engineers, *Manufacturing Processes and Materials*. 4th ed. 2000.

Dial indicators are used for many kinds of measuring and gauging operations. One example is that of inspecting a workpiece, as illustrated in Figure 11.3. They also serve to check machines and tools, alignments, and cutter runout. Dial indicators are often incorporated in special gauges in measuring instruments, as exemplified by the indicating micrometer of Figure 8.8 in. Chapter 8.

Vernier Height Gauge

The *vernier height gauge* is similar to a vernier caliper except the fixed jaw has been replaced by a fixed base, and the sliding jaw may have a scriber attached to it for layout work, or a dial indicator for measuring or comparing operations. See Chapter 8 for more information on vernier height gauges.

ANGLE MEASUREMENT INSTRUMENTS

The unit standard of angular measurement is the degree. The measurement and inspection of angular dimensions are somewhat more difficult than linear measures and may require instruments of some complexity if a great deal of angular precision is required.

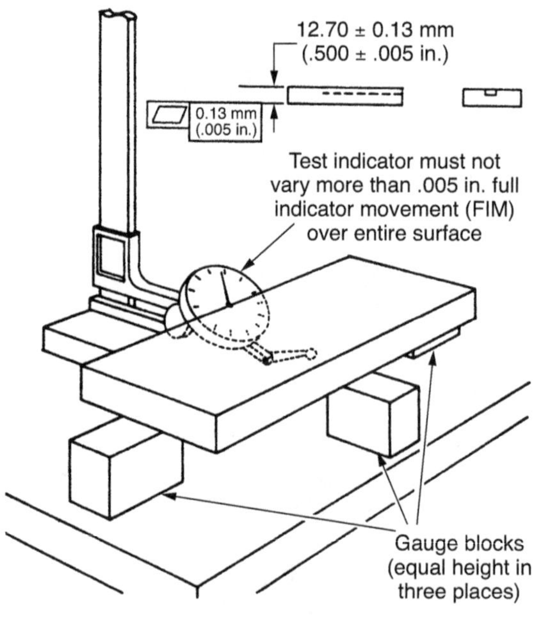

Figure 11.3 An application of dial indicators for inspecting flatness by placing the workpiece on gauge blocks and checking full indicator movement (FIM).
Reprinted by permission from Society of Manufacturing Engineers, *Manufacturing Processes and Materials*. 4th ed. 2000.

Sine Bar

The *sine bar* is a hardened, precision-ground, and lapped tool for measuring angles in different settings. It is a relatively simple device for precision measuring and checking of angles. It consists of an accurately ground, flat-steel straightedge with precisely affixed round buttons that are a definite distance apart and of identical diameters.

Sine Bar Application. Figure 11.4 illustrates one method of applying a sine bar in the determination of the angle α on the conical surface of the part located on the surface plate.

For precise results, a sine bar must be used on true surfaces. In Figure 11.4:

- The center-to-center distance of the sine bar buttons = 127 mm (5 in.)
- The distances A and B are determined by means of gauge blocks or a vernier height gauge:

$$A = 25.400 \text{ mm } (1.0000 \text{ in.})$$
$$B = 89.794 \text{ mm } (3.5352 \text{ in.})$$

Thus,

$$\text{sine } \alpha = \frac{(89.794 - 25.400)}{127.00} = 0.50704$$

From trigonometric tables, the angle α is 30° 28′.

Sine Tables. *Sine tables* are a further development of the sine bar. The sine table has a larger working surface and is much more robust than the sine bar. It is more suitable for larger, heavier parts.

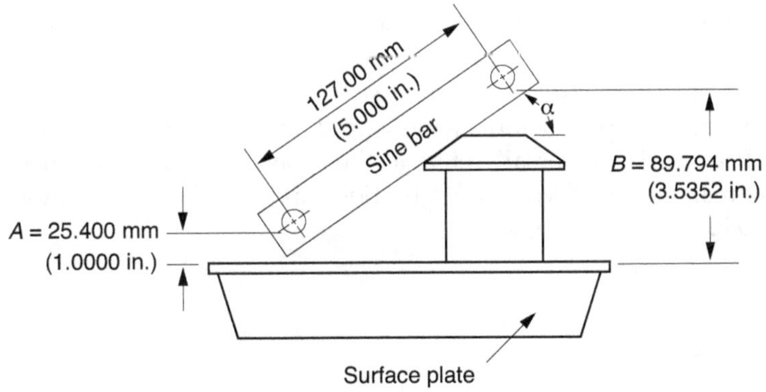

Figure 11.4 Application of a sine bar.
Reprinted by permission from Society of Manufacturing Engineers, *Manufacturing Processes and Materials*. 4th ed. 2000.

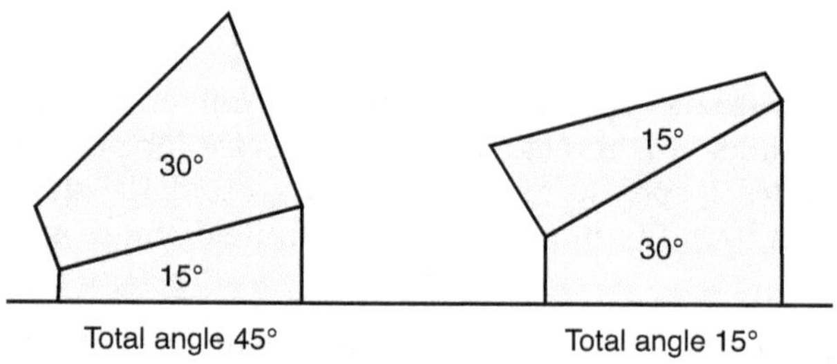

Figure 11.5 Addition and subtraction of angle blocks.

Angle Blocks

Angle blocks are simply hardened blocks further lapped to precise angles and can be wrung together like gauge blocks. Unlike gauge blocks, angle blocks can be added or subtracted, as shown in Figure 11.5.

Protractors

Protractors are used to measure angles. A *simple protractor* consists of a graduated head (in degrees) along a semicircle. A blade is connected on the center pin. By rotating the blade on the pin, any angle from 0° to 180° can be measured.

Combination Sets. Combination protractor and depth gauges combine a movable graduated depth gauge with a protractor head.

The *combination set* consists of a center head, protractor, and square with a 45° surface, all of which are used individually in conjunction with a steel rule. The heads are mounted on the rule and clamped in any position along its length by means of a lock screw. The center head is used to scribe bisecting diameters of the end of a cylindrical piece to locate the center of the piece. The protractor reads directly in degrees. Both the square head and the protractor may contain a small spirit level.

A *bevel protractor* utilizes a vernier scale to show angles as small as 5 minutes. The universal bevel protractor is widely used for angle measurement in many applications. The word "universal" refers to the capacity of the instrument to be adaptable to a great variety of work configurations and angular interrelations (Farago and Curtis 1994).

Chapter 12
Specialized Inspection Equipment

1. ***Measuring mass.*** Describe and apply weights, balances, and scales. (Apply)
2. ***Measuring finish.*** Describe and apply profilometers (e.g., optical and stylus). (Apply)
3. ***Measuring shape and profile.*** Understand and describe mechanical comparators, roundness testers, precision spindles, and profile tracers. (Understand)
4. ***Optical equipment.*** Describe and apply optical comparators and microscopes. (Apply)
5. ***Software-based measurement systems.*** Define and use digital cameras, vision inspection systems (white light / blue light), and other digital systems for product inspection. Recognize software limitations with regard to locating functional datums, target points and areas, hole positions, and the basic operation of the x, y, and z axes. (Apply)
6. ***Measuring inclination.*** Define and describe the measurement of the slope or slant of various equipment (mechanical/laser). (Understand)

Body of Knowledge II.E

MEASURING MASS

Weights, Balances, and Scales

Weight is a force, resulting from the action of gravity, that measures how hard an object presses down on a scale. Weight and mass are two fundamentally different quantities, though the weight of an object can provide an indication of the quantity that we are actually measuring—mass.

Measuring Weight and Mass

SI units (International System of Units) are used to measure physical quantities. The SI unit of mass is the kilogram, which is also used in the everyday language of measuring weights. The SI unit of weight is kg m/sec^2 (note that the weight is the mass times the gravitational acceleration [m/sec^2]). In the United States, the pound can be either a unit of weight (force) or a unit of mass.

In the international standard ISO 80000–4:2006, describing the basic physical quantities and units in mechanics as part of the international standard ISO/IEC 80000, the definition of *weight* is given as (F = m.g). We can convert between weight and mass. F (or W) is the force due to gravity (weight), *m* is the mass of the object, and *g* is the local acceleration of free fall (which equals approximately 9.8 m/s^2 or 32.2 ft/s^2).

Balances and Scales

Numerous types of balances and scales are used to cover a wide range of weight for different applications, such as laboratory, industrial, and research. They offer different capacities, resolutions, requirements, and configurations. Examples include:

- Lab balances
- Analytical balances
- Precision balances
- Industrial scales
- Bench scales
- Counting scales
- Mechanical balances
- Spring scales
- Jewelry scales

There are several national and international organizations that establish standards for weights and measures:

- The National Institute for Standards and Technology (NIST) is the nation's standards-defining authority. NIST Handbook 44 (*Specifications, Tolerances, and Other Technical Requirements for Weighing and Measuring Devices*) sets forth the minimum requirements for standards used primarily to test commercial or legal-for-trade weighing devices for compliance. NIST Special Publication 881 (*Guide for the Use of the International System of Units (SI)*) is also a good source.
- The American Society for Testing and Materials (ASTM) establishes test standards for materials, products, systems, and services for a wide range of industries. ASTM developed E617–97 (*Standard Specification for Laboratory Weights and Precision Mass Standards*) to cover various classes of weights and mass standards used in laboratories.

- The International Organization of Legal Metrology (OIML) is an intergovernmental treaty organization. OIML has two grades of membership: member states (countries that actively participate in technical activities) and corresponding members (countries that join the OIML as observers). It was established in 1955 to promote the global harmonization of legal metrology procedures. OIML has since developed a worldwide technical structure, providing metrological guidelines for the elaboration of national and regional requirements concerning the manufacture and use of measuring instruments for legal metrology applications.
- The International Organization for Standardization (ISO) is the world's largest developer and publisher of international standards. ISO is a network of the national standards institutes of 164 countries, with one member per country, and a central secretariat in Geneva, Switzerland, that coordinates the system.

Surface Metrology

Surface metrology is the measurement of the difference between what a surface actually is and what it is intended to be. *Surface measurement*, however, is involved with the relationship of a surface on the workpiece to a reference that is not actually on the workpiece. The most common aspect of surface metrology is the measurement of surface roughness as an average deviation from a mean centerline (Bosch 1984).

The quality of surface finish is commonly specified along with linear and geometric dimensions. This is becoming more common as product demands increase, because surface quality often determines how well a part performs. Heat-exchanger tubes transfer heat better when their surfaces are slightly rough rather than highly finished. Brake drums and clutch plates also work better with some degree of surface roughness. On the other hand, bearing surfaces for high-speed engines wear-in excessively and fail sooner if not highly finished, but still need certain surface textures to hold lubricants. Thus, there is a need to control all surface features, not just roughness alone.

Surface Characteristics. The American National Standards Institute (ANSI) has provided a set of standard terms and symbols to define such basic surface characteristics as profile, roughness, waviness, flaws, and lay. A *profile* is defined as the contour of any section through a surface. *Roughness* refers to relatively finely spaced surface irregularities such as might be produced by the action of a cutting tool or grinding wheel during a machining operation. *Waviness* consists of those surface irregularities that are of greater spacing than roughness. Waviness may be caused by vibrations, machine or work deflections, warping, and so on. *Flaws* are surface irregularities or imperfections that occur at infrequent intervals and at random locations. Such imperfections as scratches, ridges, holes, cracks, pits, checks, and so on, are included in this category. *Lay* is defined as the direction of the predominant surface pattern. Figure 12.1 illustrates these characteristics.

Surface Quality Specifications. Figure 12.1c includes standard symbols to specify surface quality. Roughness is most commonly specified and expressed in units of micrometers (μm), nanometers (nm), or microinches (μin.). According to

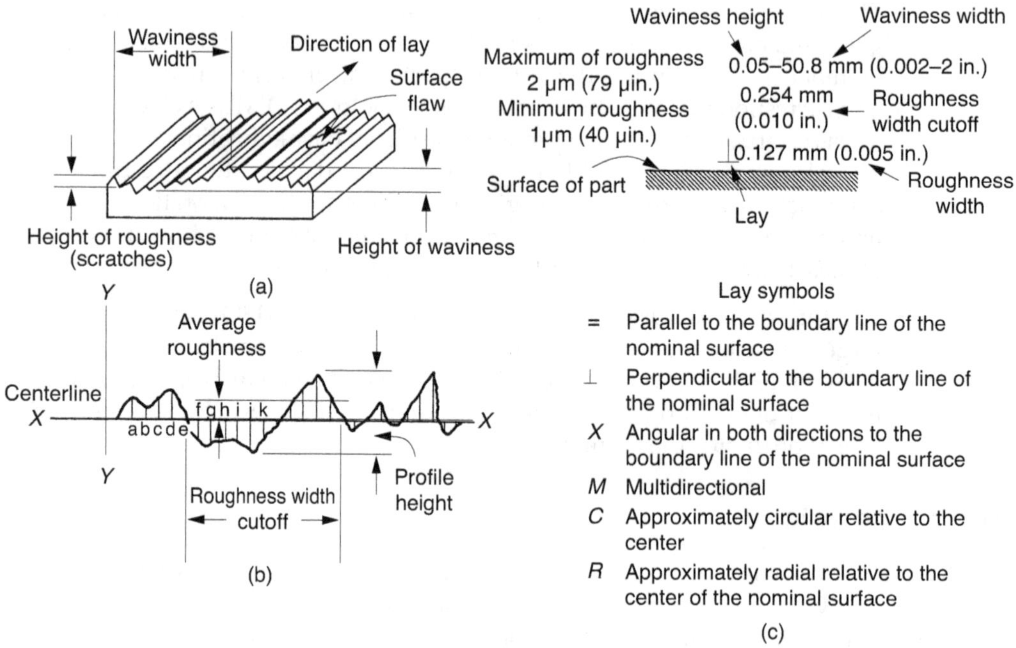

Figure 12.1 (a) Typical surface highly magnified, (b) profile of surface roughness, (c) surface quality specifications.
Reprinted by permission from Society of Manufacturing Engineers, *Manufacturing Processes and Materials*. 5th ed. 2015.

ANSI/ASME B46.1-1985, the standard measure of surface roughness adopted by the United States and approximately 25 countries around the world is the arithmetic average roughness R_a (formerly AA or CLA). R_a represents the arithmetic average deviation of the ordinates of profile height increments of the surface from the centerline of that surface. An approximation of the average roughness may be obtained by:

$$R_{a+} = \frac{y_a + y_b + y_c + \cdots + y_n}{n}$$

where

R_{a+} = Approximation of the average roughness
$y_a \ldots y_n$ = Absolute values of the surface profile coordinates
n = Number of sample measurements

The longest length along the centerline over which the measurements are made is the roughness-width cutoff or sampling length. In many cases, the maximum peak-to-valley height on a surface (R_y) is about four to five times greater than the average surface roughness as measured by R_a. This may present a problem for precision parts with small dimensional tolerances. For example, a flat surface on a part with an R_a of 0.4 μm (16 μin.) might very well have a peak-to-valley height (R_y) of 1.6 μm (64 μin.) or greater. If the tolerance on that dimension is 0.0025 mm (0.0001 in.), then the 0.4 μm (16 μin.) surface finish represents nearly two-thirds of the permissible tolerance. The root-mean-square (rms) is used to designate surface roughness. If measurements are made (plus or minus) from the centerline like the ones in Figure 12.1b, the rms average is:

$$\left[\left(\sum y_i^2\right)/n\right]^{\frac{1}{2}}$$

where

y_i = Measurement taken from the centerline
n = Number of measurements

RMS values are about 11% larger than R_a figures. Some surface roughness measuring instruments have a scale with numbers labeled "rms" and a scale calibrated in R_a values. On such instruments, the number called "rms" is a root-mean-square value only when the profile is sinusoidal. Waviness height alone may be specified, or it may be accompanied by a width specification. Thus, in Figure 12.1c, the specification 0.05–50.8 mm (0.002–2 in.) means that no waves over 0.05 mm (0.002 in.) high are allowed in any 50.8 mm (2 in.) of length. If no width specification is given, it is usually implied that the waviness height specified must be held over the full length of the work. Other specifications in Figure 12.1c are less common.

MEASURING FINISH

Waviness and roughness are measured separately. Waviness may be measured by sensitive dial indicators. A method of detecting gross waviness is to coat a surface with a high-gloss film, such as mineral oil, and then reflect it in a regular pattern, such as a wire grid. Waviness is revealed by irregularities or discontinuities in the reflected lines.

Many optical methods have been developed to evaluate surface roughness. Some are based on interferometry. One method of interference contrast makes different levels stand out from each other by lighting the surface with two out-of-phase rays. Another method projects a thin ribbon of light at 45° onto a surface. This appears in a microscope as a wavy line depicting the surface irregularities. For a method of replication, a plastic film is pressed against a surface to take its imprint. The film may then be plated with a thin silver deposit for microscopic examination, or it may be sectioned and magnified. These are laboratory methods and are economical in manufacturing only where other means are not feasible, such as on a surface inaccessible to a probe.

Except for extremely fine surface finishes that require laboratory measurement, most manufacturers measure surface texture at or near the workplace. A variety of instruments, called *surface finish gauges*, are commercially available as either hand-held or table-mounted devices. These require only moderate skill, and roughness measurements are displayed on a dial, digital readout, chart, or a digital output for statistical process control (SPC) depending on the type of instrument used. Most of these instruments employ a diamond-tipped stylus that is moved across the surface of the part to sense the point-to-point roughness of that surface. As illustrated in Figure 12.2, there are two basic types of gauges: *skid* or *skidless*. The skid-type shown in Figure 12.2a has a hinged probe that rides the work surface in proximity to a fairly large skid that also contacts the work surface. The skid-type instruments usually have

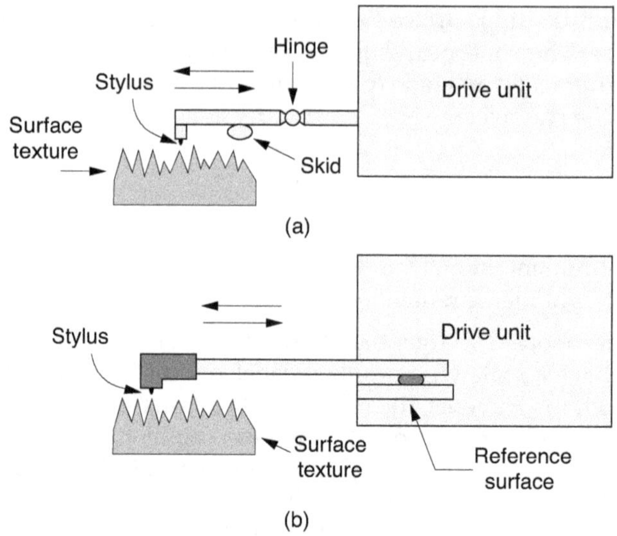

Figure 12.2 (a) Skid-type or average surface finish measuring gauge, (b) skidless or profiling gauge.
Reprinted by permission from Society of Manufacturing Engineers, *Manufacturing Processes and Materials*. 5th ed. 2015.

inductive transducers and are used predominantly for averaging measurements of surface roughness, but not waviness. The skid filters out waviness. Most portable or handheld instruments are the skid-type, and they are reasonably accurate for roughness measurements in the range of 0.30–0.51 μm (12–20 μin.) R_a.

The skidless type of instrument illustrated in Figure 12.2b has a built-in reference surface that permits the probe to sense both long- and short-wavelength variations in surface conditions. Thus, these can be used to measure both waviness and roughness, as well as surface inclination (straightness). These instruments are often referred to as *profiling gauges,* and they usually generate a profile chart on paper or on a computer screen. The international standard for the assessment of surface texture, ISO/R 468 (ISO Recommendations on Roughness Measurements, ISO/R 468, 1878, 1880, TC 57), defines three parameters: R_a (CLA), R_z, and R_{Max}, all measured relative to a straight mean line:

1. R_a (centerline average) value is the arithmetic mean of the departures of a profile from the mean line. It is normally determined as the mean result of several consecutive sample lengths L.
2. R_z (10-point height) is the average distance between the five height peaks and the five deepest valleys within the sampling length, and is measured perpendicular to it.
3. R_{Max} is the maximum peak-to-valley height within the sampling length.

Other parameters of surface measurement are defined as follows:

1. R_{tm} is the average value of R_{Max} for five consecutive sampling lengths.
2. R_p is the maximum profile height from the mean line within the sampling length. R_{pm} is the mean value of R_p determined over five sampling lengths.
3. PC (peak count) is the number of peak/valley pairs per inch projecting through a band of width *b* centered about the mean line.

MEASURING SHAPE AND PROFILE

Most of the recent applications of shape and profile measurements employ comparators and other devices for measuring roundness.

Comparators

Comparators normally employ dial indicators for their operation and come in different varieties: mechanical, optical, electronic, and pneumatic. *Optical projectors,* also known as *optical comparators,* employ a system in which light rays are directed against the object and then reflected through a projection lens onto a screen. The projections are large enough to accurately measure small configurations of objects. Optical comparators are explained in more detail in a later section.

Measurement of Roundness

Geometrically, a part can be said to be round in a given cross section if there exists within the section a point from which all points on the periphery are equidistant. In practice, however, the radius of nominally round parts tends to vary from point to point. Thus, the problem found by the metrologist is one of displaying and assessing these variations, and correctly interpreting the results.

Although many methods have been used for roundness measurement, only those that provide valid radial-deviation data lend themselves to standardization and consistent, accurate measurement of all out-of-roundness conditions. For this reason, current industry, national, and international standards primarily cover measurements taken with precision spindle-type instruments with the data recorded on a polar chart. Precision spindle instruments are those in which the spindle supports and rotates the part, with the gauge tip remaining stationary, and those in which the spindle rotates the gauge tip about the part, which remains stationary (Drews 1978).

OPTICAL EQUIPMENT

Optical inspection instruments fall into four basic categories: optical comparators, video systems, microscopes, and laser systems (Klepp 1998).

Optical comparators use traditional optics to magnify and project an object's image onto a glass screen. This type of optical inspection instrument is by far the most widely used and is also the least expensive method of optical noncontact inspection. It also offers the greatest versatility because lightweight parts, as well as ones weighing hundreds of pounds, can be inspected on many available instruments. The optical edge finder, developed later, eliminates the operator's subjectivity from the measurement and allows the system to be fully automated.

Optical comparators are available in numerous styles and configurations from domestic as well as international manufacturers. Typical options include projection screen sizes from 10 in. to 80 in., horizontal or vertical light path configurations, profile and surface illumination systems, various stage travel options, magnifications from 5× to 200×, and digital readout options such as a basic two-axis display and automatic computer numeric-controlled systems. Accuracy and repeatability of instruments currently available can vary depending on the feature being inspected but usually fall within ±0.00010 in. under certain conditions.

Inspection microscopes use traditional optics to magnify a desired detail. Many inspection microscopes today are coupled with a video system and offer manual optical inspection as well as automated video inspection. These microscopes generally work best when inspecting lightweight and/or flat parts.

Laser inspection instruments, the latest development in optical inspection, offer the greatest accuracy of any optical inspection instrument. Accuracy and repeatability within ±0.0000010 in. can be expected under certain conditions. This type of instrument utilizes a reflected laser beam to accurately determine distances by using time-delay calculations. Extremely specialized, laser inspection instruments require fixturing to locate the laser device as well as substantial setup time to align the instrument. This measurement method is best suited for specialized production inspection or calibration applications.

Optical Tooling

Telescopes and accessories to establish precisely straight, parallel, perpendicular, or angled lines are called *optical tooling*. Two of many applications are shown in Figure 12.3a and Figure 12.3b. One is to check the straightness and truth of ways of a machine tool bed at various places along the length. The other is to establish reference lanes for measurements on a major aircraft or missile component. Such methods are especially necessary for large structures. Accuracy of 1 part in 200,000 is regularly realized; this means that a point at a distance of 2.5 m (100 in.) can be located within 13 μm (0.0005 in.). Common optical tooling procedures are autocollimation, autoreflection, planizing, leveling, and plumbing.

Autocollimation is done using a telescope with an internal light that projects a beam through the crosshairs to a target mirror, as indicated in Figure 12.3a. If the mirror face is truly perpendicular to the line of sight, the crosshair image will be reflected back on itself. The amount the reflected image deviates from the actual reticle image is an indication of the tilt in the target. A target may have a cross-line pattern for alignment with the line of sight. An autocollimated image is not clear for distances over 15.2 m (50 ft), and thus a somewhat less accurate method must be used. This is *autoreflection*, with an optical flat containing a cross-line pattern mounted on the end of the illuminated telescope and focused to twice the distance of the target mirror. If the mirror is perpendicular to the line of sight, the pattern of the flat is reflected in coincidence with the crosshairs in the telescope.

Planizing involves fixing planes at 90° to other planes or with a line of sight. This may be done from accurately placed rails on which transits are mounted in a tooling dock, as indicated in Figure 12.3b. A *transit* is a telescope mounted to swing in a plane perpendicular to a horizontal axis. Square lines also may be

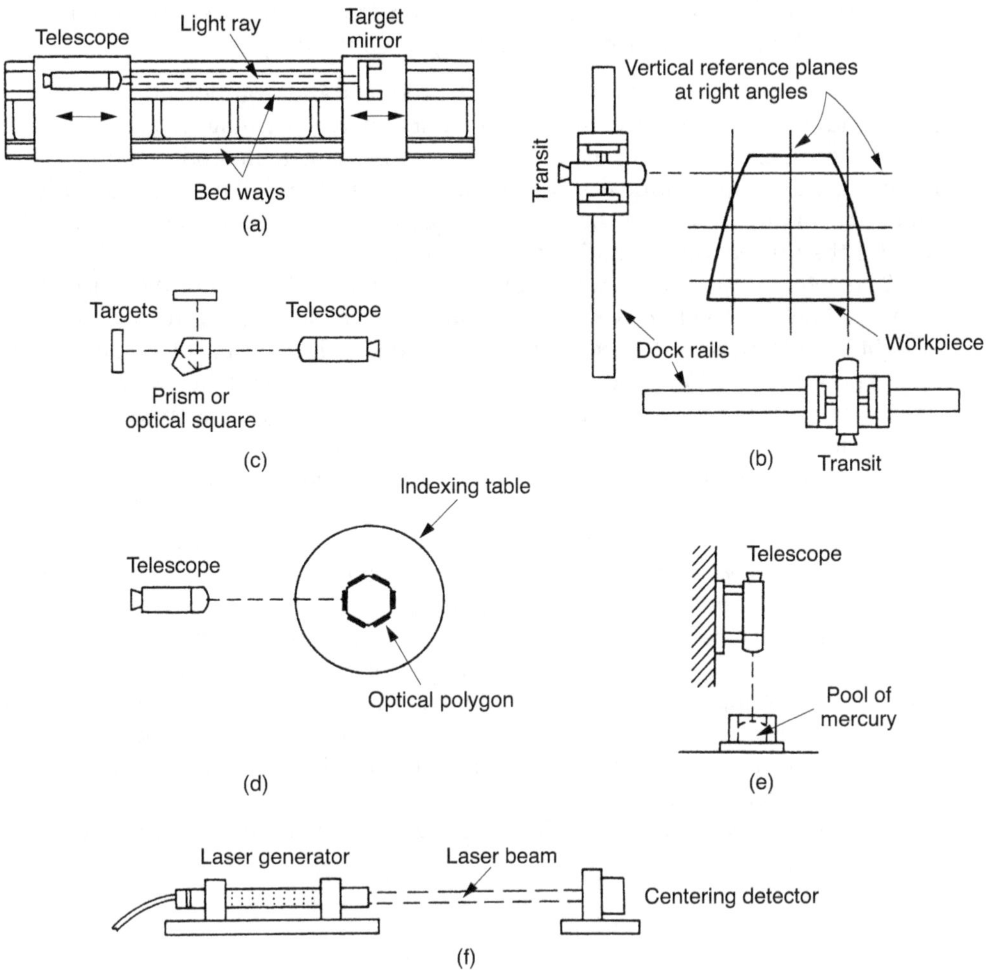

Figure 12.3 Optical tooling.
Reprinted by permission from Society of Manufacturing Engineers, *Manufacturing Processes and Materials*. 5th ed. 2015.

established with an optical square or planizing prism mounted on or in front of a telescope, as depicted in Figure 12.3c. Angles may be precisely set by autocollimating on the precisely located faces of an optical polygon, as in Figure 12.3d.

Leveling establishes a horizontal line of sight or plane. This may be done with a telescope fitted with a precision spirit level to fix a horizontal line of sight. A transit or sight level set in this manner may be swiveled around a vertical axis to generate a horizontal plane. *Plumbing*, as shown in Figure 12.3e, consists of autocollimating a telescope from the surface of a pool of mercury to establish a vertical axis.

An advanced step in optical tooling is the use of the intense light beam of a laser. A centering detector, as shown in Figure 12.3f, has four photocells equally spaced (top and bottom and on each side) around a point. Their output is measured and becomes equalized within the device, and is centered with the beam. This provides a means to obtain alignment with a straight line. Squareness may be established by passing a laser beam through an optical square.

Optical Comparators. Many industrial products and component parts are so small and of such complex configuration that they require magnification for accurate discernment. For this purpose, a number of measuring and gauging instruments using various optical systems (such as the toolmakers' microscope, the binocular microscope, and the optical projecting comparator) find wide application for the inspection of small parts and tools.

The *optical projecting comparator* projects a magnified image of the object being measured onto a screen. A workpiece is staged on a table to cast a shadow in a beam of light in *diascopic projection*, as shown in Figure 12.4. The outline of the part is magnified and displayed on a screen. In *episcopic projection*, the light rays are directed against the side of the object and then reflected back through the projection lens.

Optical projection provides a means to check complex parts quickly to small tolerances. Commonly, a translucent drawing is placed over the screen with lines drawn to scale for the contour of the part, the limits of the outline, or critical features such as angles. For instance, the outline of a part can be compared with a drawing on the screen, allowing deviations in the whole contour to be quickly seen. A fixture or stage may be supplied for a part to mount all pieces in the same way in rapid succession. The table can be adjusted in coordinate directions by micrometer screws or servomotors to 2 μm (0.0001 in.). Table positions can be determined from the micrometer readings or from digital readout devices. Thus, a part can be displaced to measure precisely how far a line is from a specified position. In addition, the screen can be rotated to a vernier scale to measure angular deviations in minutes or small fractions of a degree. Magnifications for commercial comparators range from 5× to as much as 500×. For example, at 250× magnification, 0.0020 mm (0.0008 in.) on a part

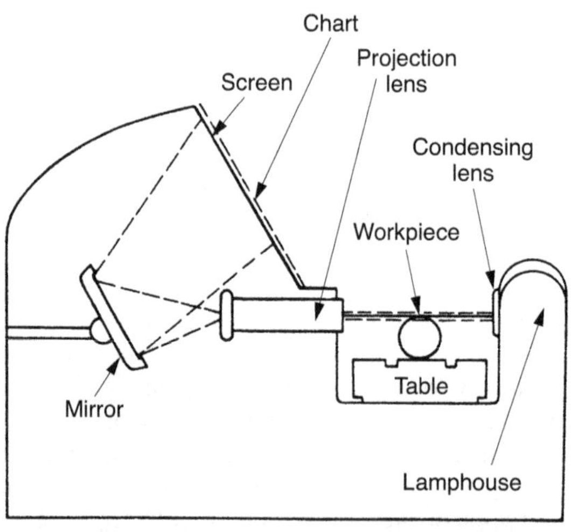

Figure 12.4 Optical comparator system.
Reprinted by permission from Society of Manufacturing Engineers, *Manufacturing Processes and Materials*. 5th ed. 2015.

becomes 0.500 mm (0.0197 in.) on the screen, which is readily discernible. The horizontal optical comparator shown in Figure 12.5 is table mounted and has a 356 mm (14 in.) diameter viewing screen. The designation "horizontal" means that the lens system is mounted horizontally, as illustrated in Figure 12.4. Comparators are also commercially available with a vertical lens configuration to facilitate the staging of thin parts.

One of the features of the comparator shown in Figure 12.5 is a computerized digital readout (DRO) located on top of the machine. The DRO has a two-axis digital display for establishing measurements in the x–y plane. In addition, a 12-character alphanumeric readout displays help messages, setup options, and the results of calculations. A fiber-optic edge-sensing device is also shown extending

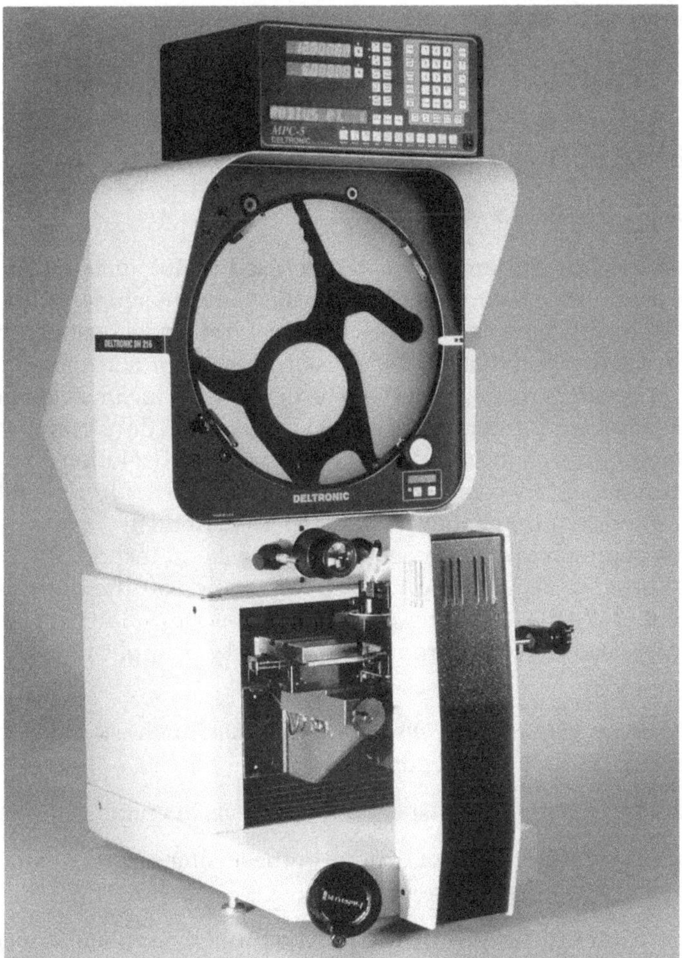

Figure 12.5 Horizontal optical comparator with a 356 mm (14 in.) viewing screen, digital readout, and edge-sensing device.
Source: Deltronic Corporation.

down the upper left portion of the screen. This device permits the DRO to precisely indicate the edges of a part. A 16-key external keypad mounted on the lower base provides the option of using the dedicated keys as they are identified or redefining any or all of those keys to execute any of 20 different programs containing up to 100 keystrokes each. The keypad includes a joystick capable of *x*-, *y*-, and *z*-axis control.

Another feature of the comparator shown in Figure 12.5 is an electric screen protractor that reads angles directly to either a minute or 0.01°. The angular setting of the protractor is displayed on an LED readout at the bottom right of the screen. The machine has built-in provisions for either diascopic projection (contour illumination) or episcopic projection (surface illumination) via a high-intensity tungsten-halogen light source. Lens changing is facilitated by the use of quick-change bayonet-type lens holders. Seven different lens magnifications are available, ranging from 5× to 100×, all with an optical focusing range of 76 mm (3 in.). This comparator includes the edge-sensing device, the computerized DRO, and one lens.

More discussion on optical flats and optical tooling can be found in Chapter 8.

SOFTWARE-BASED MEASUREMENT SYSTEMS

Machine Vision Systems

Typically, a machine vision system can be used to automate go/no-go inspection decisions, assembly verification, part location and machine guidance, gauging/dimensional measurements, feedback control loops, and a host of other tasks. It is a common misperception that machine vision systems provide generic optical detection and processing capabilities. While every system includes essential functions, most customers require some level of customization in development and should be cautious of vendors claiming to have one-size-fits-all solutions. Systems perform best in their own tightly controlled, highly specialized environment (Fabel 1997).

Application requirements vary drastically by industry, but a number of components are common to every machine vision system. Technology is rapidly evolving in all these areas, creating new opportunities on the manufacturing floor (Fabel 1997):

Machine vision systems are used in a variety of applications, including (Heil and Daigle 2004):

- *Pharmaceutical packaging inspection.* The machine vision systems are typically 21 CFR Part 11 compliant.
- Automotive lightbulb inspection that meets Six Sigma standards.
- Inspection of fill levels of opaque bottles using smart cameras in concert with X-ray technology.
- Inspection to ensure Teflon coating is properly applied to medical inhalant canisters.

Machine vision systems range from small, integrated units, or *smart cameras,* to PC-controlled units with frame-grabbers, to high-speed multiprocessor systems. At the high end, recent improvements include enhanced communications capabilities, permitting use with Ethernet, PC, robots, and Fieldbus systems with transfer of data and images to an enterprise file server or display for operator use. Dramatic changes also have been made in small integrated units, also called smart cameras, that include the basic camera with lens and camera functions to control speed and exposure, a microprocessor, and software (Axelrod 2010).

Other Digital Inspection Systems

As laser tracker technology has become more economically accessible, the laser trackers themselves have also become smaller, lighter, faster, and increasingly user-friendly. Options such as battery operation and IP54 (ingress protection) have allowed today's laser trackers to go where people never thought possible. As demand for this technology continues to expand, its availability has expanded to even the smallest job shops. Laser trackers are demanding a second look from a larger audience.

A laser tracker is a portable coordinate measurement machine that uses a laser beam to accurately measure and inspect the features of an object in 3-D space. It excels in terms of accuracy, reliability, and durability for large-scale metrology applications (Martin 2015). Handheld scanning, probing, and machine control products have increased the laser tracker's flexibility, and the number of possible applications has grown exponentially.

There are several triggering events that can justify making the investment in a new vision system. One is a desire to increase the throughput of an inspection system, which is set by the camera frame rate. A new requirement for color vision can also prompt movement to a new camera system. The need to replace failing components or ones that have become obsolete can be another triggering event. Virtually all of the new development in vision is based on digital camera systems, so a replacement for an obsolete analog camera that offers equivalent or better performance may be difficult to find. A digital camera quickly becomes the obvious choice.

Another benefit of digital inspection is the ability gained to augment camera operation with customized preprocessing. This preprocessing can represent unique value-added elements to the vision system's capabilities, or free the host PC from some imaging tasks to manage additional work. Digital cameras afford developers an opportunity to either lower the requirements for (and the cost of) the host PC's performance or increase system capabilities without requiring a new PC.

With all of the options and flexibility available, digital inspection makes sense. With the right camera, machine vision systems can increase productivity, simplify maintenance, and lower their total cost of ownership (Ahearn 2015).

MEASURING INCLINATION

Various companies have instruments that are designed for measuring slopes or inclination. Among these is the MSEM Inclinometer. An *inclinometer,* or *tilt sensor,* is an instrument used for measuring slope, tilt, or inclination by using gravity. An inclinometer creates an artificial plane defined at zero degrees, and measures the change in angle with respect to this plane. These devices go by many names, including tilt sensor, level sensor, bubble level, angle indicator, clinometer, protractor, slope meter, tilt meter, tilt/slope alarm, pitch/roll indicator, or level gauge. All refer to the same device (2GIG Engineering n.d.).

Chapter 13

Calibration

1. ***Calibration systems.*** Describe the principles and purpose of a calibration system, including the importance of establishing calibration intervals and uncertainty. Identify and use basic tracking and identification methods such as logs, stickers, radio frequency identifications (RFID), barcodes, and other identification codes to control calibration equipment. (Apply)
2. ***Calibration standards and equipment traceability.*** Understand and describe the hierarchy of standards, from working standards through international standards, and the documentation process of a measurement device traceable to the international standards. Recognize measurement results before and after an adjustment or repair is made. (Understand)
3. ***Gauge calibration environment.*** Understand the effects that environmental conditions have on the calibration process, such as temperature, humidity, vibration, and cleanliness of the gauge. (Understand)
4. ***Out-of-calibration effects.*** Describe the effects that out-of-calibration instruments can have on product acceptance and the actions to take in response to this situation. (Apply)

Body of Knowledge II.F

CALIBRATION SYSTEMS

The purpose of *calibration* is to ensure that various types of measurement and process equipment accurately and consistently perform as designed and intended. Further, it is also to ensure that equipment accuracy and consistency remain correlated with known quantities or values that are commonly referred to as *standards*. The basic principle of calibration, then, refers to the process of aligning measurement and process equipment performance with known quantities or values as specified in standards.

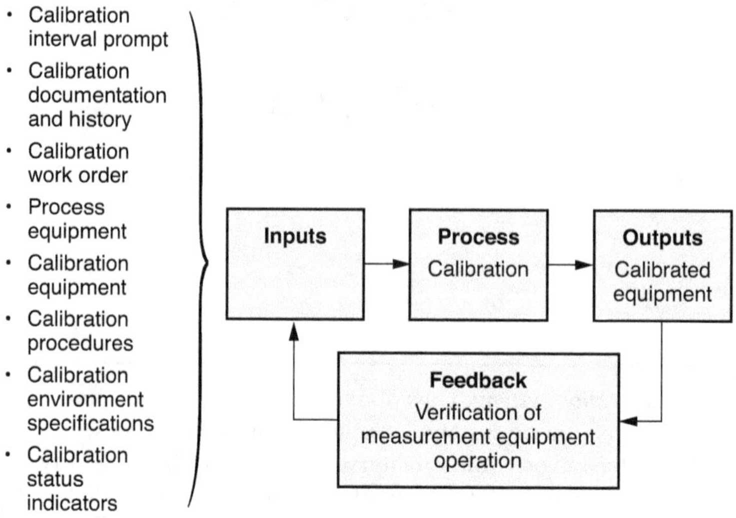

Figure 13.1 The calibration system.

A calibration system, like any type of system, is composed of inputs, processes, outputs, and feedback (see Figure 13.1).

Calibration Interval Prompt

The calibration process is initiated in one of two ways: (1) a piece of malfunctioning equipment is submitted for repair and calibration, or (2) a piece of equipment is identified as being in need of calibration in accordance with an established interval.

A calibration interval is an interval based on time, such as weekly, monthly, quarterly, semiannually, annually, or biannually. A calibration interval may also be an interval based on cycles of operation, such as every 1000 uses. A calibration interval is established for equipment identified as being influenced by, or characteristic of, any of the following (Vogt 1980):

1. Regulatory or oversight control
2. Importance in process operation
3. Manufacturer guidelines or requirements
4. Historical performance accuracy and consistency

The duration of a calibration interval may be shortened at the discretion of an equipment owner, calibration laboratory manager, or other authorized individual within an organization. A number of techniques are in use to establish calibration intervals initially and adjust them later. These methods include setting the same interval for all equipment in the user's inventory, the same interval for families of instruments (for example, oscilloscopes, digital voltmeters [DVMs], and gauge blocks), and the same interval for a given manufacturer and model number. These initial intervals are then adjusted for the entire inventory, for individual families, or for manufacturer and model numbers, respectively, based on analyses or history. A study conducted for the National Institute of Standards and Technology (NIST) in connection with a review of government laboratory practices identifies these and other methods (Vogt 1980). It is generally not possible or advisable to lengthen the duration of a calibration interval without a detailed analysis of equipment performance and, in the case of regulatory control, authorization by an agent of the cognizant department or agency.

The actual prompt initiating the calibration process may be in either electronic or hard copy form. Electronic information systems supporting modern industrial operations most commonly track equipment on established calibration intervals and prompt the appropriate individuals when calibration is required. Paper-based information systems supporting industrial operations for calibration require the establishment of a "tickler" file system. A *tickler file* is most commonly some form of card index wherein each piece of equipment with a calibration interval has its own card, and cards are sorted such that equipment in need of calibration moves toward the front of the index, and the appropriate individuals physically pull the cards as the calibration interval requires.

Uncertainty

In gauge calibration work, it is not uncommon to encounter situations where the typical uncertainty for a measurement is 50% of the tolerance for a given element when the gauge was new. Typical of this situation are pitch diameter measurements where the tolerances were generated long before measurement uncertainty became well known in the gauge making industry. When NIST's uncertainty falls in this range, you are not going to do better elsewhere, so you'll have to review your acceptance criteria to balance them against reality (Cox 2010).

Tracking and Identification Methods

The objective of identification and tracking systems has expanded in terms of usage and overall function, thus, the use of the captured data becomes the real differentiating factor. The time of big data has arrived, and the way you capture and identify information has become all the more important. Using the right technology to identify your products in production and shipping can really help (Anderson 2015).

Object-specific information can be saved, updated, and called up at any time on the transponder without having to be connected to a central system. This means that objects equipped with RFID transponders can be tracked along the entire logistics or production chain. For example, load carriers, which often circulate in large numbers, can be located and tracked, and are therefore guaranteed to find the way back to their own inventory (Anderson 2015).

The use of RFID is also beneficial when there are harsh ambient conditions, extreme temperatures, or heavy mechanical loads on the objects to be identified. Optical technologies always require a line of sight in order to detect the code, and are therefore more sensitive to wear or contamination, so often there is less maintenance with an RFID-based system (Anderson 2015).

So, whether you are using the information to help keep the food supply safe, reduce the proliferation of counterfeit medication, or just make sure your shipment gets to Grandma's house, the technology used can play a big role in making sure the data are accurate (Anderson 2015).

Calibration Documentation and History

Each piece of equipment with a calibration interval generally has repair, maintenance, and calibration documentation. Such documentation forms a critically important component in the traceability and history of the equipment. The information contained in the calibration documentation and history includes, but is not limited to, the calibration interval, identification of calibration procedures, a summary of actions taken during calibration, a summary of failures/discrepancies, a summary of parts replaced, identification of calibration technicians who have worked on the equipment, and perhaps any special comments or warnings, such as any particular equipment function that may not have been calibrated, or functions that are inoperable.

Calibration Equipment

Calibration equipment refers to equipment used during, or in support of, the calibration process. Calibration equipment may be assemblies or subassemblies needed to evaluate the performance of any given piece of equipment. Calibration equipment usually encompasses, but is certainly not limited to, things such as oscilloscopes, electrical meters, flow meters, temperature gauges, special jigs and fixtures, and associated clips, leads, and wires.

It is important to note that any or all of the equipment needed to support calibration processes may have its own calibration interval and, in fact, may be one of the many pieces of equipment with its own calibration requirements.

Calibration Procedures

Calibration procedures are the step-by-step instructions that describe the calibration process for a given piece of equipment. Such procedures are not normally stored with the equipment. While the equipment is installed for normal process use, the procedures are kept in the calibration laboratory.

Calibration procedures identify all process equipment models or configurations to which the procedures apply. The procedures also provide important safety warnings and identify any specifications for the calibration environment.

It is important to note that the calibration procedures are step-by-step instructions, not guidelines. Therefore, calibration technicians are not at liberty to deviate from the instructions for any reason unless there is a documented cause for process improvement and/or equipment traceability.

CALIBRATION STANDARDS AND EQUIPMENT TRACEABILITY

Calibration standards are known, highly accurate, and verifiable quantities used as the basis of comparison in calibration processes. Virtually all industrialized nations maintain a set of calibration standards for the measurement of various quantities and phenomena.

NIST is the custodian of measurement standards in the United States. NIST was established by an act of Congress in 1901, although the need for such a body had been noted by the founders of the Constitution. NIST has two main facilities and laboratories in Gaithersburg, Maryland, and Boulder, Colorado, where research into the phenomenon of measurement, the properties of materials, and calibration of reference standards is carried out.

There are several levels of calibration standards arranged in a hierarchy. At the highest level are international standards, which serve as the basis of trade between nations. At the lowest level are transfer standards, which serve as the basis of trade between organizations. Calibration standards at the lowest levels in the hierarchy are used to support day-to-day operations by technicians and shop floor operators. Calibration standards in the middle of the hierarchy are generally used by personnel working in calibration laboratories who are dedicated to calibration processes. Calibration standards at the highest levels in the hierarchy are generally used by calibration specialists and government officials.

Figure 13.2 presents a hierarchy of calibration standards.

- *International standards.* A standard recognized by an international agreement to serve internationally as the basis for fixing the value of all other standards of the quantity concerned.
- *National standards.* A standard recognized by an official national decision to serve in a country as a basis for fixing the value of all other standards of the quantity concerned. Generally, a national standard in a country is also a primary standard to which other standards are traceable.
- *Primary standards.* A standard designed or widely acknowledged as having the highest metrological quality and whose value is accepted without reference to other standards of the same quantity. National standards are generally primary standards.

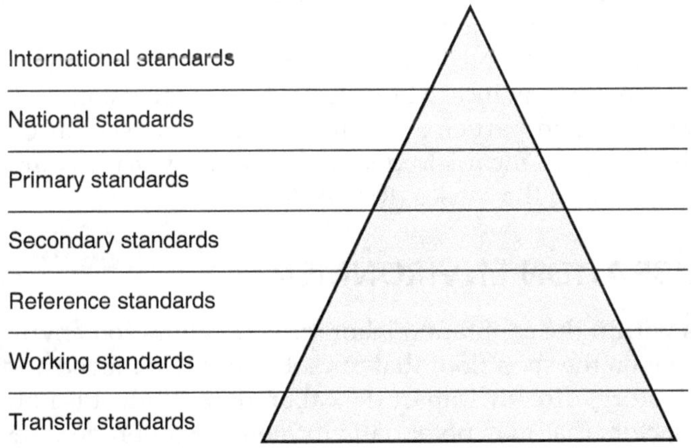

Figure 13.2 Calibration standards hierarchy.
Source: Used by permission from Bucher (2004, 2006).

- *Secondary standards.* A standard whose value is based on comparisons with some primary standard. Note that a secondary standard, once its value is established, can become a primary standard for some other user.
- *Reference standard.* A standard with the highest metrological quality available at a given location from which the measurements made at that location are derived.
- *Working standard.* A measurement standard not specifically reserved as a reference standard that is intended to verify measurement equipment of lower accuracy.
- *Transfer standard.* A standard that is the same as a reference standard except that it is used to transfer a measurement parameter from one organization to another for traceability purposes.

To maintain accuracy, standards in industrialized nations must be traceable to a single source, usually the country's national standards. Since the national laboratories of industrialized nations maintain close connections with the International Bureau of Weights and Measures, there is assurance that items produced from calibration standards in one country will be consistent with items produced from calibration standards in other countries.

Equipment Traceability

Traceability is a process intended to quantify a laboratory's measurement uncertainty in relationship to the national standards. Traceability is based on analyses of error contributions present in each of the measurement transfers: the calibration of the laboratory's reference standards by the NIST, the measurements made in the calibration transfers within the laboratory, and the measurements made on a product. Evidence of traceability is normally required. Such traceability may be as simple as retention of certificates and reports of calibration or as complex as reproduction of the analyses demonstrating the uncertainties claimed for the measurements (Rice 1976).

A laboratory that maintains its own reference standards (that is, it relies on no laboratory other than the NIST for calibration of its standards) must continually monitor its own performance. Measurements on check standards, intercomparisons of standards, and participation in measurement assurance programs sponsored by the NIST are means to quantify laboratory error sources as well as to provide indications of the causes (Rice 1976).

GAUGE CALIBRATION ENVIRONMENT

Conditions within the calibration laboratory or calibration environment (such as for equipment on the shop floor that cannot be easily removed and taken to a calibration laboratory) directly impact the calibration. While it is not possible to control all the factors that may potentially influence a calibration, several factors are controlled for nearly all calibrations, such as cleanliness, temperature, humidity, and vibration. Any additional factor that is known or believed to impact a calibration (for example, pressure or flow rates in certain types of sensors) may require additional controls.

OUT-OF-CALIBRATION EFFECTS

The effects of using out-of-calibration equipment are the same as the type I and type II errors (also known as producer's and consumer's risks) discussed in Chapters 15 and 18 of this book. In essence, the effects of out-of-calibration equipment cause stakeholders to believe that (1) the equipment is calibrated and functioning properly when it is not, or (2) the equipment appears to fail calibration when it is, in fact, functioning correctly.

Using out-of-calibration equipment in production or service delivery operations can cause a number of difficulties for manufacturers and service providers. In a best-case scenario, once discovered, out-of-calibration equipment functions correctly and exceptions reporting must document the out-of-calibration incident, a *corrective action* plan must be developed/initiated, and product or service quality must be systematically verified—all of which is wasteful, potentially compromises customer confidence and goodwill, and costs the company time and money. In a worst-case scenario, once discovered, out-of-calibration equipment does not function correctly and containment of product or service delivery must be initiated, *material control* and segregation procedures must be employed, a comprehensive evaluation of product/service performance must be conducted, material review board action is required, a root cause analysis is required, warranty or recall may become necessary, and a company is exposed to legal/regulatory action.

At a minimum, procedures for dealing with an out-of-calibration event or discovery include the following:

1. Identification of the out-of-calibration condition (as described earlier in terms of type I or type II errors).
2. Determination of the magnitude of the condition (that is, how far out of calibration was the equipment?).
3. Assessment of when the out-of-calibration condition occurred.
4. Quantification of the amount of product/service delivery produced/delivered during the out-of-calibration condition.
5. Evaluation of product or service delivery status (that is, was any product produced or any service provided to customers during the out-of-calibration condition?).
6. Identification of who is authorized to manage the containment efforts.

Chapter 14

Measurement System Analysis (MSA)

Define and understand the following elements of MSA. (Understand)

1. Bias
2. Stability
3. Precision
4. Accuracy
5. Linearity
6. Repeatability and reproducibility (R&R) studies

Body of Knowledge II.G

Measurement system analysis (MSA) is used to understand and quantify the variability associated with measurements and measurement systems. It should be noted that a discussion of MSA should come *after* the mastery of statistical content provided in Chapter 7 (Basic Statistics and Applications). However, MSA is being discussed prior to the discussion of basic statistics and applications in this book to preserve the order of topics as presented in the Certified Quality Inspector Body of Knowledge.

The third edition of *Measurement Systems Analysis Reference Manual,* from the Automotive Industry Action Group (AIAG), defines *measurement* as "the assignment of numbers [or values] to material things to represent the relations among them with respect to particular properties" (AIAG 2002). Similarly, the manual defines a *measurement system* as "the collection of instruments or gauges, standards, operations, methods, fixtures, software, personnel, environment and assumptions used to quantify a unit of measure or fix assessment to the feature characteristic being measured; the complete process used to obtain measurements" (AIAG 2002).

These definitions will serve as the basis of our discussion. However, before we begin that discussion we must first provide a clarification. The term "MSA" is commonly used interchangeably with the term "gauge repeatability and reproducibility" (gauge R&R). MSA is a more comprehensive analysis quantifying variability components from gauge stability, gauge bias, gauge linearity, gauge repeatability, and reproducibility (i.e., variability from operator measurements). Gauge R&R is, in fact, a subset or component of MSA. The reason underlying the mistaken interchange of the terms "MSA" and "gauge R&R" is that many, if not most, problems with measurement systems are detected and corrected with the gauge R&R procedure without having to continue the more comprehensive analysis of MSA. For a detailed discussion of analysis techniques for gauge stability, gauge bias, and gauge linearity, readers are referred to the third edition of the AIAG *Measurement Systems Analysis Reference Manual.*

PROCESS VARIABILITY

All processes, no matter how well they are designed and executed, possess the component of variability. It is only the degree of variability that varies from process to process. The fact that variability exists in virtually any and all processes cannot be overemphasized. The total variability of a process consists of two components: parts variability and gauge variability. Further, note that variability in a process can be quantified by a probability distribution and can be expressed as:

$$\sigma^2_{\text{Total}} = \sigma^2_{\text{Parts}} + \sigma^2_{\text{Gauge}} \tag{14.1}$$

VARIABILITY IN THE MEASUREMENT PROCESS

Extending our conversation from the previous section, it is critically important that we understand that the methods, procedures, tools, and equipment we use to make measurements constitute an independent process that creates, and is susceptible to, its own variation. This means there are two sources or components of variation in each measurement, as illustrated by Equation 14.1 and Figure 14.1.

It is important to ensure that the component of variability associated with the measurement system does not consume an excessive amount of variability as is allowed in the process specification. Therefore, quantifying measurement system variability is the purpose of MSA.

EVALUATING MEASUREMENT SYSTEM PERFORMANCE

In any process, we usually focus on bringing the process under statistical control. In other words, we work to reduce the overall variation in the process. In this chapter, we focus entirely on MSA, which constitutes an important part of the total process variation. As mentioned earlier, the total process variation can be divided into two major categories: variation due to parts and variation due to gauge or measurement system. Part-to-part variation may be due to the environment, methods, materials, machines or some combination thereof, and other factors.

The variation due to the measurement system mainly consists of two major components: the instrument being used for taking measurements and the operators who use the instrument.

In the industrial world, these components are usually referred to as repeatability and reproducibility, respectively. Repeatability and reproducibility may be considered as the major indicators of measurement system performance. We will discuss repeatability and reproducibility in more detail later.

Since repeatability refers to the variation generated by the instrument (i.e., measurement equipment or gauge), it is referred to as equipment variation (EV).

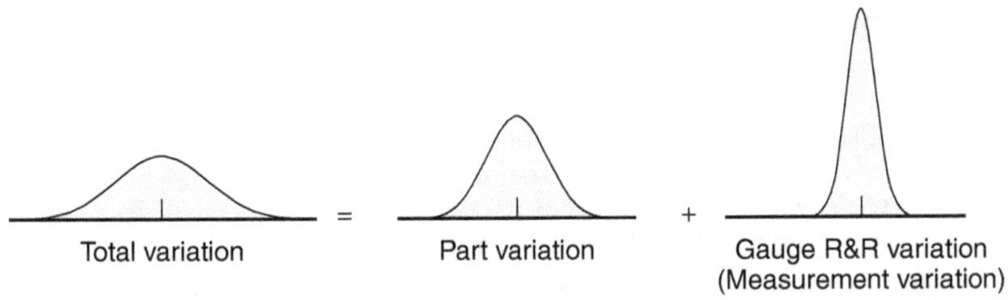

Figure 14.1 Components of total variation.

Reproducibility refers to variation generated by operators using measurement instruments, and it is referred to as operator (i.e., appraiser) variation (AV). The study of gauge R&R is usually referred to as *gauge R&R study*.

The *Measurement Systems Analysis Reference Manual* (AIAG 2002) prescribes several methods for conducting gauge R&R. In this book we will study two methods: the range-based method and the analysis of variance (ANOVA) method. Since ANOVA is an advanced statistical technique that is not covered in this book, we will not discuss it in detail. We will focus on explaining and interpreting the results of our example that we will work out using computer software.

GAUGE REPEATABILITY AND REPRODUCIBILITY STUDIES

The range-based method has been presented by various authors including IBM (1986) and Barrentine (2003). Before we discuss the details of this method, we will define the terms *measurement capability index*, K_1-factor, and K_2-factor, as well as other terms useful in understanding MSA.

MSA is a technique for collecting and analyzing data to evaluate the effectiveness of the gauge. To collect data, we randomly select a few parts and a number of operators (usually three or more, but as a general rule, the more the better). Each operator takes multiple measurements (at least two) on each part, and all the parts are measured in random order. These measurements are also known as trials. Using the terminology of control charts, the measurements on each part or the number of trials constitutes a rational subgroup, and the number of parts times the number of operators constitutes the number of subgroups or samples. Then $\bar{R}$ is defined as the average of the ranges of trials within the same operator, and $\bar{\bar{R}}$ is defined as the average of the $\bar{R}$'s among the operators.

The *measurement capability index* (MCI) is a measurement that quantifies our belief that the gauge is reliable enough to support the decisions we make under the existing circumstances. The MCI relates to the following characteristics that are key to any measurement system:

- Bias
- Precision
- Accuracy

- Stability
- Linearity
- Repeatability and reproducibility (R&R)

A measurement system is *biased* if it gives a measurement value that consistently overstates or understates the true value of the measurement. *Precision* refers to how close to each other any two or more measurements of a part are. Note that precision is independent of accuracy. A measurement may be very precise but not very accurate (see Figure 14.2). The characteristic *precision* is further subdivided into two categories: repeatability and reproducibility. *Repeatability* measures the preciseness of the observations taken under the same conditions, which is achieved by computing the variance of such observations. For example, we say a gauge possesses the characteristic of repeatability if an operator obtains similar observations when measuring the same part again and again. *Reproducibility* measures the preciseness of the observations taken by different operators when measuring the same part. For example, we say a gauge possesses the characteristic of reproducibility if various operators obtain similar observations when measuring the same part again and again.

Accuracy of a measurement system is the closeness of the average of measurements taken to the true value. The distinction between precision and accuracy is explained well in Figure 14.2.

Stability is defined by the total variation in measurements obtained with a measurement system on the same master or same parts when measuring a single characteristic over an extended period of time. The smaller the total variation, the more stable the measurement system.

Linearity is the difference between the true value (master measurement) and the average of the observed measurements of the same part that has the same distribution over the entire measurement range. Linearity is best explained by the diagram in Figure 14.3.

The following example illustrates gauge repeatability and reproducibility study.

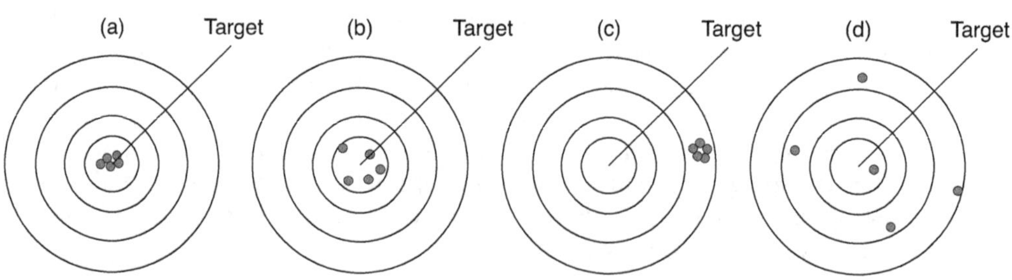

Figure 14.2 (a) Accurate and precise, (b) accurate but not precise, (c) not accurate but precise, (d) neither accurate nor precise.

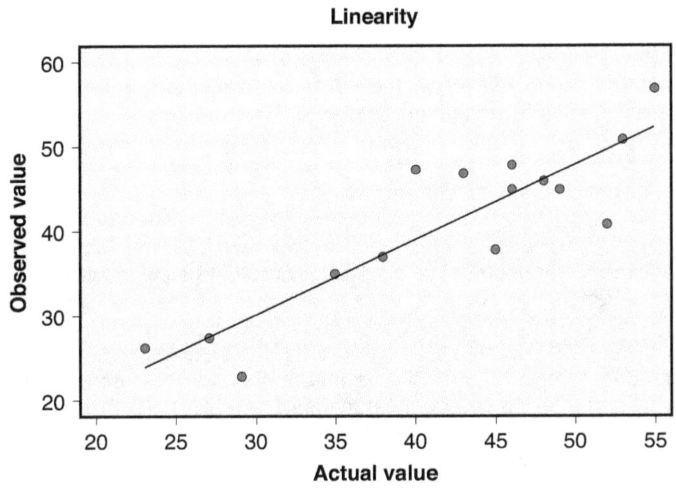

Figure 14.3 Diagram showing the linear relationship between the actual and the observed values.

Example 14.1

A manufacturer of bolts used in automotive applications has installed a new measuring gauge. In order to perform the MSA on the new gauge, the quality manager randomly selected three operators from the Department of Quality Control, who decided to take a random sample of 10 bolts. Each operator takes 3 measurements on each bolt, which is randomly selected. The data obtained are shown in Table 14.1.

Table 14.1

Data on an experiment involving three operators, ten bolts, and three measurements (in mm) on each bolt by each operator.

Bolt (part)	Operator 1			Operator 2			Operator 3		
number	Trial 1	Trial 2	Trial 3	Trial 1	Trial 2	Trial 3	Trial 1	Trial 2	Trial 3
1	26	22	26	21	23	21	24	22	26
2	28	26	28	24	29	26	24	25	24
3	28	31	28	28	27	28	32	30	27
4	35	33	31	35	31	30	34	35	31
5	37	35	38	36	38	35	35	34	35
6	40	38	40	40	38	40	36	37	38
7	39	42	41	40	39	43	43	41	43
8	42	43	46	42	46	42	43	44	45
9	50	52	50	53	52	53	49	53	49
10	28	31	28	28	27	28	32	30	27
	$\bar{R}_1 = 3.0$ $\bar{x}_1 = 35.40$			$\bar{R}_2 = 2.8$ $\bar{x}_2 = 34.77$			$\bar{R}_3 = 3.0$ $\bar{x}_3 = 34.93$		

(continued)

Solution

We first discuss gauge R&R using the range-based method, the approach used by IBM (1986), Barrentine (2003), and others.

Step 1:

Verify that gauge calibration is current.

Step 2:

Identify operators. Three operators are typically used in gauge studies; however, the more operators the better.

Step 3:

Select a random sample of parts and have each operator measure all parts. One operator measures all parts, taking several measurements on each part, then the second operator takes measurements, the third operator takes measurements, and so on. All parts are measured in random order.

Step 4:

Calculate the sample mean, intertrial range for each sample, and average range for each operator. The sample means and average ranges are provided in Table 14.1

Step 5:

Calculate the range of sample means ($R_{\bar{x}}$), that is

$$\begin{aligned} R_{\bar{x}} &= \max(\bar{x}_i) - \min(\bar{x}_i) \\ &= 35.40 - 34.77 = 0.63 \end{aligned}$$

Step 6:
Calculate the average range ($\bar{R}$) for operators

$$\begin{aligned} \bar{R} &= \frac{\bar{R}_1 + \bar{R}_2 + \bar{R}_3}{3} \\ &= \frac{3.00 + 2.80 + 3.00}{3} = 2.93 \end{aligned}$$

Step 7:
Calculate repeatability (this value is also referred to as equipment variation (EV) and the estimate of the standard deviation of repeatability $(\sigma_{\text{Repeatability}})$

$$\text{Repeatability } (EV) = \bar{\bar{R}} \times K_1 = 2.93 \times 3.05 = 8.94$$

where from Appendix F with number of trials $r=3$, $n=\#$ of parts, $\times \#$ of operators $=30$, we have $K_1 = 3.05$.

$$\hat{\sigma}_{\text{Repeatability}} = \frac{EV}{5.15} = \frac{8.94}{5.15} = 1.74$$

Step 8:
Calculate reproducibility (this value is also referred to as appraiser or operator variation (AV) and the estimate of the standard deviation of reproducibility [$\sigma_{\text{Reproducibility}}$])

$$\text{Reproducibility (AV)} = \sqrt{(R_{\bar{x}} \times K_2)^2 - \left[(EV)^2 / m \times r\right]}$$

where from Appendix G for three operators, we have $K_2 = 2.70$, # of parts $= m = 10$, and # of trials $r = 3$. Thus, we have:

$$\begin{aligned}\text{Reproducibility (AV)} &= \sqrt{(0.63 \times 2.70)^2 - \left[(8.94)^2 / 10 \times 3\right]} \\ &= 0.48\end{aligned}$$

Note that if the number under the radical is negative, then AV is zero.

The estimate of the standard deviation of reproducibility $\left(\sigma_{\text{Reproducibility}}\right)$ is:

$$\hat{\sigma}_{\text{Reproducibility}} = \frac{0.48}{5.15} = 0.09$$

With this method, since reproducibility is calculated by ignoring the interaction term, the standard deviation of reproducibility may merely be looked upon as the operator standard deviation.

Step 9:
Calculate gauge R&R (i.e., repeatability and reproducibility) and the estimate of gauge R&R standard deviation:

$$\begin{aligned}\text{Gauge R\&R} &= \sqrt{(\text{Repeatability})^2 + (\text{Reproducibility})^2} \\ &= \sqrt{(8.94)^2 + (.48)^2} \\ &= 8.95\end{aligned}$$

The estimate of gauge R&R standard deviation is given by

$$\hat{\sigma}_{\text{Gauge}} = \frac{8.95}{5.15} = 1.74$$

$$\sigma^2_{\text{Total}} = \sigma^2_{\text{Parts}} + \sigma^2_{\text{Gauge}}$$

where σ^2_{Total}, σ^2_{Parts} and σ^2_{Gauge} are the variance of total variability, parts variability, and gauge variability, respectively.

MEASUREMENT CAPABILITY INDICES

As the process capability index (PCI) quantifies the ability of a process to produce products of desired quality, the MCI quantifies the ability of a measurement system to provide accurate measurements. In other words, MCI evaluates the adequacy of the measurement system. There are various MCIs in use. We discuss here the other two most commonly used MCIs: percent of process variation and process specification. The MCI should always be used in conjunction with the PCI (Barrentine 2003).

MCI as a Percentage of Process Variation (MCI_{pv})

$$MCI_{pv} = 100 \times \frac{\hat{\sigma}_{\text{Gauge R\&R}}}{\hat{\sigma}_{\text{Total}}} \tag{14.2}$$

The criteria for assessment of this index (Barrentine 2003) are usually as follows:

1. $\leq 20\%$ implies the measurement system is good.

2. $>20\%, \leq 30\%$ implies the measurement system is marginal.
3. $>30\%$ implies the measurement system is unacceptable.

MCI as a Percentage of Process Specification (MCI_{ps})

$$MCI_{ps} = \frac{6 \times \hat{\sigma}_{\text{Gauge R\&R}}}{USL - LSL} \tag{14.3}$$

The criteria for assessment of MCI_{ps} are similar to MCI_{pv} (Barrentine 2003), that is,

1. $\leq 20\%$ implies the measurement system is good.
2. $>20\%, \leq 30\%$ implies the measurement system is marginal.
3. $>30\%$ implies the measurement system is unacceptable.

Section III

Inspection and Test

Chapter 15

Blueprints; Drawings; Geometric Dimensioning & Tolerancing (GD&T); and Model Based Definitions

1. ***Blueprints, engineering drawings, and model based definitions.*** Define and interpret various sections of technical drawings, such as title blocks, tolerances, and change or revision blocks, as well as notes, scale, and size details. Use this information to conduct inspections. (Apply)
2. ***Terminology and symbols.*** Define and interpret drawing views and details for product specifications or other controlling documents. Define and use various terms and symbols from the ASME Y14.5 Standard. (Analyze)
3. ***Position and bonus tolerances.*** Calculate position and bonus tolerances from various drawings. (Analyze)
4. ***Part alignment and datum structure.*** Determine part alignment and setup using the datum structure. (Analyze)

Body of Knowledge III.A

INTRODUCTION

Geometric dimensioning and tolerancing (GD&T) is defined as the language of symbols used in mechanical drawing to efficiently and accurately communicate geometry requirements for features on parts and assemblies (ASME 1994). GD&T is used in a wide variety of manufacturing industries nationally and internationally to ensure successful and accurate communication between the design and the production departments. It is the language that enables design engineers to express what they mean on a drawing, thus improving product designs and lowering costs. Process engineers and manufacturing use the language to interpret the design intent and to determine the best manufacturing approach. Quality control and inspection use the GD&T language to determine proper setup and part verification (ASME 1994).

Understanding how to implement and translate GD&T drawings will aid in improving product design and effectively communicate design requirements for suppliers and manufacturing (Effective Training 2012).

Blueprints and Engineering Drawings, and Model Based Definitions

An *engineering drawing* is a type of drawing that is technical in nature, used to fully and clearly define requirements for engineered items. The purpose of engineering drawings is to accurately and clearly capture all the geometric features of a product or a component that will convey the required information for the production team.

Engineering drawings are often referred to as *blueprints,* a name that reflects the way those prints were once made using chemical material to print blue lines on white paper. Although design engineers now use computers and other modern production processes that use different colors, they still call engineering drawings "blueprints" (Ware 1999).

Figure 15.1 depicts an example of a blueprint done using today's technology. Engineering drawings can be used in manufacturing, construction, operation, installation, and maintenance, among other applications. They are made in a way that is understandable for all users. When these drawings are used by two or more countries, it is essential that the drawing prepared in one country be easily and correctly interpreted in the other(s). These countries don't have to speak the same language or use the same units of measure. The drawing can be understood and translated based on the international standards provided by ISO/TC 10 technical drawing standards (ISO 2002).

Sections of Technical Drawings

In technical drawings, the following sections can be identified (Dirks 1995–2000):

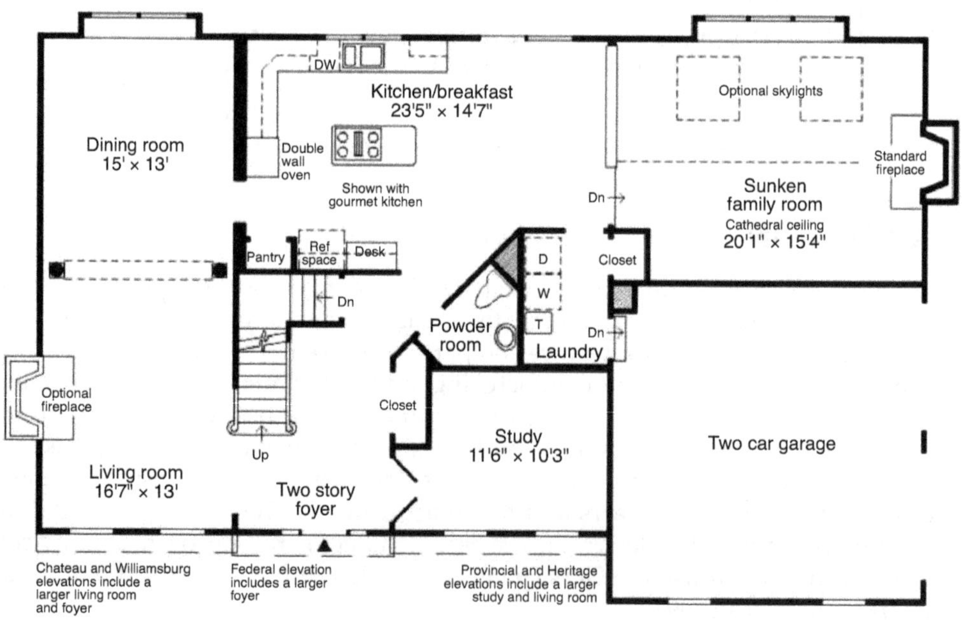

Figure 15.1 Blueprint for a house floor plan.

- Title block
- Notes
- Scale
- Revision block
- Tolerance

Title Block. The title block includes the border and the various sections for providing quality, administrative, and technical information, i.e., all the information that enables the drawing to be interpreted, identified, and archived. The title block is usually located in the lower right corner, as shown in Figure 15.2. This block must contain the title of the project or drawing, a description of the drawing, the scale, the name of the designer and/or draftsman, and the date. It might also contain information about the company, registration or ID number, and previous revisions. The title block should also include boxes for the legal signatures of the originator and other persons involved in production of the drawing to the required quality. The drawing sheet size should be in accordance with BS EN ISO 5457 TD *Sizes and layout of drawing sheets* (Dirks 1995–2000).

Notes. Notes are placed on drawings to give additional information to clarify an object on the blueprint. The notes area is not part of the title block. Figure 15.2 shows the placement of notes on engineering drawings. Notes might include legends and symbols or specially marked places on the blueprint (Dirks 1995–2000).

Scale. The scale block in the title block of the blueprint shows the size of the drawing compared with the actual size of the item. The scale may be shown as an enlargement or a reduction of the actual size. For example, a scale of 1/100 means that 1 unit on the drawing is equal to 100 units in real (actual) size (see Figure 15.3).

Revision Block. The revision block is located in the upper-right corner of the blueprint or technical drawing. All modifications to the drawing are documented in this block. The most important modifications needed in the revision block are the revision date, revision number, revision symbol, change description, change authorization, and change code (MTAG 2001) (see Figure 15.4).

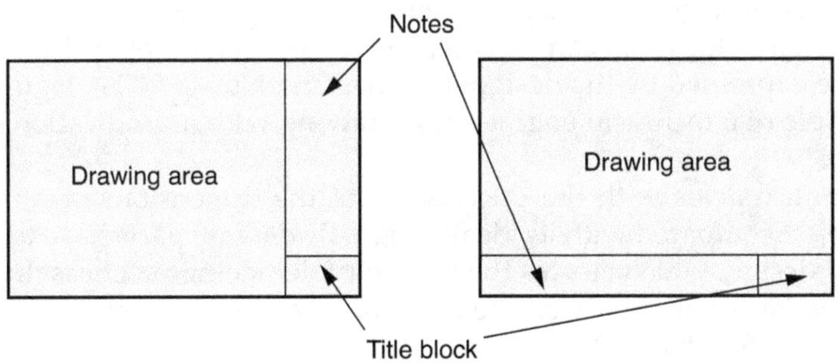

Figure 15.2 Placement of the title block and notes on engineering drawings.
Source: Fox (2012).

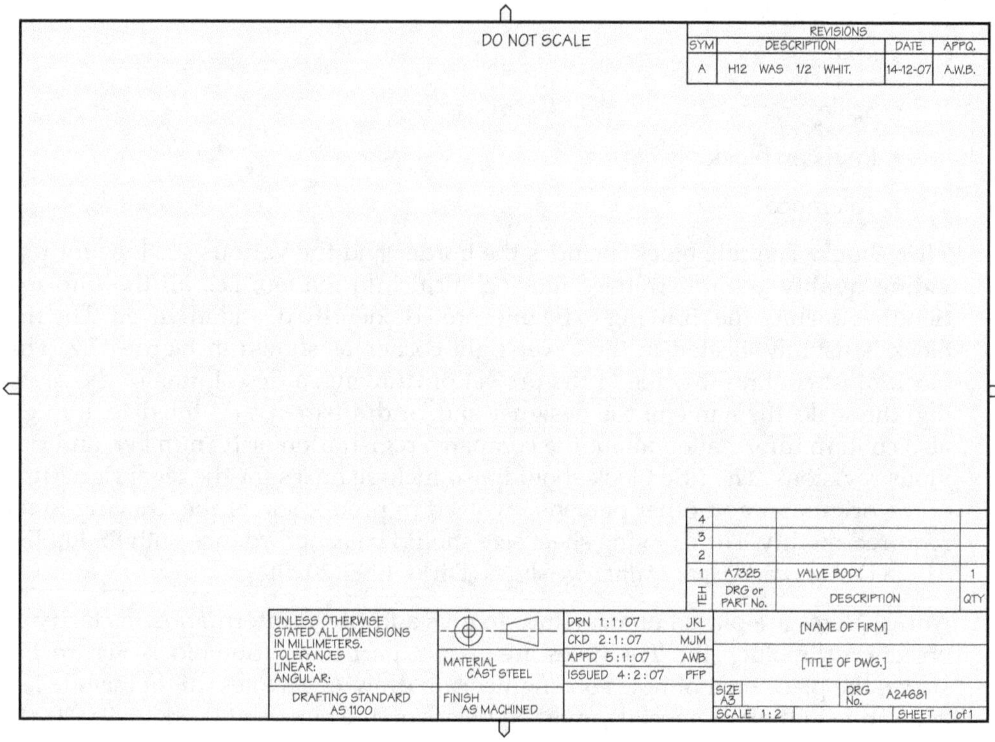

Figure 15.3 Example of title and notes blocks on an engineering drawing.
Source: Digital Design (2012).

Tolerance. ASME Y14.5 (1994) defines *tolerance* as "the total amount by which a specific dimension may vary." Tolerance is the range of allowed values, which is normally specified by a certain standard. Tolerance is one of the main sections that appear in the engineering drawing and shows the variation of each dimension (Pyzdek and Berger 2003; ASTM 1996). The drawing includes the allowed boundaries of a generic shape in terms of its geometry and of the allowed dimensional values that can be accepted for a variation over the exact or nominal value expressed by the designer (Cañas and Novak 2006). Figure 15.5 shows an example of a technical engineering drawing with an indication of geometric tolerance.

Tolerance deals with the uncertainty of the dimensions of a real manufactured object compared with its ideal design. By default, parts have to deviate from the ideal design, which creates the need for tolerancing practices. In the drawing world, tolerances are noted on the drawing per standard notations such as those in ASME Y14.5 or ISO 1101.

There are two main classes of tolerances: dimensional and geometric. Geometric tolerances are the more complex of these. Geometric tolerances provide more flexible means for controlling shape than do dimensional tolerances. They achieve this by enabling tolerances to be defined independently of explicit dimensions. This enables tolerances to be specified that are more closely related to the

13	14	15	16		
Rev No	Revision note		Date	Signature	Checked
A	SECTION ON A–A DELETED		29/12/03	R.B.	All other

Figure 15.4 Example of a revision block.
Source: Roymech, UK (2007).

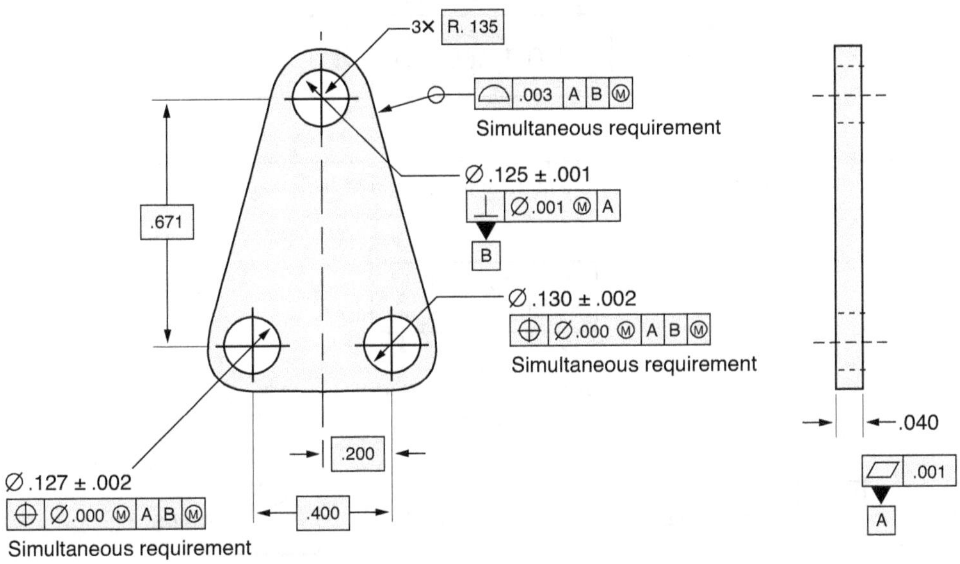

Figure 15.5 Example of a technical engineering drawing with an indication of geometric tolerances.
Source: Cañas and Novak (2006).

functional requirements of the design, such as strength and fit, by specifying item features and characteristics (Lamit and Kitto 1994).

From the previous discussion we can understand that a geometric tolerance describes a constraint on the acceptable deviation of a manufactured object from the ideal design. Tolerances are applied to the geometric features of a part, such as faces and holes.

There are several subtypes of the geometric tolerance entity that are not mutually exclusive. For example, tolerances that reference datums are of type *geometric tolerance with datum reference*. Tolerances that include a modifier such as maximum material condition are of type *modified geometric tolerance*. Many typical engineering tolerances combine these two types (Briggs and Hendrix 2003).

Consider an example of a location tolerance on a hole whose visual representation as a control frame is shown in Figure 15.6. Reading the control frame left to right, the location tolerance is a position tolerance, locating a cylindrical feature, with a tolerance value of 0.1, with a modifier condition of maximum material condition, referencing datums A, B, and C, in that order (Briggs and Hendrix 2003). Figure 15.7 provides the features that can be represented by geometric tolerancing. Figure 15.8 provides a simple 2-D example wherein a position tolerance ⊕ is imposed on hole H. The exact ("true") position of H is defined in Figure 15.8a by two basic (boxed, signifying "exact") dimensions that are anchored to part features A and B; the labels on these features designate them as datums. The left-to-right ordering of A and B in the ⊕ "callout block" sets the datum precedence: A is primary and B is secondary. The position tolerance ("<tolspec>" in Figure 15.8a) is a small disk-shaped zone defining the allowable variability of H's position (Briggs and Hendrix 2003).

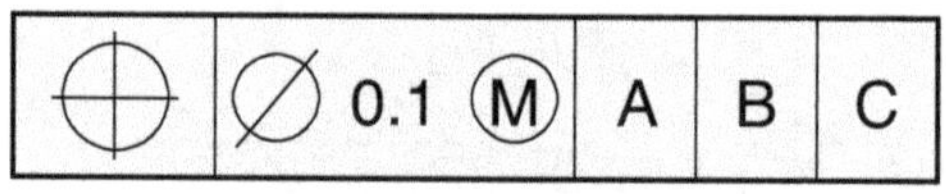

Figure 15.6 Visual representation of the control frame of a hole.

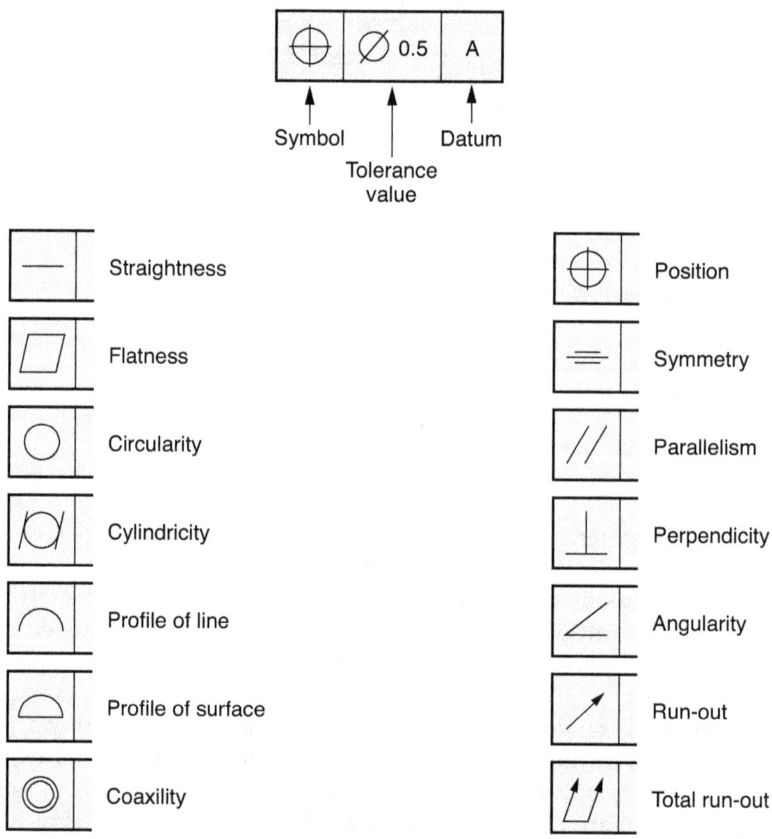

Figure 15.7 Features that can be specified by geometric tolerancing.
Source: Roymech, UK (2007).

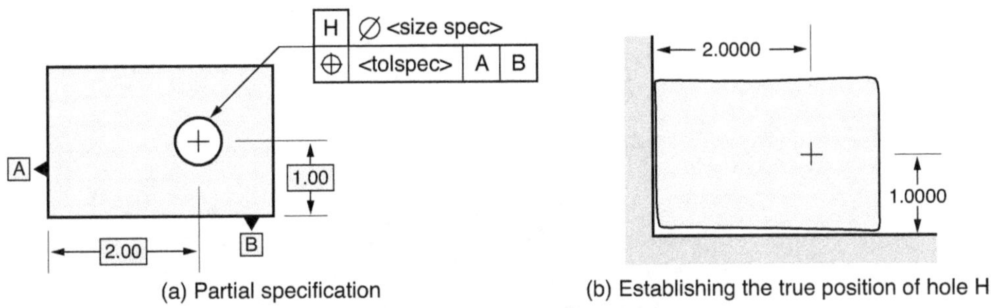

Figure 15.8 Simple 2-D example of position tolerance.

Model Based Definition

Model based definition (MBD) is not a new concept in the aerospace manufacturing industry. While it has been applied to product definition and manufacturing for years, its use as a fully implemented process for quality control has been realized only recently. Many companies are modernizing quality assurance processes to embrace MBD because prime manufacturers in aerospace are demanding it. Advancements in both software and hardware technologies are making it easier to put all the key elements in place, giving aerospace manufacturers the last piece of the puzzle for an enterprise-wide system of digital product definition. With MBD, an organization eliminates ambiguities, and creates many opportunities for cost reductions and improvements in lean principles (Knoche 2006).

Terminology and Symbols

The *Y14.3-2003 Multiview and Sectional View Drawings* standard published by ASME establishes the requirements for creating orthographic views for item description. The standard includes the multi-view system of drawing, selection, and arrangement of orthographic views, auxiliary views, sectional views, details, and conventional drawing practices. This allows engineers to view different angles and look at more detailed pictures for the item (ASME 2003). The different views and symbols can be interpreted to extract the required information and the specifications for that item. For example, orthographic views show a top view, a front view, and a right side view. One view is not sufficient to describe the item in detail, so additional views are required. The number of orthographic views needed to describe an object fully depends on the complexity of the object. For example, a simple metal washer can be fully described using only one orthographic view; however, an extremely complex object may require as many as six views (top, front, left side, right side, back, and bottom). Most objects can be sufficiently described using three views: top, front, and right side, as shown in Figure 15.9.

The symbols on the drawing have different representations based on the ASME standards. For example, each line style has its own interpretation. Figure 15.10 shows the different line symbols that can be used in an engineering drawing (Gunderson 1994).

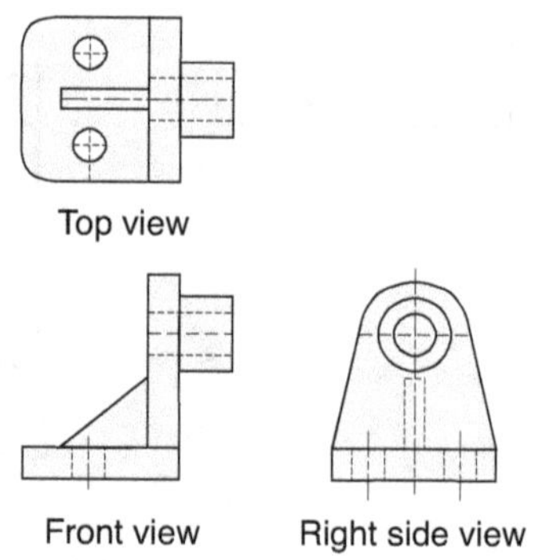

Figure 15.9 Top, front, and right side views of an item.

Table 15.1 Other feature control symbols.

Symbol	Modifier
Ⓕ	Free state
Ⓛ	Least material condition
Ⓜ	Maximum material condition
Ⓟ	Projected tolerance zone
Ⓢ	Regardless of feature size
Ⓣ	Tangent plane
Ⓤ	Unilateral

Other symbols that are used in a feature control frame to specify a feature's description, tolerance, modifier, and datum(s) are shown in Table 15.1.

Position and Bonus Tolerances

Bonus tolerance is an additional tolerance used for geometric control. Whenever a geometric tolerance is applied to a feature of size, and it contains a maximum material condition (MMC) or a least material condition (LMC) modifier in the tolerance portion of the feature control frame, a bonus tolerance is permissible (Effective Training 2012).

Tolerance of position control is a geometric tolerance that defines the location tolerance of a feature of size from its true position (Effective Training 2012).

If a position was identified exactly on a drawing, the tolerance is applied at LMC or MMC, or is applied regardless of feature size. If LMC or MMC is listed, we can calculate a wider tolerance, or bonus, depending on the actual size of the measured feature. When using the regardless-of-feature-size approach, we are not allowed to consider bonus tolerance, and the center of the measured feature must lie within the tolerance zone specified. For example, the symbols in Figure 15.11 on a drawing indicate that the diameter of a tolerance zone is 0.001″ and the true point is in the middle.

Line standards			
Name	**Convention**	**Description and application**	**Example**
Center lines		Thin lines made up of long and short dashes alternately spaced and consistent in length Used to indicate symmetry about an axis and location of centers	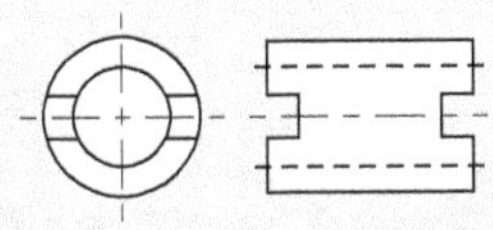
Visible lines		Heavy unbroken lines Used to indicate visible edges of an object	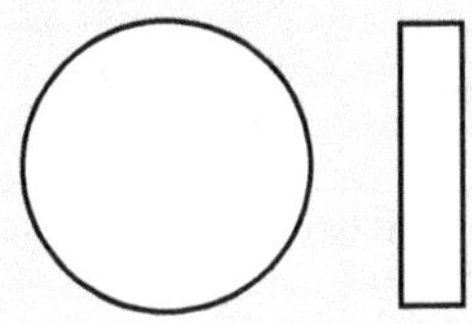
Hidden lines		Medium lines with short, evenly spaced dashes Used to indicate concealed edges	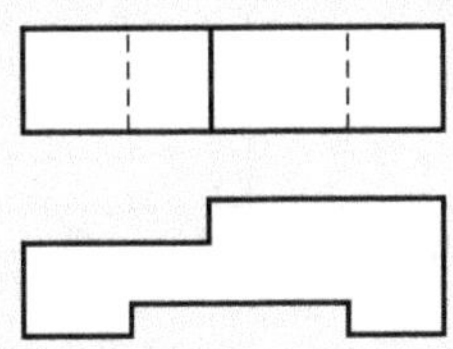
Extension lines		Thin unbroken lines Used to indicate extent of dimensions	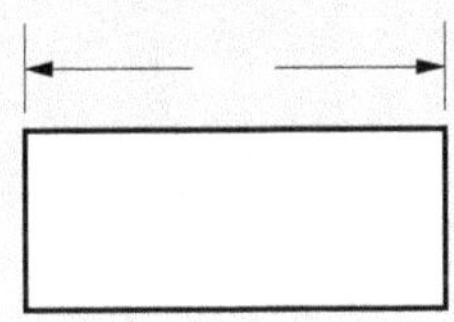
Dimension lines		Thin lines terminated with arrowheads at each end Used to indicate distance measured	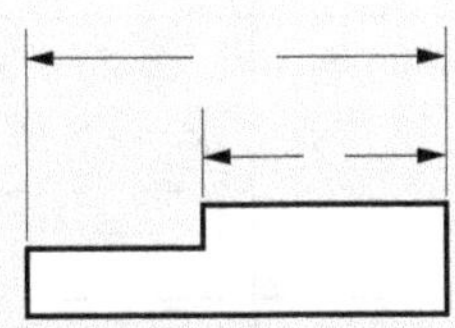
Leader		Thin line terminated with arrowhead or dot at one end Used to indicate a part, dimension, or other reference	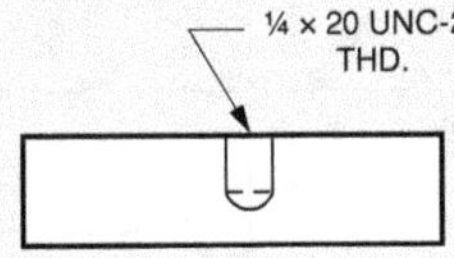

Figure 15.10 Engineering drawing line types and styles.
Source: Gunderson (1994).

Line standards			
Name	**Convention**	**Description and application**	**Example**
Break (long)	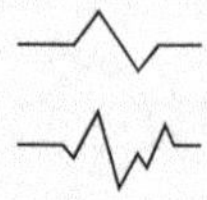	Thin solid ruled lines with freehand zig-zags Used to reduce size of drawing required to delineate object and reduce detail	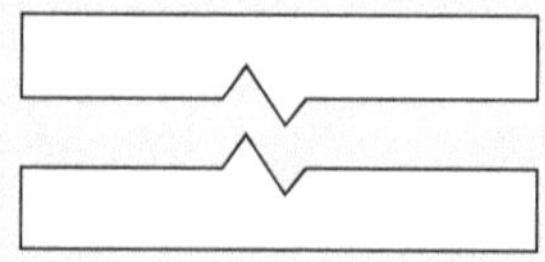
Break (short)		Thick solid freehand lines Used to indicate a short break	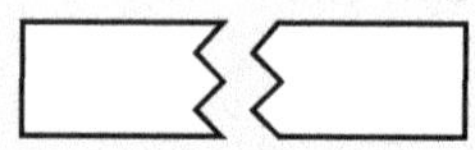
Phantom or datum line		Medium series of one long dash and two short dashes evenly spaced ending with long dash Used to indicate alternate position of parts, repeated detail, or a datum plane	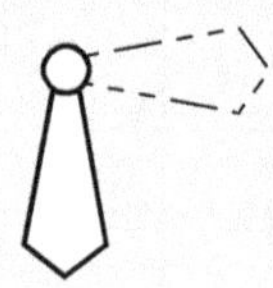
Stitch line		Medium line of short dashes evenly spaced and labeled Used to indicate stitching or sewing	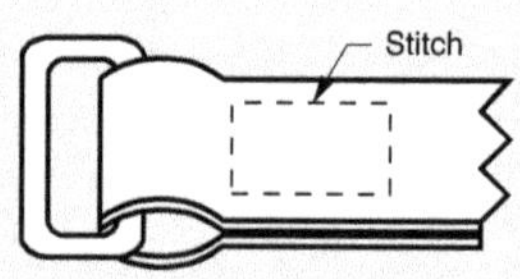
Cutting-plane line		Used to designate where an imaginary cutting took place	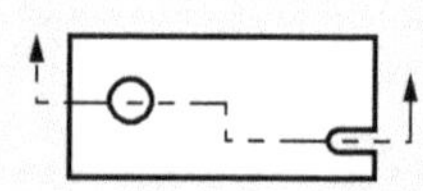
Viewing-plane line	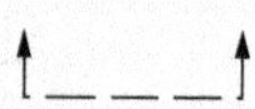	Used to indicate direction of sight when a partial view is used	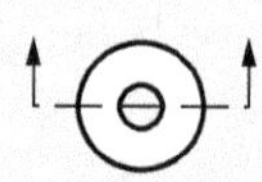
Section lines		Used to indicate the surface in the section view imagined to have been cut along the cutting-plane line	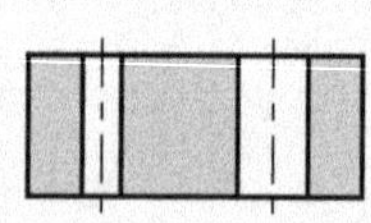
Chain line		Used to indicate that a surface or zone is to receive additional treatment or considerations	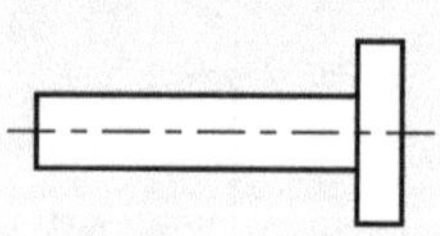

Figure 15.10 *(continued)*

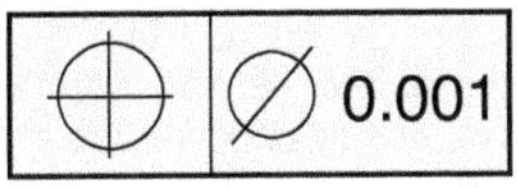

Figure 15.11 Example of symbols on an engineering drawing.

The diameter is a tolerance of size. However, the location also affects the fit and function of the feature to its mating part(s). Bonus tolerances allow the tolerance amount not used in the diameter to be applied to the location. This allows the maximum use of tolerances while keeping the fit and function intact (Clark 2012).

Example

Imagine we're trying to assemble a 5" long plug into a 4" deep bore. If the diameter of the bore is 2.0000" and the diameter of the plug is 1.9999", the plug would have to be straight within (or probably below) 0.0001" to ensure it would fit and function. On the contrary, if our plug diameter were 1.9995", it could have a straightness deviation of almost 0.0005" and still fit. This is an example of form, as opposed to location, but the concept is very similar (Clark 2012).

In our example, we'll use an inside feature (see Figure 15.12). Our drawing states we have a diameter tolerance of 1.5000" +0.0005" (so our MMC is 1.5000"). In addition, we have a positional location tolerance of 0.001" applied at MMC (Clark 2012).

The intent of positional location when applied at LMC or MMC is to ensure an acceptable amount of clearance between the part being made and its mating part. If the size of our inside feature is at MMC (smallest diameter), we need to be within 0.001" of the location for the mating part to fit (our part) properly. The larger we make our inside diameter, the more "room" we have for the location to be outside the 0.001" location tolerance zone, with the clearance remaining acceptable. Whatever amount above MMC (above our minimum) our diameter becomes is applied as a bonus to the tolerance of the positional location. If the diameter of our feature was 1.5005", the five-tenths we are above MMC would be applied to the location tolerance, which then becomes 0.0015" (Clark 2012).

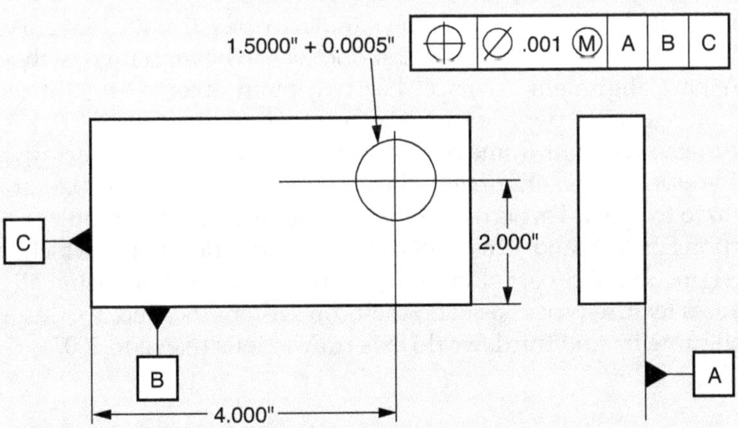

Figure 15.12 Form tolerance example.

(continued)

> If we look at the diameter tolerance and the location tolerance as two sides of a balanced equation, whatever we add to one side we also add to the other side to keep the two sides in balance (Clark 2012):
>
> 1.5000″ (Diameter at MMC) = 0.001″ (Location)
>
> But if we were to add:
>
> 1.5000″ + 0.0003″ (Diameter) = 0.001″ + 0.0003″ (Location)
>
> 1.5003″ (Diameter) = 0.0013″ (Location)

PART ALIGNMENT AND DATUM STRUCTURE

Datums are points, lines, planes, cylinders, axes, and so on, from which the location, or geometric relationship, of other part features may be established or related.

Datum targets define the setup or fixture that will align the part to the datum reference frame. Datum targets do the same thing that a surface plate, angle plate, gauge pin, or a stack of gauge blocks does. They are all used to establish the datum points, axes, and planes. It would be like tolerancing the datum reference framework. If a fixture is built to establish the datum targets, there will be a drawing of the fixture, and this fixture drawing will be toleranced. But, rule of thumb (Y14.43) says the tolerances will usually be a factor of 10 better than the part tolerances (Tec-Ease 2012).

Part alignment is crucial for achieving accurate machine measurements. Knowing the software package that comes with a machine is not enough; operators of these machines must have good metrology skills, understand basic geometry, and, at a minimum, know how to read a blueprint and perform geometric dimensioning and tolerancing.

Incorrectly defining or failing to define a part alignment is the single most influential factor in gauge repeatability errors and incorrect measurements.

For example, consider that a perfectly round 1 in. diameter hole on a flat plate would show nearly 0.002 in. of form (roundness) error if measured as a circle and the part were 5° skewed from being parallel to the current CMM axis. Never assume that because a part sits on a CMM surface, it can be measured without alignment.

A part alignment accomplishes two main objectives: (1) it defines the part function—usually depicted on the print via datum—and (2) it establishes the coordinate system or frame of reference for physically measuring the part.

To define a part alignment, the operator locks all three axes of the coordinate system to the part. This is done via machine software and requires establishing the directions of *X*, *Y*, and *Z*, as well as their intersection point, which can be taken as the origin. To build part alignment, we first do a level or spatial alignment, which separates feature types based on their dimensions. Second, we do a rotation or planar alignment. And third, we do axis translations (Nevado 2007).

Chapter 16

Sampling

> Define and utilize the following terms concepts to sampling. (Apply)
>
> 1. Acceptance quality limit (AQL)
> 2. Random sampling
> 3. Lot and sample size
> 4. Acceptance number
> 5. Sampling plans
>
> Body of Knowledge III.B

INTRODUCTION

Acceptance sampling is a method for inspecting the product. Inspection can be done with screening (also called *sorting* or *100% inspection*), in which all units are inspected, or with sampling. *Acceptance sampling* is the process of inspecting a portion of the product in a lot for making a decision regarding classification of the entire lot as either conforming or nonconforming to quality specifications.

Whether inspection is done with screening or with sampling, the results can be used for the following purposes:

1. To distinguish between good lots and bad lots using acceptance sampling plans (as in incoming material inspection and final product inspection).
2. To distinguish between good products and bad products.
3. To determine the status of process control and whether the process is changing. This is usually done in conjunction with control charts.
4. To evaluate process capability. In this case, inspection is used to determine whether the process exhibits excessive variation and whether it is approaching or exceeding the specification limits.
5. To determine process adjustment. Based on inspection results of process output—as depicted by a histogram, for example, the process mean may require adjustment, and/or process variation may need to be reduced.

A process might require adjustment even though all the units produced to date conform to the quality standards agreed on with the customer.

6. To rate the accuracy of inspectors or inspection equipment by comparing the inspection results with corresponding standards. An inspection operation can result in two types of error: classification of a conforming unit as nonconforming and classification of a nonconforming unit as conforming. The probability of both types of error can be easily estimated using probability theory and other statistical methods.
7. To serve as a mechanism for evaluating vendors in terms of their products' quality. Vendors that consistently deliver high-quality products can receive preferred status involving reduced inspection and priority in bidding for new contracts, while vendors that do not stand up to quality requirements can be warned or discontinued altogether. This type of procedure is known as *vendor qualification* or *vendor certification*.

ADVANTAGES

Sampling provides the economic advantage of lower inspection costs due to fewer units being inspected. In addition, the time required to inspect a *sample* is substantially less than that required for the entire lot, and there is less damage to the product due to reduced handling. Most inspectors find that selection and inspection of a random sample is less tedious and monotonous than inspection of the complete lot. Another advantage of sampling inspection is related to the supplier/customer relationship. By inspecting a small fraction of the lot and forcing the supplier to screen 100% in case of lot rejection (which is the case for rectifying inspection), the customer emphasizes that the supplier must be concerned about quality. On the other hand, the variability inherent in sampling results in sampling errors: rejection of lots of conforming quality and acceptance of lots of nonconforming quality.

Acceptance sampling is most appropriate when inspection costs are high and when 100% inspection is monotonous and can cause inspector fatigue and boredom, resulting in degraded performance and increased error rates. Obviously, sampling is the only choice available for destructive inspection. Rectifying sampling is a form of acceptance sampling. Sample units detected as nonconforming are discarded from the lot, replaced with conforming units, or repaired. Rejected lots are subject to 100% screening, which can involve discarding, replacing, or repairing units detected as nonconforming.

In certain situations it is preferable to inspect 100% of the product. This would be the case for critical or complex products, where the cost of making the wrong decision would be too high. Screening is appropriate when the fraction nonconforming is extremely high. In this case, most of the lots would be rejected under acceptance sampling, and those accepted would be so as a result of statistical variations rather than better quality. Screening is also appropriate when the fraction nonconforming is not known and an estimate based on a large sample is needed.

SAMPLING CONCEPTS

Sampling may be performed according to the type of quality characteristics to be inspected. There are three major categories of sampling plans: sampling plans for attributes, sampling plans for variables, and special sampling plans.

Random Sampling

The simple random sampling design consists of selecting n (sample size) sampling units in such a way that each sampling unit has the same chance of being selected.

The Operating Characteristic Curve

No matter which type of attribute sampling plan is being considered, the most important evaluation tool is the operating characteristic (OC) curve.

The *OC curve* allows a sampling plan to be almost completely evaluated at a glance, giving a pictorial view of the probabilities of accepting lots submitted at varying levels of percent defective. The OC curve illustrates the risks involved in acceptance sampling. Figure 16.1 shows an OC curve for a sample size *(n)* of 50 drawn from an infinite lot size, with an acceptance number *(c)* of 3.

As can be seen by the OC curve, if the lot were 100% to specifications, the probability of acceptance P_a would also be 100%. But if the lot were 13.4% defective, there would be a 10% probability of acceptance.

Acceptance Sampling by Attributes

Acceptance sampling by attributes is generally used for two purposes: (1) protection against accepting lots from a continuing process whose average quality deteriorates beyond an acceptance quality limit, and (2) protection against isolated lots

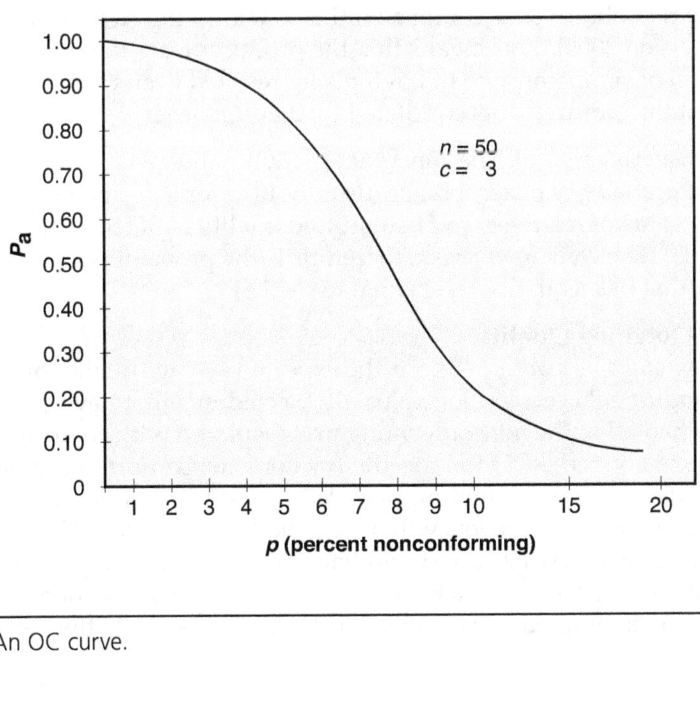

Figure 16.1 An OC curve.

that may have levels of nonconformances greater than can be considered acceptable. The most commonly used form of acceptance sampling plan is sampling by attributes. The most widely used *standard* of all attribute plans, although not necessarily the best, is ANSI/ASQ Z1.4-2008. The following sections provide more detail on the characteristics of acceptance sampling and discuss military standards in acceptance sampling.

Acceptance Quality Limit

Acceptance quality limit (AQL) is defined as the quality level that is the worst tolerable process average when a continuing series of lots is submitted for acceptance sampling. This means that a lot that has a fraction defective equal to the AQL has a high probability (generally around 0.95, although it may vary) of being accepted. As a result, plans that are based on AQL, such as ANSI/ASQ Z1.4-2008, favor the producer in getting lots accepted that are in the vicinity of the AQL for fraction defective in a lot.

Lot Tolerance Percent Defective

The *lot tolerance percent defective (LTPD)*, expressed in percent defective, is the poorest quality in an individual lot that should be accepted. The LTPD has a low probability of acceptance. In many sampling plans, the LTPD is the percent defective having a 10% probability of acceptance.

Producer's and Consumer's Risks

There are two risks involved in using acceptance sampling plans: (1) producer's risk and (2) consumer's risk. These risks correspond with type 1 and type 2 errors in hypothesis testing and are defined as follows:

- *Producer's risk* (α). The producer's risk for any given sampling plan is the probability of rejecting a lot that is within the AQL (ASQ Statistics Division 2005). This means that the producer faces the possibility (at level of significance α) of having a lot rejected even though the lot has met the requirements stipulated by the AQL level.
- *Consumer's risk* (β). The consumer's risk for any given sampling plan is the probability of acceptance (usually 10%) for a designated numerical value of relatively poor submitted quality (ASQ Statistics Division 2005). The consumer's risk, therefore, is the probability of accepting a lot that has a quality level equal to the LTPD.

Average Outgoing Quality

The *average outgoing quality (AOQ)* is the expected average quality of outgoing products, including all accepted lots, plus all rejected lots that have been sorted 100% and have had all of the nonconforming units replaced with conforming units.

There is a given AOQ for specific fractions nonconforming of submitted lots sampled under a given sampling plan. When the fraction nonconforming is very low, a large majority of the lots will be accepted as submitted. The few lots that are rejected will be sorted 100% and will have all nonconforming units replaced with conforming units. Thus, the AOQ will always be less than the submitted quality. As the quality of submitted lots declines in relation to the AQL, the percentage of lots rejected increases in proportion to accepted lots. As these rejected lots are sorted and combined with accepted lots, an AOQ lower than the average fraction of nonconformances of submitted lots emerges. Therefore, when the level of quality of incoming lots is good, the AOQ is good; when the incoming quality is bad and most lots are rejected and sorted, the result is also good.

Average Outgoing Quality Limit

The AOQ is a variable dependent on the quality level of incoming lots. When the AOQ is plotted for all possible levels of incoming quality, a curve as shown in Figure 16.2 results. The *average outgoing quality limit (AOQL)* is the highest value on the AOQ curve.

Lot Size, Sample Size, and Acceptance Number

For any single sampling plan, the plan is completely described by the lot size, sample size, and acceptance number.

TYPES OF ATTRIBUTE SAMPLING PLANS

There are several types of attribute sampling plans in use, with the most common being single, double, multiple, and sequential sampling plans. The type of sampling plan used is determined by ease of use and administration, general quality level of incoming lots, average sample number, and so on.

Single Sampling Plans

When single sampling plans are used, the decision to either accept or reject the lot is based on the results of the inspection of a single sample of n items from a submitted lot. In Figure 16.2, the OC curve and AOQ curve were calculated for a single sampling plan where $n=50$ and $c=3$. Single sampling plans have the advantage of ease of administration, but due to the unchanging sample size, they do not

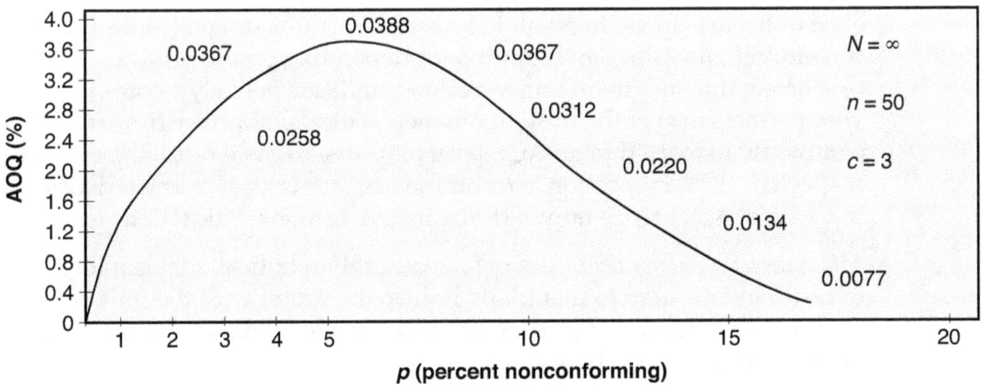

Figure 16.2 AOQ curve for $N=\infty$, $n=50$, $c=3$.

take advantage of potential cost savings of reduced inspection when incoming quality is either excellent or poor.

Double Sampling Plans

When using double sampling plans, a smaller first sample is taken from the submitted lot, and one of three decisions is made: (1) accept the lot, (2) reject the lot, or (3) draw another sample. If a second sample is to be drawn, the lot will either be accepted or rejected after the second sample. Double sampling plans have the advantage of a lower total sample size when the incoming quality is either excellent or poor, because the lot is either accepted or rejected on the first sample.

Multiple Sampling Plans

Multiple sampling plans work in the same way as double sampling with an increase in the number of samples to be taken up to seven, according to ANSI/ASQ Z1.4-2008. In the same manner that double sampling is performed, acceptance or rejection of submitted lots may be reached before the seventh sample depending on the acceptance/rejection criteria established for the plan.

SAMPLING STANDARDS AND PLANS

ANSI/ASQ Z1.4-2008

ANSI/ASQ Z1.4-2008 is probably the most commonly used standard for attribute sampling plans. The wide recognition and acceptance of the plan could be due to government contracts stipulating the standard, rather than its statistical importance. Producers submitting products at a nonconformance level within AQL have a high probability of having the lot accepted by the customer.

When using ANSI/ASQ Z1.4-2008, the characteristics under consideration should be classified. The general classifications are critical, major, and minor defects, and are described as follows:

- *Critical defect.* A *critical defect* is a defect that judgment and experience indicate is likely to result in hazardous or unsafe conditions for the individuals using, maintaining, or depending on the product, or a defect that judgment and experience indicate is likely to prevent performance of the unit. In practice, critical characteristics are commonly inspected to an AQL level of 0.40%–0.65% if not 100% inspected. 100% inspection is recommended for critical characteristics if possible. Acceptance numbers are always zero for critical defects.
- *Major defect.* A *major defect* is a defect, other than critical, that is likely to result in failure or to materially reduce the usability of the unit of product for its intended purpose. In practice, AQL levels for major defects are generally about 1%.
- *Minor defect.* A *minor defect* is a defect that is not likely to materially reduce the usability of the unit of product for its intended purpose. In practice, AQL levels for minor defects generally range from 1.5% to 2.5%.

Levels of Inspection

There are seven levels of inspection used in ANSI/ASQ Z1.4-2008: reduced inspection, normal inspection, tightened inspection, and four levels of special inspection. The special inspection levels should only be used when small sample sizes are necessary and large risks can be tolerated. When using ANSI/ASQ Z1.4-2008, a set of switching rules must be followed as to the use of reduced, normal, and tightened inspection.

The following guidelines are taken from ANSI/ASQ Z1.4-2008:

- *Initiation of inspection.* Normal inspection Level II will be used at the start of inspection unless otherwise directed by the responsible authority.
- *Continuation of inspection.* Normal, tightened, or reduced inspection shall continue unchanged for each class of defect or defectives on successive lots or batches except where the following switching procedures require change. The switching procedures shall be applied to each class of defects or defectives independently.

Switching Procedures. Switching rules are graphically shown in Figure 16.3.

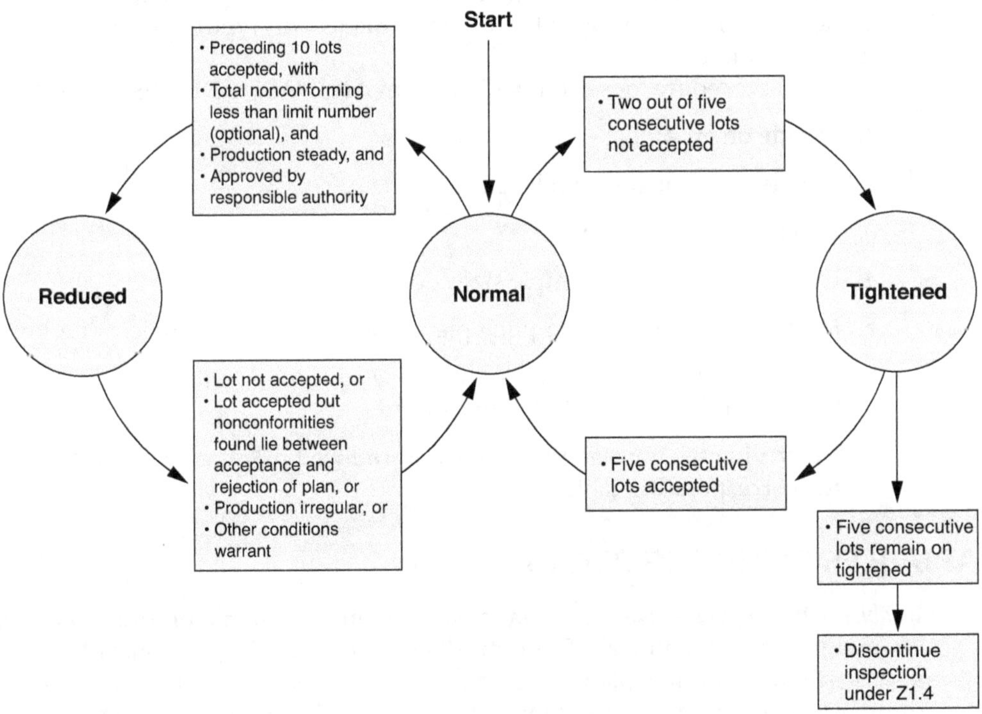

Figure 16.3 Switching rules for normal, tightened, and reduced inspection. Types of sampling.

Types of Sampling

ANSI/ASQ Z1.4-2008 allows for three types of sampling plans:

1. Single sampling
2. Double sampling
3. Multiple sampling

The choice of the type of plan depends on many variables. Single sampling is the easiest to administer and perform but usually results in the largest average total inspection. Double sampling results in a lower average total inspection than single sampling but requires more decisions to be made, such as:

- Accept the lot after first sample.
- Reject the lot after first sample.
- Take a second sample.
- Accept the lot after second sample.
- Reject the lot after second sample.

Multiple sampling plans further reduce the average total inspection but also increase the number of decisions to be made. As many as seven samples may be required before a decision to accept or reject the lot can be made. This type of plan requires the most administration.

A general procedure for selecting plans from ANSI/ASQ Z1.4-2008 is as follows:

1. Decide on an AQL.
2. Decide on the inspection level.
3. Determine the lot size.
4. Find the appropriate sample size code letter.
5. Determine the type of sampling plan to be used: single, double, or multiple.
6. Using the selected AQL and sample size code letter, enter the appropriate table to find the desired plan to be used.
7. Determine the normal, tightened, and reduced plans as required from the corresponding tables.

VARIABLES SAMPLING PLANS

Variables sampling plans use the actual measurements of sample products for decision making rather than classifying products as conforming or nonconforming, as in attribute sampling plans. Variables sampling plans are more complex in administration than attribute plans, thus they require more skill. They provide some benefits, however, over attribute plans:

1. The protection is equal to an attribute sampling plan with a much smaller sample size. There are several types of variables sampling plans in use, three of these being (1) α known, (2) α unknown but can be esti-

mated using sample standard deviation s, and (3) α unknown and the range R is used as an estimator. If an attribute sampling plan sample size is determined, the variables plans previously listed can be compared as a percentage to the attribute plan.

Plan	Sample size (percent)
Attribute	100
σ unknown, range method	60
σ unknown, s estimated from sample	40
σ known	15

2. Variables sampling plans allow the determination of how close to nominal or a specification limit the process is performing. Attribute plans either accept or reject a lot; variables plans give information on how well or poorly the process is performing.

Variables sampling plans, such as ANSI/ASQ Z1.9-2008, have some disadvantages and limitations:

1. The assumption of normality of the population from which the samples are being drawn.
2. Different characteristics on the same parts will have different averages and dispersions, resulting in a separate sampling plan for each characteristic.
3. Variables plans are more complex in administration.
4. Variables gauging is generally more expensive than attribute gauging.

ANSI/ASQ Z1.9-2008

The most common standard for variables sampling plans is ANSI/ASQ Z1.9-2008, which has plans for (1) variability known, (2) variability unknown—standard deviation method, and (3) variability unknown—range method. Using the aforementioned methods, this sampling plan can be used to test for a single specification limit, a double (or bilateral) specification limit, estimation of the process average, and estimation of the dispersion of the parent population.

Figure 16.4 summarizes the structure and organization of ANSI/ASQ Z1.9-2008.

SEQUENTIAL SAMPLING PLANS

When tests are either destructive in nature or costly, it may be advantageous to use sequential sampling plans popularized by Wald (1973). These plans have the advantage of greatly reduced sample sizes while giving good protection.

To determine a sequential sampling plan, the following parameters must be defined:

α = Producer's risk

AQL = Acceptance quality limit = p_1

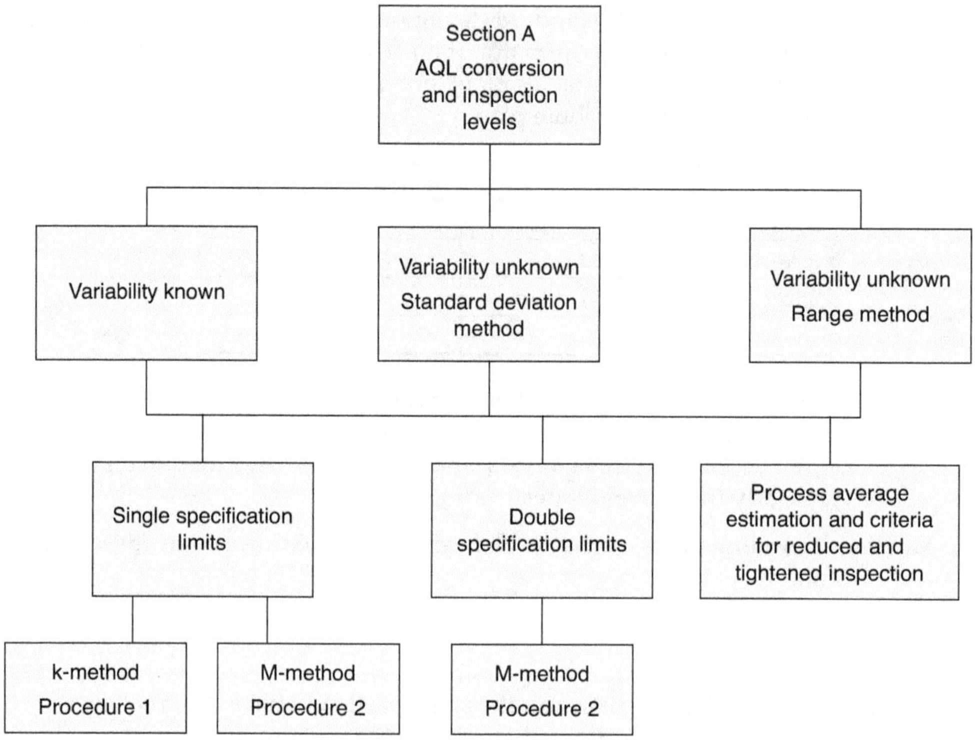

Figure 16.4 Structure and organization of ANSI/ASQ Z1.9-2008.

β = Consumer's risk

RQL = Rejectable (or unacceptable) quality level = p_2

Figure 16.5 shows the accept, reject, and continue testing areas for a sequential sampling plan. The y-axis represents the number of nonconforming items in the sample, and the x-axis scales the number of units inspected.

CONTINUOUS SAMPLING PLANS

Many production processes do not produce lots, thus lot-by-lot acceptance sampling plans, discussed earlier, cannot be applied. In cases such as these, continuous sampling plans are developed. In continuous sampling plans, 100% inspection and sampling inspection are alternately applied. The most recent standard for developing continuous sampling plans is the military standard MIL-STD-1235B.

MIL-STD-1235B

MIL-STD-1235B uses the same parameters, i and f, as previously defined. The standard includes CSP-1, CSP-2, CSP-F, CSP-T, and CSP-V plans.

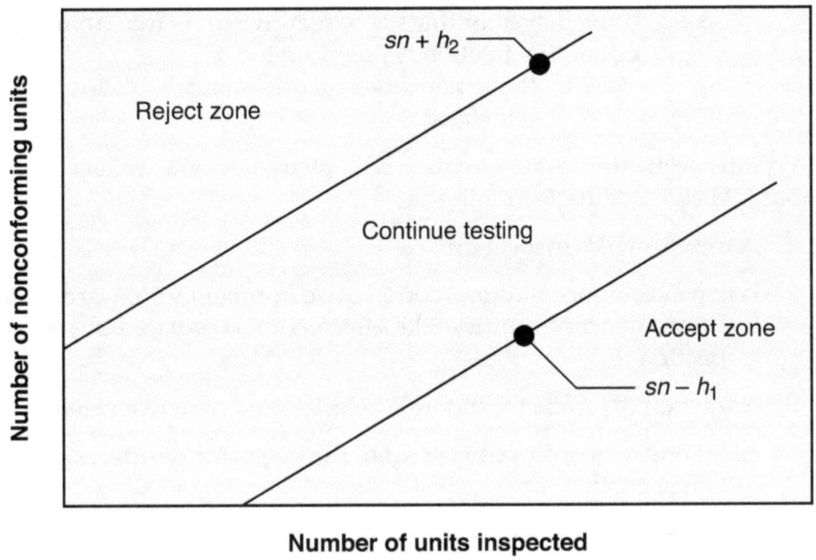

Figure 16.5 Decision areas for a sequential sampling plan.

CSP-1 and CSP-2 plans operate in the same way as Dodge's CSP-1 and CSP-2 plans, but they are selected based on a sample size code letter and an AQL value as a quality index. The sample size code letter is selected based on the number of units in the production interval.

CSP-F plans work the same way as CSP-1 plans, providing alternate sequences of 100% and sampling inspection procedures, but the difference is that AOQL and the number of units in the production interval are used in this case to characterize the plans. Once AOQL and *f* values are selected, go to the corresponding table to read *i*, the clearance number. CSP-F is a single-level continuous sampling scheme.

CSP-T plans include the provision of reduced sampling frequency once the product shows superior quality.

The CSP-T plan works as follows:

1. Start with 100% inspection.
2. When *i* (clearance number) consecutive number of units are found free from nonconformities, 100% inspection is replaced by sampling inspection.
3. A fraction of *f* units is randomly selected and then inspected.
   - 3.1. If one nonconformity is found, the inspector reinstates 100% inspection.
   - 3.2. If the inspector finds *i* consecutive units free from nonconformities, the fraction *f* is reduced to *f*/2.
     - 3.2.1. If one nonconformity is found, the inspector switches back to 100% inspection.

3.2.2. If the inspector finds i consecutive units free from nonconformities, the fraction f is reduced to $f/4$.
3.2.2.1. If one nonconformity is found, 100% inspection is reinstated.

CSP-V plans work the same way as CSP-T plans but with reduced i instead of reduced f. The procedure is as follows:

1. Start with 100% inspection.
2. When i (clearance number) consecutive number of units are found free from nonconformities, 100% inspection is replaced by sampling inspection.
3. A fraction of f units is randomly selected and then inspected.

3.1. If one nonconformity is found, the inspector reinstates 100% inspection.

3.2. If the inspector finds i consecutive units free from nonconformities, the inspection continues with inspecting the same fraction f.

3.2.1. If one nonconformity is found, the inspector switches back to 100% inspection.

3.2.2. If the inspector finds i/3, the sampling inspection continues with the same fraction f.

Chapter 17

Inspection Planning and Processes

1. ***Inspection types.*** Define and distinguish between inspection types, such as incoming material, first-article (first-piece), in-process, and final. (Analyze)
2. ***Inspection errors.*** Identify potential inspection errors and their sources, such as bias, fatigue, flinching, distraction, poor time management, setup, and cosine error when using indicators. (Apply)
3. ***Product traceability.*** Identify methods to trace products and materials, such as age control, shelf life, first-in first-out (FIFO), barcoding, date codes, and lot and part numbering. (Apply)
4. ***Identification of nonconforming material.*** Describe and apply various methods of identifying nonconforming material, such as tagging, labeling, and segregating. (Apply)
5. ***Level of severity.*** Define and describe levels of severity i.e., (critical, major, and minor) and apply them to product features and defects. (Apply)
6. ***Disposition of nonconforming material.*** Describe disposition methods, such as rework, reprocess, reinspect, scrap, and customer waiver as determined by a material review board (MRB) or other authority. (Understand)

Body of Knowledge III.C

INTRODUCTION

Inspection is the evaluation of product quality by comparing the results of measuring one or several product characteristics with applicable standards. From this definition, it is evident that the inspection function involves a number of tasks (Raz 1992):

1. Measurement, which could be on a qualitative or quantitative scale. The objective is to make a judgment about a product's conformance to specifications.

2. Comparison of the measurement results to certain standards that reflect the intended use of the product by the customer and the various production costs. If the product is found to be nonconforming, a decision as to whether nonconforming products are fit for use may be reached.
3. Decision making regarding the disposition of the unit inspected and, under sampling inspection, regarding the lot from which the sample was drawn.
4. Corrective action(s) in order to improve the quality of the product and/or process based on the aggregate results of inspection over a number of units.

Inspection decisions for most products are shown in Figure 17.1.

INSPECTION PLANNING

Inspection planning includes the determination of the location of inspection and/or quality control methods and procedures at the various points in the production process. It also involves the determination of the types of inspections to be carried out and the quality levels, identification of critical characteristics to be inspected, and classification of defects.

The location of inspection stations can be determined based on the following considerations (Raz 1992):

- Inspect incoming materials to prevent the entry of defective components into the production system. This step could be eliminated if the suppliers provide sufficient evidence of the use of process control techniques to maintain product quality.

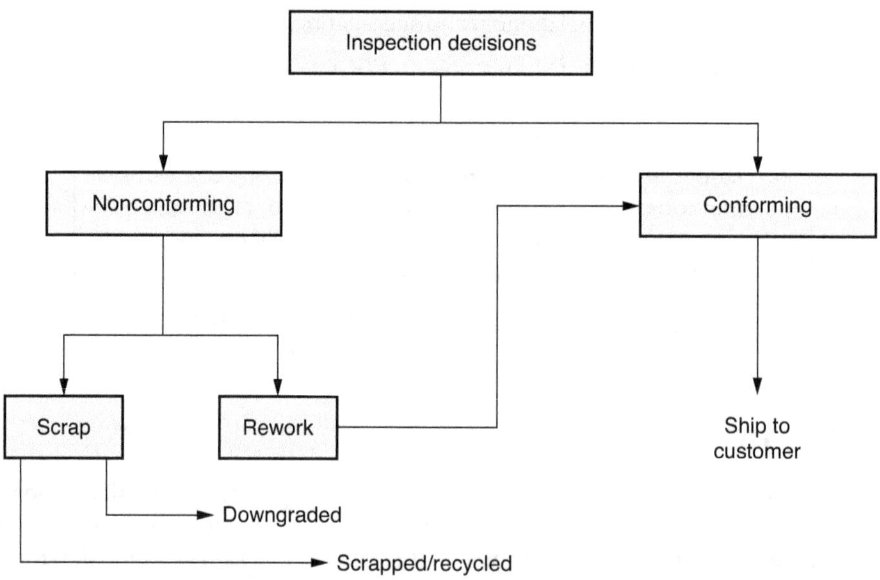

Figure 17.1 Inspection decisions.

- Inspect prior to costly operations to avoid further investment in an already nonconforming product.
- Inspect prior to processing operations that may mask defects; for example, surface finish should be inspected prior to painting.
- Inspect prior to processing operations that may cause an increase in repair costs. For example, inspect and test circuit boards prior to assembly into their enclosures.
- Inspect following operations known to have a relatively high defect rate.
- Inspect final or finished goods before moving the product to another department or plant prior to shipping to the customer.
- Inspect the first few units of each new batch in order to verify that the setup is correct.

The basic tool for choosing the location of inspection stations is the *flowchart*. The most common locations are as follows (Gryna 2001):

- At receipt of goods from suppliers, usually called "incoming inspection" or "supplier inspection."
- Following the setup of a process to provide added assurance against producing a defective batch. In some cases this "setup approval" also becomes approval of the batch.
- During the running of critical or costly operations, usually called "process inspection."
- Prior to delivery of goods from one department to another, usually called "lot approval" or "tollgate inspection."
- Prior to shipping completed products to storage or customers, usually called "finished goods inspection."
- Before performing a costly, irreversible operation, for example, pouring a melt of steel.
- At natural "peepholes" in the process. A flowchart example is shown in Figure 17.2.

INSPECTION TYPES

Incoming Material Inspection

Incoming material inspection is performed when one company purchases materials or supplies from another company. This is considered a producer–supplier relationship, where the acceptance or rejection decisions on submitted lots are based on a certain criterion agreed on between both companies. The purpose of incoming inspection is to distinguish between good lots and bad lots. The

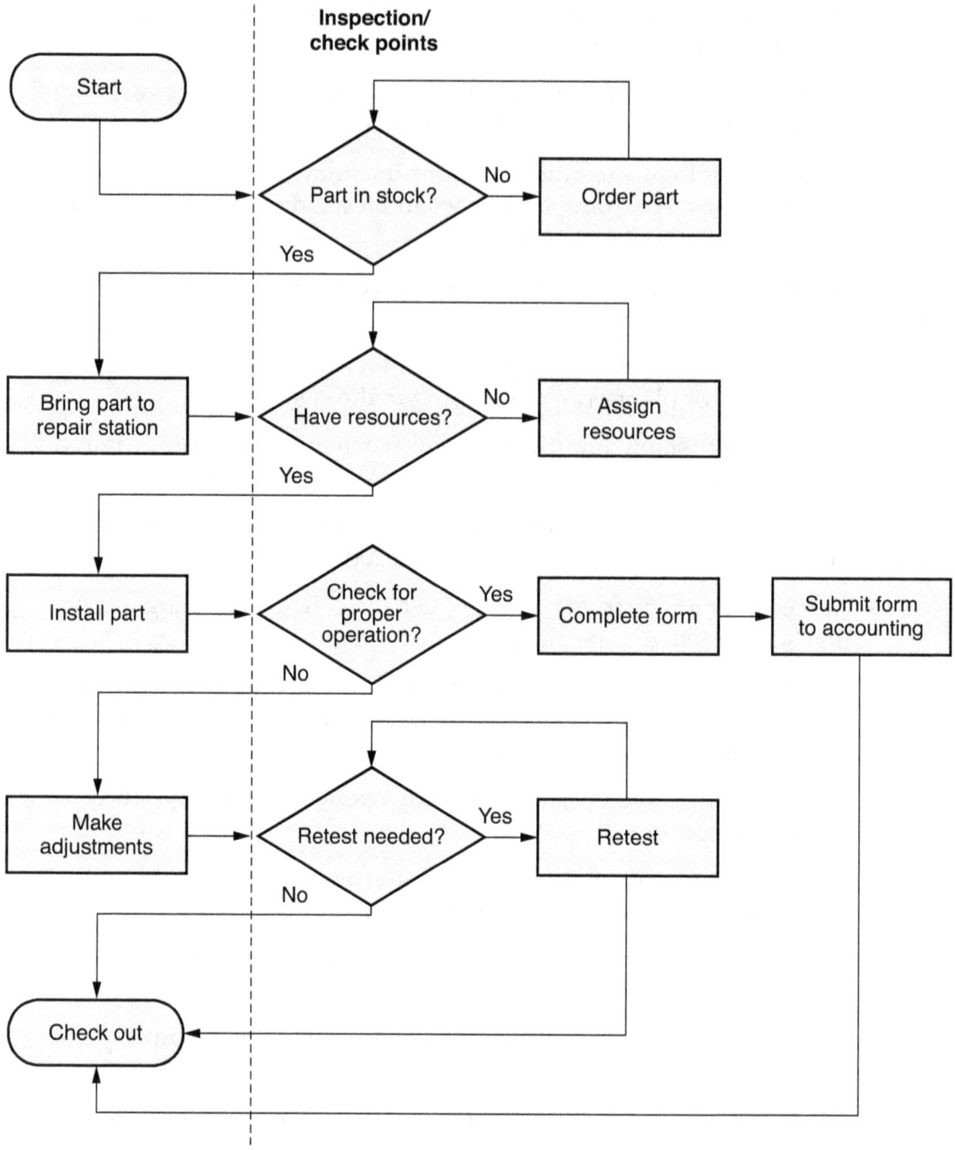

Figure 17.2 An example of a flowchart of a repair job.

decision that follows an incoming inspection procedure falls into one of the following categories:

1. Accept the lot; that is, the lot conforms to specifications
2. Reject the lot; that is, the lot is considered nonconforming
3. The lot should go to a screening inspection (100% inspection)

First-Piece Inspection

In first-piece inspection, the operator is likely to carry out the inspection. The purpose is to make sure that all procedures and execution of work are properly employed according to a specified work standard or set of specifications.

In-Process Inspection

The purpose of in-process inspection is to determine the appropriateness of control techniques employed by the company to decide on the degree of conformance to a system of standard requirements. In many cases, sampling is employed. The results of in-process inspection are as follows:

1. Products are inspected and then classified into classes or grades depending on the degree of conformity to standards.
2. Nonconforming products may be classified as scrap or rework. In some cases, the cost of reworking a product is more costly than producing it.
3. Nonconforming products may be analyzed to determine the root cause of the problem and the appropriate follow-up corrective action.

Final Inspection

Finished product inspection is carried out to confirm that the product conforms to the customer's specifications. It may include assembly, installation, and packaging. In all cases final inspection is thorough and complete and may be intensive and costly depending on the accuracy and severity of incoming inspection and in-process inspection.

Inspection planning also includes the preparation of a list of characteristics to be inspected. The following guidelines may prove helpful (Raz 1992):

- Inspect characteristics that affect the performance of the product. To the extent possible, product testing should be done under conditions that simulate actual use.
- Select characteristics that can be measured objectively, to the extent possible.
- Provide a seriousness classification to improve consistency for characteristics that are evaluated subjectively.
- Inspect characteristics that can be related to a specific production process in order to simultaneously obtain information about the process.

A detailed inspection plan should be prepared and approved by the customer and the production, engineering, and manufacturing departments prior to the start of full-scale production. The inspection plan should include the following items (Raz 1992):

- The location of each inspection station in the sequence of production operations

- The type of inspection or test to be carried out, including a description of the environment, equipment, and procedures
- Accuracy requirements from the measurements
- The conformance criteria, normally based on product specifications
- The sample size and procedure for drawing a sample in the case of sampling inspection
- The lot size and the criteria for lot acceptance, if applicable
- The disposition of nonconforming units (for example, repair, scrap, or salvage) and of rejected lots (for example, screen or return to vendor)
- The criteria for initiating a review of the process, vendor, or inspector

INSPECTION ERRORS

Errors are bound to happen in any system that has a combination of interrelated factors that make up an outcome. This is true in inspection, as the inspection process involves many factors that combine to determine the outcome of the inspection, often a decision. Major contributors to error sources are the following (Darmody 1967):

1. Human error or errors due to the physiology of the inspector
2. Inspection equipment errors due to accuracy, instrument selection, methodology, and so on
3. Errors due to the environment that surrounds the inspection location
4. Procedural errors due to the introduction of new processes, equipment, and so on

Inspection involves a measurement, which is comparing the actual value of the measured item, component, or material with a standard, and making the decision as a result of such comparison. Measurement by itself has different error sources that are often unavoidable.

Error in measurement is the difference between the indicated value and the true value of a measured quantity. The true value of a quantity to be measured is seldom known. Errors are classified as one of two types:

1. Random errors
2. Systematic errors

Random errors are accidental in nature. They fluctuate in a way that cannot be predicted from the detailed employment of the measuring system or from knowledge of its functioning. Sources of error such as hysteresis, ambient influences, or variations in the workpiece are typical but not all-inclusive in the random category.

Systematic errors are those not usually detected by repetition of the measurement operations. An error resulting from either faulty calibration of a local standard or a defect in contact configuration of an internal measuring system is typical but not completely inclusive in the systematic class of errors. It is important to know all the sources of error in a measuring system rather than merely to be aware of the details of their classification. Analysis of the causes of errors is helpful in attaining the necessary knowledge of achieved accuracy (Darmody 1967).

There are many different sources of errors that influence the precision of a measuring process in a variety of ways according to the individual situation in which such errors arise. The permutation of error sources and their effects, therefore, is quite considerable. In general, these errors can be classified under one of three main headings (Rashed and Hamouda 1974):

1. Process environment
2. Equipment limitation
3. Operator fallibility

These factors constitute an interrelated three-element system for the measuring process.

The requirement of any precision measuring instrument is that it be able to represent, as accurately as possible, the dimension it measures. This necessitates that the instrument itself have a high degree of inherent accuracy. Small inaccuracies will exist, however, due to the tolerances permitted in the instrument's manufacture. These inaccuracies will influence the degree of precision attainable in its application.

The areas in which operator fallibility arises can be grouped as follows:

1. Identification of the measuring situation
2. Analysis of alternative methods
3. Selection of equipment
4. Application (or measurement)

The identification of measuring situations becomes increasingly complex in modern metrology. As parts become smaller and more precise, greater attention has to be paid to geometric qualities such as roundness, concentricity, straightness, parallelism, and squareness. Deficiencies in these qualities may consume all of the permitted design tolerance, so that a simple dimensional check becomes grossly insufficient.

Operators have to be knowledgeable about what they have to measure and how satisfactorily the requirements of the situation will be met by the measuring instrument. Correct identification of the measuring situation will eliminate those methods unsuitable for the situation. Proper measuring equipment can therefore be selected from a smaller range of measuring process alternatives. Method analysis can then be applied to these alternatives to determine which best suits the situation. This usually involves examining each method for different characteristics and evaluating the relative accuracies of the different methods. Figure 17.3 depicts the factors that affect the performance of the measuring process.

Other common error sources in inspection include flinching, bias, distraction, and fatigue. *Flinching* is an unintended human observation in measurement that

Operator fallibility

- Identification of the situation
- Analysis of alternative methods
- Selection of equipment
- Application/measurement

Measuring equipment

- Sensitivity
- Accuracy and precision
- Consistency
- Repeatability
- Gauge or instrument wear

Environmental conditions

- Temperature
- Vibration
- Structural instability
- Humidity
- Factors of atmospheric pollution
- Atmospheric pressure
- Gravity

Figure 17.3 Factors affecting the measuring process.

lies on the border between manipulation and approximation. Inspectors readily miscount a dial division or two to get a subconsciously desired reading (Kennedy et al. 1987).

Poor time management is another source of inspection error that is not commonly highlighted in many reference books.

Inspector Qualifications and Training

Basic requirements for inspection personnel include the following (Raz 1992):

1. The ability to perform the relevant measurements
2. Understanding of product specifications to the point of being capable of determining product quality
3. Basic mathematical skills for recording and analyzing data
4. Basic understanding of statistical concepts needed for sampling inspection and process characterization
5. Knowledge of measurements and measurement technology
6. Understanding of company's inspection policies, inspection procedure, products, materials, and processes

Training refers to the formal procedures used to improve job-related capability. Training programs for inspection personnel should be designed to address three main generic aspects:

1. *Attitude.* This includes developing a genuine concern for the product and for the customer, as well as fostering a positive self-image of the inspection function. To a significant extent, attitude is affected by the leadership of management and supervisory staff.
2. *Knowledge.* This includes not only knowledge directly related to the inspection function but also knowledge of the various production processes, materials, equipment, procedures, and so on.
3. *Skills.* This category refers to mastering the performance of the technical activities that are part of the inspector's job.

The Characteristics of a Quality Inspector

The quality inspector is involved in many aspects of product quality, including:

- Evaluating hardware documentation
- Performing laboratory procedures
- Inspecting products against specified standards
- Measuring process performance
- Recording data
- Preparing formal reports

Because of this, the quality inspector needs technical expertise in the field where they perform given tasks. This knowledge should be related to the processes where inspection is applied. The quality inspector should be aware of all applicable standards and regulations, and know how to systematically perform feature comparisons to prescribed requirements. The quality inspector should also be familiar with the products and processes, and how to use the inspection monitoring and measuring instruments. A quality inspector is required to understand and apply a variety of tools and techniques, including technical math, metrology, inspection and test techniques, and quality assurance and improvement methodologies. These are the essential features that determine the technical competency of the profession.

The level of excellence within the inspection field requires a specific personal skill set that distinguishes average from outstanding inspectors. The quality inspector must be attentive, objective, and detail oriented. This profession requires a high level of responsibility and accountability when performing quality inspection duties. Quality inspection activities are best performed by specialists who are accurate and error-free, analytical, consistent, confident, reliable, and organized.

Since quality inspectors need to work in cross-functional teams, solid interpersonal skills are essential. This interpersonal skill set includes verbal and nonverbal communication, listening, problem solving, and decision making (*Quality Magazine* 2016).

Product Traceability

Traceability is an explicit part of the ISO 9001 and ISO/TS 16949 standards. See paragraph 7.5.3 in. ANSI/ISO/ASQ Q9001–2008, for example. Traceability is like a pedigree for a dog breed or the provenance of a painting. It allows one to learn the history of an item. Commodity products such as nuts and bolts have limited need for traceability. Complex products such as automobiles must have multiple paths that trace back through many levels and many different sources. Sensitive material such as pharmaceuticals and food products must be traceable at all times. Even in the case of nuts and bolts, however, wise manufacturers will keep different lots segregated and identified as long as it is economically possible.

The ISO 9001 standard requires product identification and traceability, where appropriate, for recall of nonconforming product, hazardous product, or product in conflict with laws, regulations, or statutes. Product identification must be provided when required by a customer. Properly identified items must have a unique number and are tracked by their location in the process. Differences between items and lots must be distinguishable.

Traceability, that is, the ability to preserve the identity of the product and its origins, should begin when the process is first designed. Today, appropriate software and database designs are available. Training of workers may be required to create the proper climate and means to accomplish this. Frank M. Gryna (1988) listed four reasons why traceability is needed:

1. To assure that only materials and components of adequate quality enter the final product, for example, sterility of drug materials, adequate metallurgical composition, and heat treatment of structural components.
2. To assure positive identification to avoid mix-up of products that otherwise look alike.
3. To permit recall of suspect product on a precise basis. Lacking traceability programs, huge recalls of automobiles and other products have been required in the past. The number of defectives in the recalled set was often quite small.
4. To localize causes of failure and take remedial action at minimal cost.

There are other uses of traceability, such as in inventory control and scheduling. Some of these uses also affect quality. For example, use of materials on a first-in, first-out basis reduces the risk of quality deterioration of perishable materials.

Factors to Consider

- What is the cost of the product? A more expensive product requires more accountability over time, and thus better traceability.
- How long will the product last? If it is going to be around a long time, there is more concern about its origin, as new discoveries of chemical characteristics and environmental effects are often made. The discovery that asbestos is a carcinogen after its routine use for decades is a good example.

- Will the product be built into another product?
- Does the product contain items or materials that have not been thoroughly evaluated over a long period of time?
- Is there a significant possible health hazard associated with the product?
- Are field modifications often required, with different replacement items required on different models? (Automobiles are a prime example.)

Ten items to consider in a traceability program are:

1. Product category
2. Product life
3. Unit cost
4. Recall or modification in the field
5. Product complexity
6. Level of downstream traceability
7. Documents providing traceability
8. Type of identification
9. Coded versus uncoded identification
10. Method of identification—tags, nameplates, ink stamps, other means

The use of a tracing code (Feigenbaum 1991) is required for efficient operation. This code is established at the beginning of material flow, and a traceability flowchart is established.

The major activities on the flowchart include:

1. Critical component selection and listing by part number.
2. Vendor part coding (recording vendor name and date of receipt).
3. Coding internally manufactured parts, subassembly, assembly, and storage in a daily tally. At the end of the assembly line, each shipping container is date coded. This sequential coding procedure provides sufficient data to tie critical components to specific dates of receiving inspection, manufacturing, and final assembly.
4. Computerized shipping records, including date codes, customer name, and destination. Correlation of these data with tracing code numbers results in very effective traceability of critical components.

Other methods that are commonly used to trace products and materials include age control, shelf life, first-in first-out (FIFO), barcoding, date codes, and lot and part numbering.

IDENTIFICATION OF NONCONFORMING MATERIAL

The only reason to identify anything is to be able to trace it. The only reason to trace it is to be able to find out something about it later. These are two significant reasons in today's highly technological and litigious society. Without product traceability, many manufacturers would be exposed to unacceptable risk.

Principles of Identification

Modern technology has produced a wide array of identification methods. The physical application of markings and subsequent tracking by means of scanners and sensors provide many options. It is necessary to maintain records not only of items produced and their identification but also of how the record-keeping system itself is operated and modified. After all, information storage and retrieval is a rapidly changing field.

Mechanics

To illustrate the mechanics of product identification, consider the case of the Sauer Danfoss Company in Ames, Iowa. This company makes moderately complex mechanical products that require 100% testing and periodic design modifications. It recently improved its materials management system by creating a multifunctional task team of four people. The team collected data for two and a half years and finally decided to scrap the existing system for tracking material, which was dependent on manual entry into paper "move tags" and then manual keying into a computer database. Determination of current status required frequent physical counts of all items.

The team switched the company's system over to bar code and RFID (radio frequency identification) technologies. Now, whenever an item of hardware moves, it is automatically accounted for, either by a bar code scanner or by an RFID receiver. A sophisticated database system automatically processes each scan. The database maintains a variety of characteristics about each unit, including:

- Model number
- Unit number
- Date produced
- Result of test
- Date of test
- Rework record

Product identification is vital when producing complex products, and often unnecessary for mundane commodities. However, an example of the lack of sufficient product identification and control is the Starlink seed corn problem of 2000. Starlink was a form of seed corn that was approved for growing animal feed but not for human consumption. Inadequate controls were put into place when the seed corn was sold to farmers, and as a result the animal feed corn was inextricably intermixed with human consumption corn at grain elevators throughout the Midwest. At the time of delivering the corn, neither the farmers nor the grain elevator operators realized there was a problem. Soon, consumer groups were testing products made out of this corn, and the use of the unacceptable corn was detected. A great outcry resulted, and many losses were incurred as the intermixed corn had to be converted to animal feed.

Lennox Industries in Marshalltown, Iowa, uses a 10-digit alphanumeric product identification code. This set of 10 digits allows traceability to a diverse set of factors, including the date of fabrication, the supplier of each subsystem, the product model, and the date of final assembly. Several things must be considered when setting up such a code:

- The amount of liability exposure.
- The number of levels of components and subcomponents.
- The process design must incorporate the ability to trace products back to their point of creation and installation.

LEVELS OF SEVERITY

In certain types of products, more than one defect could be present, and a relatively small number of minor defects could be acceptable to the customer. Product quality in such cases may be judged by the total number of defects or the number of defects per unit. Control charts for attributes are a tool that may be used for this purpose. In such cases, the objective of inspection is to determine the number of defects or nonconformities present rather than to classify units as conforming or nonconforming.

A *nonconformity* is defined as the failure of a quality characteristic to meet its intended level or state, occurring with severity sufficient to cause the product not to meet a specification. A *defect* is a nonconformity severe enough to cause the product not to satisfy normal usage requirements. Thus, the difference between a nonconformity and a defect is based mainly on perspective. The former is defined according to specifications, while the latter is defined according to fitness for use. The numerical result generated by inspection consists of the number of defects or nonconformities for each product unit. Often, it is possible to classify the different types of defects according to their severity and then assign a weight to each class based on the importance of the affected quality characteristic regarding the product specifications. The selection of the weights should reflect the relative importance of the various defect categories and their likelihood of causing product failure or customer dissatisfaction. A typical seriousness classification includes four levels of defect seriousness:

1. A *critical* defect may lead directly to severe injury or catastrophic economic loss.
2. A *serious* defect may lead to injury or significant economic loss.
3. A *major* defect may cause major problems during normal use. A major defect will likely result in reduced usability of the product.
4. A *minor* defect may cause minor problems during normal use.

Training programs should be aimed at new inspection personnel as well as those with on-the-job experience. On-the-job training by a supervisor or a training coordinator could be the most prevalent and least costly approach to training. Other training methods include the following:

- Classroom instruction by in-house experts or outside consultants. Best suited for theoretical subjects such as basic mathematics, statistics, experimental design, and computer use.
- Self-study using audiovisual programs and self-instruction training manuals. This method allows the inspectors to study at their own pace and during convenient times.
- Outside programs offered by professional organizations and their local chapters, and by universities and community colleges through their extension divisions.

CONTROL OF NONCONFORMING MATERIALS

ISO 9001:2008 provides guidelines for control of nonconforming products (or materials) that are considered systematic courses of action in many organizations. The following is a summary of these guidelines.

Control of Nonconforming Product

- Product that does not conform to requirements is identified and controlled to prevent unintended use or delivery.
- Documented procedure defining the controls, responsibilities, and authorities.
- Shall deal with nonconforming product by one or more of the following:
  - Take action to eliminate nonconformity
  - Authorize use, release, or acceptance by concession of relevant authority (may be customer)
  - Take action to preclude its original intended use or application
- Record nature of conformities and actions taken, including any concessions obtained.
- When nonconforming product is corrected, it is subject to reverification to requirements.
- When detected after delivery, take action appropriate to effects, or potential effects, of the nonconformity.
- Should empower people at any stage in a process to report nonconformities.
- Should effectively and efficiently control nonconforming product identification, segregation, and disposition.

- Should establish effective and efficient process to provide review and disposition.
- Review should be conducted by authorized, competent people with authority to disposition and take appropriate corrective action.
- Trends or patterns should be reviewed and should be considered for improvement.

Corrective Action

- Take action to eliminate the cause of nonconformities in order to prevent recurrence.
- Corrective actions shall be appropriate to the effects of the nonconformities encountered.
- Documented procedure to define requirements for:
  - Reviewing nonconformities (including customer complaints)
  - Determining the causes of nonconformities
  - Evaluating the need for action to ensure that nonconformities do not recur
  - Determining and implementing action needed
  - Records of the results of action taken (4.2.4)
  - Reviewing corrective action taken
- Should ensure the corrective action is used as improvement tool
- Should include evaluation of the significance of problems and potential impact on:
  - Operating cost
  - Costs of nonconformity
  - Product performance
  - Dependability
  - Safety
  - Customer satisfaction
- Corrective action should be focused on eliminating causes to avoid recurrence. Examples of sources for corrective action:
  - Customer complaints
  - Nonconformity reports
  - Internal audit reports
  - Outputs from management review

- Outputs from data analysis
- Organization's people
- Process measurements

Preventive Action

- Determine action to eliminate causes of potential nonconformities to prevent their occurrence.
- Preventive actions shall be appropriate to the effects of the potential problems.
- Documented procedure to define requirements for:
  - Determining potential nonconformities and their causes
  - Evaluating the need for action to prevent occurrence
  - Determining and implementing action needed
  - Records of results of action taken (4.2.4)
  - Reviewing preventive action
- Should plan to mitigate the effects of loss to organization to maintain performance of processes and products.
- Loss prevention in the form of planning should be applied to realization and support processes, activities, and product to ensure satisfaction of interested parties.
- Should be based on data from appropriate methods, including evaluation of data trends and criticality relative to the performance of the organization and its products. Data can be generated from:
  - Risk analysis tools, such as fault modes and effect analysis
  - Customer needs and expectations
  - Management review output
  - Outputs from data analysis
  - Satisfaction measurements
  - Results of self-assessment
- Results of the evaluation of the effectiveness and efficiency of loss prevention plans should be an output from management review.
- Should be used as input for the modification plans and as input to the improvement processes.

DISPOSITION OF NONCONFORMING MATERIAL

Material Segregation

There are two major situations that demand disposition of nonconforming products. The first is when a product fails to pass inspection or a test, and a decision regarding it must be made. This is the function of the material review board. The second situation, considerably more serious, is when a problem develops after the product is out of the plant, on store shelves, in dealer showrooms, and in use by customers. Now a product recall may be required. In view of the very negative aspects of product recall, all the prior work concerning product traceability and product integrity will pay off quite handsomely in organizing the recall.

Material Review Board

The *material review board* (MRB) is an appointed group of individuals with different backgrounds and expertise whose assignment is to determine what corrective actions must be taken after nonconforming parts or components are discovered. In a larger sense, the purposes of the MRB are to determine the disposition of nonconforming parts, components, and subassemblies, determine the causes of the nonconformance of these items, and take the necessary corrective actions to prevent such nonconformance from taking place in future production.

The basic function of an MRB is to (1) review material that does not conform to standards, (2) determine what its disposition should be, and (3) drive the development of effective corrective action to prevent recurrence. The MRB is a broad-based reviewing agency whose membership usually consists minimally of representatives from the following:

- *Engineering.* The cognizant designer is often the representative.
- *Quality assurance.* The representative is often from quality control engineering.
- *Customers.* The representative may be from the customer's organization (for example, the government inspector) or from marketing. In some companies the role of the MRB is solely one of judging fitness for use of nonconforming products. Bond (1983) discusses board composition, philosophy, and problem documentation.

In general, the MRB procedural steps can be summarized as follows: After a defect is discovered, verification by inspection may be needed. A complete description of any nonconformance is then initiated. A quality engineer picked by the MRB will review the facts and include the case in an appropriate tracking system. The MRB committee may then follow up with investigation and analysis. After this, the quality engineer takes the case again, recommending the appropriate corrective action(s) and steps for implementation. The term *standard repair* is common within the MRB framework. It signifies a procedure where a certain type of defect occurs time and time again. A standard repair procedure is then initiated, documented, and implemented for such situations. Minor defects are most likely to be

Table 17.1 Standards pertaining to MRB operations.

Standard	Purpose
MIL-STD-1520C	Sets "the requirements for cost-effective corrective action and disposition system for nonconforming material"
MIL-Q-9858A	Quality program requirements, Section 6.5, "Nonconforming Material," requires the contractor to establish "an effective and positive system for controlling nonconforming material"
MIL-STD-481B	Configuration control—engineering changes
MIL-I-8500	Establishes interchangeability and replaceability requirements
ANSI/ASQ Z1.4-2008	Sampling procedures and tables for inspection by attributes
ANSI/ASQ Z1.9-2008	Sampling procedures and tables for inspection by variables percent nonconforming

treated with a standard repair procedure. Minor defects, unlike major ones, may not adversely affect the integrity of the part, component, or assembly. In many cases, the MRB concludes that the lot containing nonconforming products should not be shipped as is.

The decision, concurred with by inspection personnel, may be to sort (100% inspection), downgrade, repair, rework, or scrap. A decision to ship may also be authorized by the MRB; in such cases, a unanimous decision should be reached by all members. The decision should also create factual data; thus, it is an important source of information. A successful MRB program requires that the board not only makes decisions about immediate disposition of rejected material, but also directs ongoing programs of root cause analysis to eliminate future rejections of the same type.

Several military documents are associated with the MRB concept. Table 17.1 shows a partial list.

Chapter 18

Testing Methods

Define and use the following methods in various situations.

1. ***Nondestructive testing:*** X-ray, eddy current, ultrasonic, dye penetrant, magnetic particle, optical, visual, and profile (Understand)
2. ***Destructive testing:*** tensile, force testing, and drop test (Apply)
3. ***Functionality testing:*** tension, torque, leak testing, and compression (Apply)
4. ***Hardness testing:*** Brinell, Rockwell, durometer, and micro-hardness scales (Apply)

Body of Knowledge III.D

NONDESTRUCTIVE TESTING

Definition

Nondestructive testing (NDT) is testing that does not destroy the test object. It is also referred to as *nondestructive evaluation* or *nondestructive examination (NDE).* The American Society for Nondestructive Testing (ASNT) defines NDT as "test methods used to examine or inspect a part or material or system without impairing its future usefulness" (Dettmers 2004; ASNT 2012).

NDT Applications and Objectives

NDT dates back to the early 1900s, and evolved over the years from being captive in the lab to becoming a very powerful production tool. NDT is now used worldwide to detect variation in product finish, product defects, and/or cracks. This type of testing is used in most stages of the production process, if not all. The Center for Nondestructive Evaluation at Iowa State University gives the following applications for NDT (Larson 2000):

- Screening and inspection of incoming material
- Product development
- Inspecting for in-service damage
- Verifying the proper process
- Monitoring and controlling manufacturing processes
- Verifying the proper assembly

The objectives of such testing are (ASNT 2012):

- To ensure customer satisfaction through ensuring product reliability and integrity
- To ensure the safety of the end user
- To compete with the current evolving market by maintaining reputation through high-quality products and better product design
- To minimize waste, which will lead to minimizing the overall manufacturing cost and maximizing the profit
- To ensure the flexibility and adaptation of product to market changes

Methods of NDT

The National Materials Advisory Board (NMAB) Ad Hoc Committee on Nondestructive Evaluation adopted a system that classifies methods into six major categories: visual, radiation, magnetic-electrical, mechanical vibration, thermal, and chemical-electrochemical (ASNT 2012; Wenk and McMaster 1987; McMaster and Wenk 1951). Within these categories, the most identified methods are visual, X-ray, eddy current, ultrasonic, magnetic, and liquid penetrant.

Visual Method. The visual method is the oldest and most common method of examination. It has been used in a variety of industrial and commercial applications (Aircraft NDT 2005). The examiner usually follows a procedure that might vary in complexity from simple to very complex. Visual tools also vary, from the naked eye to very complex remote-controlled equipment. Figure 18.1 shows a piece of equipment that might be used to inspect products or materials. Some tools can be visual but controlled with a robot to inspect hazardous or tight areas, such as nuclear reactors. Video cameras and camcorders are considered tools for visual inspection.

Radiation. An X-ray machine or radioactive isotope is used as a source of radiation. The radiation is directed through the test part and onto a recording medium such as film. When the film is developed, a radiograph is obtained that shows the internal condition of a part (Aircraft NDT 2005). This radiation has higher energy and a shorter wavelength than visible light, which gives it the ability to penetrate the different objects, thus revealing cracks and other object defects (Larson 2000). Figure 18.2 shows the method of inspecting using X-ray.

The different shades on the film indicate the different thicknesses showing defects existing in the object. Figure 18.3 shows an example of an object that has been inspected by X-ray.

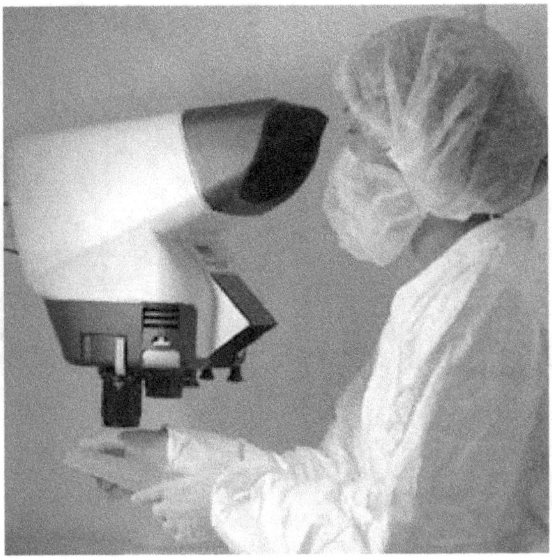

Figure 18.1 Visual inspection ensures consistency.
Source: Larson (2000).

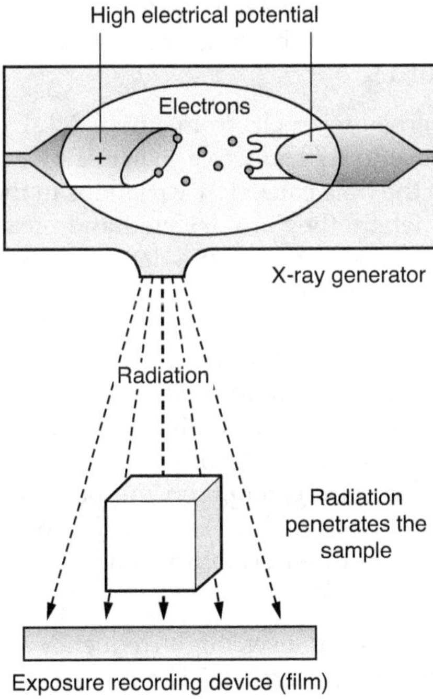

Figure 18.2 Inspection using X-ray.
Source: Larson (2000).

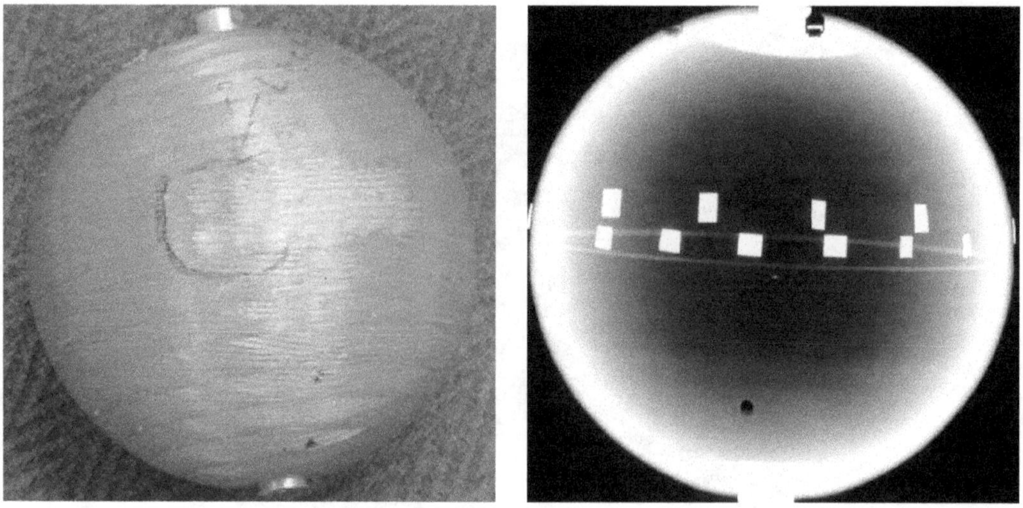

Figure 18.3 Example of an inspection using X-ray.
Source: Larson (2000).

X-ray and radiation testing in general affect the health of the people performing the test. Appropriate precautions and policies have to be followed to protect users. Radioactive waste should be considered for environmental and biological issues (NDT-ED.org 2012c).

Eddy Current. *Eddy current* is an electromagnetic NDT method in which electric currents are generated into a conductive material by an induced, varying magnetic field. Changes in the flow caused by variations in the object are reflected into a nearby coil or coils, where they are detected and measured by suitable instrumentation (Aircraft NDT 2005). Figure 18.4 depicts the eddy current method. This type of testing can be used to detect surface cracks but can also be used to make electrical conductivity and coating-thickness measurements (Larson 2000).

Eddy current can also be used for monitoring aircraft structural components and metallic structures such as aircraft fuselage. Some disadvantages of this type of testing are that it works only with conductive materials and that the surface of the material must be accessible. Other limitations are the depth of penetration, which is bounded to a certain extent, and the possibility that the finish of the product might provide a bad reading. Even with these limitations, this method has many advantages, including portability and the ability to provide immediate feedback (Soleimani et al. 2006).

Ultrasonic. Ultrasonic testing uses high-frequency sound waves to conduct examinations and make measurements. The pulse echo technique is performed by introducing sound into the test object, and reflections (echoes) are returned to a receiver from internal imperfections or from geometrical surfaces of the part. It is typically used to detect subsurface defects, or defects originating from surfaces not accessible without disassembly or removal. It can also be used to detect

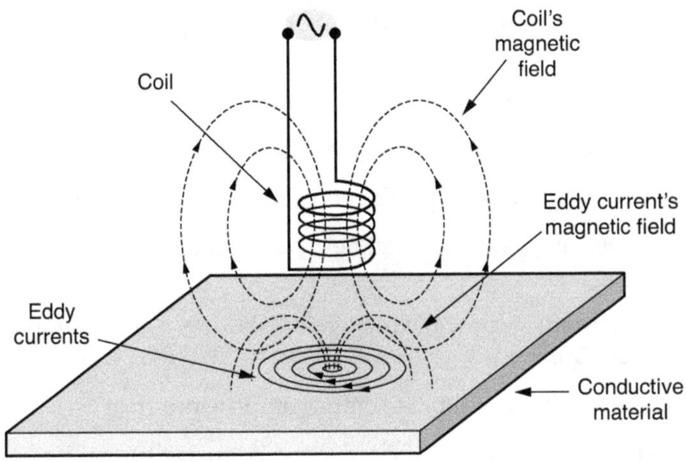

Figure 18.4 Eddy current method.
Source: Larson (2000).

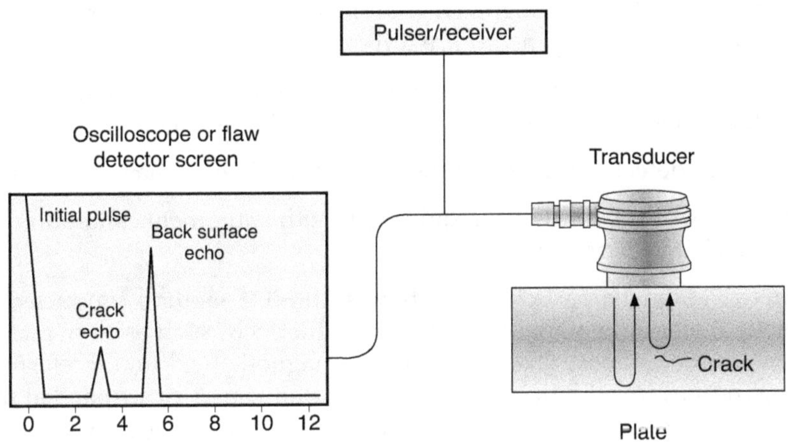

Figure 18.5 The general inspection principle for ultrasonic testing.
Sources: Aircraft NDT (2005); NDT-ED.org (2012a).

laminations, lack of fusion, and corrosion of various materials (Aircraft NDT 2005; NDT-ED.org 2012a). Figure 18.5 shows the general inspection principle.

Some often-cited advantages of ultrasonic inspection include the following (NDT-ED.org 2012a):

- It is sensitive to both surface and subsurface discontinuities.
- The depth of penetration for flaw detection or measurement is superior to other NDT methods.
- Only single-sided access is needed when the pulse echo technique is used.

- It is highly accurate in determining reflector position and estimating size and shape.
- Minimal part preparation is required.
- Electronic equipment provides instantaneous results.
- Detailed images can be produced with automated systems.
- It has other uses, such as thickness measurement, in addition to flaw detection.

Like all NDT methods, ultrasonic inspection also has its limitations, which include (NDT-ED.org 2012a):

- Surface must be accessible to transmit ultrasound.
- Skill and training necessary are more extensive than with some other methods.
- It normally requires a coupling medium to promote the transfer of sound energy into the test specimen.
- Materials that are rough, irregular in shape, very small, exceptionally thin, or not homogeneous are difficult to inspect.
- Cast iron and other coarse-grained materials are difficult to inspect due to low sound transmission and high signal noise.
- Linear defects oriented parallel to the sound beam may go undetected.
- Reference standards are required for both equipment calibration and the characterization of flaws.

Magnetic. Magnetic particle inspection is an NDT method conducted on ferromagnetic material (material that can be magnetized) such as iron, nickel, cobalt, or some of their alloys. The test induces a magnetic field into the ferrous object, coating the surface with iron particles that are either dry or suspended in a liquid. Surface imperfections will distort the magnetic field, causing an accumulation of the iron particles near imperfections, thus indicating their presence in the material (Aircraft NDT 2005). Many industries use magnetic particle inspection for determining a component's fitness for use. Some examples of these are the structural steel, automotive, petrochemical, power generation, and aero-space industries. Underwater inspection is another area where magnetic particle inspection may be used to test items such as offshore structures and underwater pipelines (NDT-ED.org 2012b). Figure 18.6 is an example of testing using magnetic particle inspection.

One disadvantage, as mentioned earlier, is that it works only on ferromagnetic material. On the other hand, it has many advantages:

- It is a simple concept compared to other concepts.
- It is considered a combination of two NDT methods—magnetic flux leakage testing and visual testing.
- It can be used underwater.

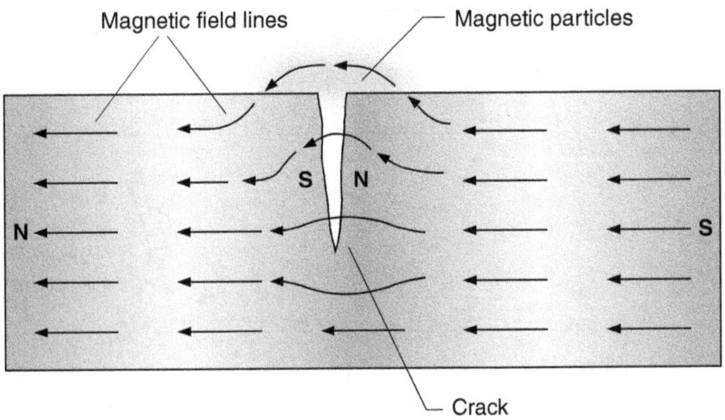

Figure 18.6 Magnetic particle inspection.
Sources: Larson (2000); NDT-ED.org (2012b).

- It is easy to apply, and the object's surface preparation is not as critical as in other NDT methods (Larson 2000; Aircraft NDT 2005; NDT-ED.org 2012b).

Liquid Penetrant. Liquid or dye penetrant inspection is considered an extension of visual inspection and is used for detecting surface-breaking flaws on any non-absorbent material's surface. For this type of test, the object must be thoroughly cleaned to remove dirt and grease. The object is then soaked in a fluorescent or visible dye solution. After a specified dwell time, the excess penetrant is removed from the surface and a developer is applied. Often, a developer is applied to help draw penetrant out of imperfections open to the surface, making them much more visible. With fluorescent penetrants, the object must be viewed in a dark room with an ultraviolet lamp, which is sometimes not practical. Some industries use red dye in aerosol cans instead of the fluorescent material. This red dye doesn't need the ultraviolet light and can be viewed with the naked eye. Liquid penetrant inspections are able to detect only surface-breaking flaws (Larson 2000; Aircraft NDT 2005; Munns 2004). Figure 18.7 shows the steps for the liquid penetrant testing.

Other nondestructive inspection methods include laser testing, leak testing, inspection, thermal/infrared testing, vibration testing and analysis, and profile testing.

DESTRUCTIVE TESTING

Definition

Destructive testing is a test that puts the sample object under certain circumstances until it actually fails. The failed pieces are then studied and compared with known standards to determine the quality of the object. This type of testing is generally much easier to carry out, more likely to yield more information, and much easier to interpret than NDT (Integrated Publishing 2012).

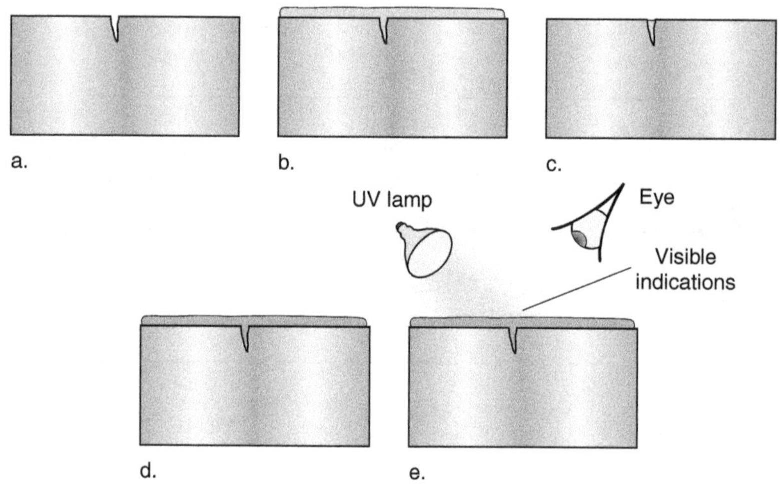

Figure 18.7 Liquid penetrant testing steps.
Source: Munns (2004).

Destructive Testing Applications and Objectives

Destructive testing usually provides a more reliable assessment for objects, but destruction of the test object usually makes this type of test costlier, than NDT, for the test object's owner. Destructive testing is also inappropriate in many circumstances, such as forensic investigation. A decision has to be made regarding the trade-off between the cost of the test and its reliability. If a nondestructive test can provide the required results, the destructive test can be waived. Destructive testing is performed on a sampling of test objects randomly drawn for the purpose of characterizing the testing reliability of the nondestructive test (Integrated Publishing 2012).

Destructive testing is used mostly in metal testing, especially to test the strength of welded parts. It is also used to determine the ability of a metal to stretch or pent. It is common in automobile manufacturing testing and other metal manufacturing industries.

Methods of Destructive Testing

Tensile Testing. A *tensile test*, also known as a *tension test*, is the method for determining the behavior of materials under axial stretch loading. Data from the test are used to determine elastic limit, stretching, modulus of elasticity, proportional limit, reduction in area, tensile strength, yield point, yield strength, and other tensile properties. This test is probably the most fundamental type of mechanical test that can be performed on a material. Tensile tests are simple, relatively inexpensive, and fully standardized. By pulling on something, you will very quickly determine how the material reacts to forces being applied in tension. As the material is being pulled, you will find its strength, along with how much it will stretch. Figure 18.8 shows a tensile test for a furnace (Integrated Publishing 2012; Instron 2012).

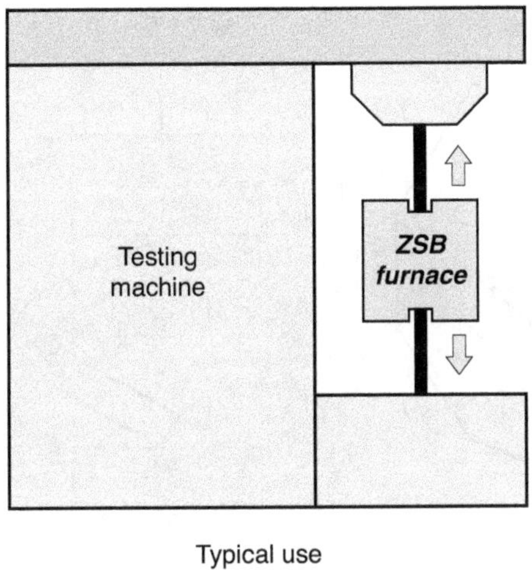

Figure 18.8 Tensile test for a furnace.
Source: Micropyretics Heaters International (2012).

The results of tensile testing are used in selecting materials for engineering applications. Tensile properties are included in material specifications to ensure quality. Tensile properties are used to predict the behavior of materials under certain forms of tension (Davis 2004).

Force Testing. Force testing includes free-bend, guided-bend, nick-break, impact, and crash testing. The *free-bend* test is designed to measure the flexibility of the weld deposit and the heat-affected area adjacent to a weld. It is also used to determine the percentage of stretching of the weld metal (Integrated Publishing 2012). Figure 18.9 shows a free-bend test and the forces applied in the test to measure the flexibility.

The *guided-bend* test is used to determine the quality of weld metal at the face and root of a welded joint. The sample object is placed across the supports of the die. A plunger, operated from above by hydraulic pressure, forces the sample into the die. To fulfill the requirements of this test, the sample object has to be bent to 180° with no cracks on the surface greater than 1/8 in. (Integrated Publishing 2012).

The *nick-break* test is useful for determining the internal quality of the weld metal. This test is used for detecting whether internal defects exist, such as slag inclusions, gas pockets, lack of fusion, and oxidized or burned metal. For example, there should not be more than six pores or gas pockets per square inch of exposed broken surface of the weld (Integrated Publishing 2012).

The *impact* test is used to check the ability of a weld to absorb energy under impact without fracturing. This is a dynamic test in which a test specimen is broken by a single blow, and the energy used in breaking the piece is measured in foot-pounds. This test compares the toughness of the weld metal with the base metal. It is useful for determining whether any of the mechanical properties of the base metal were destroyed by the welding process (Integrated Publishing 2012).

A *crash* test is usually performed in order to ensure safe design standards in crash compatibility for automobiles or related components, as shown in Figure 18.10.

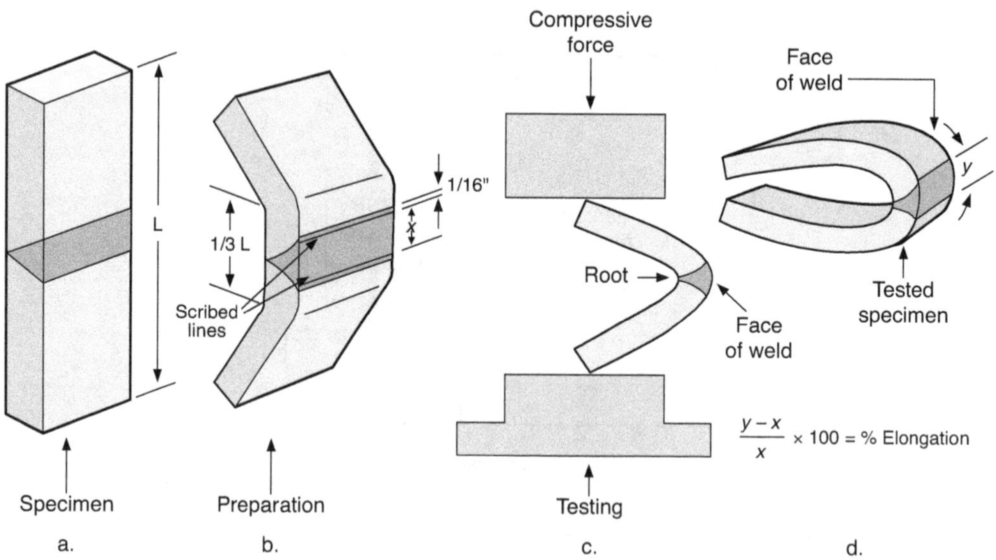

Figure 18.9 Free-bend test.
Source: Integrated Publishing (2012).

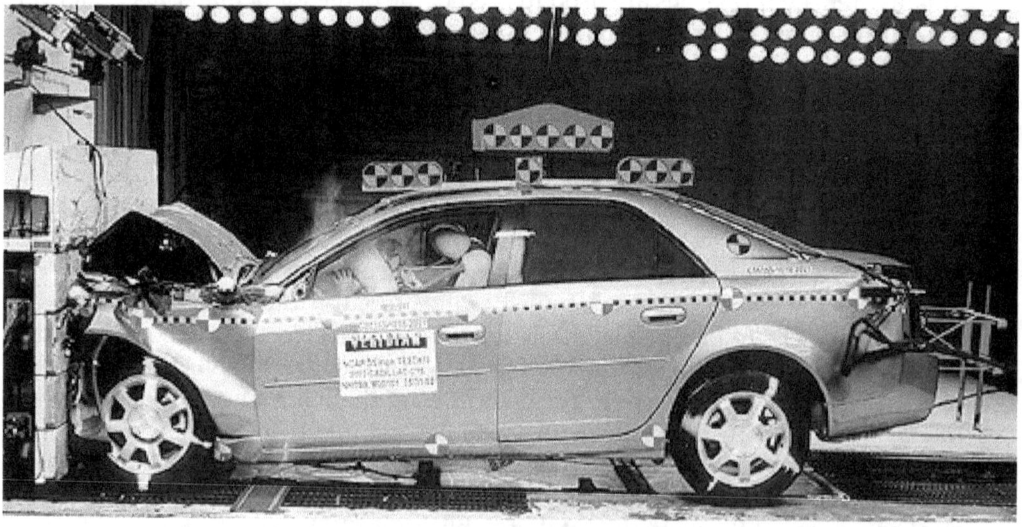

Figure 18.10 Crash testing.
Source: Integrated Publishing (2012).

Hardness Testing. *Hardness* is the property of a material that enables it to resist plastic deformation, usually by penetration. However, the term "hardness" may also refer to resistance to bending, scratching, abrasion, or cutting. Hardness is not an intrinsic material property dictated by precise definitions in terms of fundamental units of mass, length, and time. A hardness property value is the result of a defined measurement procedure. Relative hardness of minerals can be assessed by reference to the *Mohs scale*, which ranks the ability of materials to resist scratching by another material. Similar methods of relative hardness assessment are still used today.

Most hardness tests are limited in practical use and do not provide accurate numeric data or scales, particularly for modern-day metals and materials. The usual method to achieve a hardness value is to measure the depth or area of an indentation left by an indenter of a specific shape, with a specific force applied for a specific time. There are three principal standard test methods for expressing the relationship between hardness and the size of the impression: Brinell, durometer, and Rockwell (see "Hardness Testing," for more details about these types of hardness testing). For practical and calibration reasons, each of these methods is divided into a range of scales, defined by a combination of applied load and indenter geometry (England 2012).

Metallography. *Metallography* is the microscopic study of the structural characteristics of a metal or an alloy. The microscope is by far the most important tool of the metallurgist from both a scientific and a technical standpoint. It is possible to determine grain size and the size, shape, and distribution of various phases and inclusions, which have a significant effect on the mechanical properties of the metal (ASM International 2005).

FUNCTIONALITY TESTING

Definition

Functionality testing is used to verify whether a product meets the intended design specifications and functional requirements stated in the development documentation (nResults 2008).

Functionality testing can identify potential product bugs, verify the success of the fixes, and ensure that the product is within standards.

Functionality Testing Objectives

Functionality testing provides products that have only a minimum number of bugs or problems. This type of test ensures that a product will function as intended. Minimizing the bugs in the product will minimize the cost of service and maintenance after the sale, which will build the reputation of the product and make it competitive in the market.

Methods of Functionality Testing

The most common functionality testing is performed on software packages. This topic is covered in Chapter 19 Other functionality testing is being performed in manufacturing facilities to ensure that products can be exposed to the maximum tension, torque, and/or compression.

Tension. *Tension* is the act of stretching an object using an external force. Test engineers use test equipment that can apply a large amount of force on an object and measure the magnitude and duration of the force on that object. Figure 18.11 shows a tension test machine.

Torque. A *torque* is a vector that measures the tendency of a force to rotate an object about some axis (Serway and Jewett 2003). This type of test is performed on the object to measure the magnitude of torque that can be applied on that object before it breaks. The torque is calculated by multiplying the force applied to a lever by its distance from the lever's fulcrum (illustrated in Figure 18.12), which can be represented by the following formula:

$$\tau = r \times F$$

Compression. *Compression* is the act of applying axially directed pushing forces on an object, as shown in Figure 18.13. One example of such a functional test is compression exerted on a concrete block to test the ability of that block to withstand the pressure. Many concrete structures have compressive strengths in excess of 50 MPa.

The compressive strength of a material is the value of uniaxial compressive stress reached when the material fails completely.

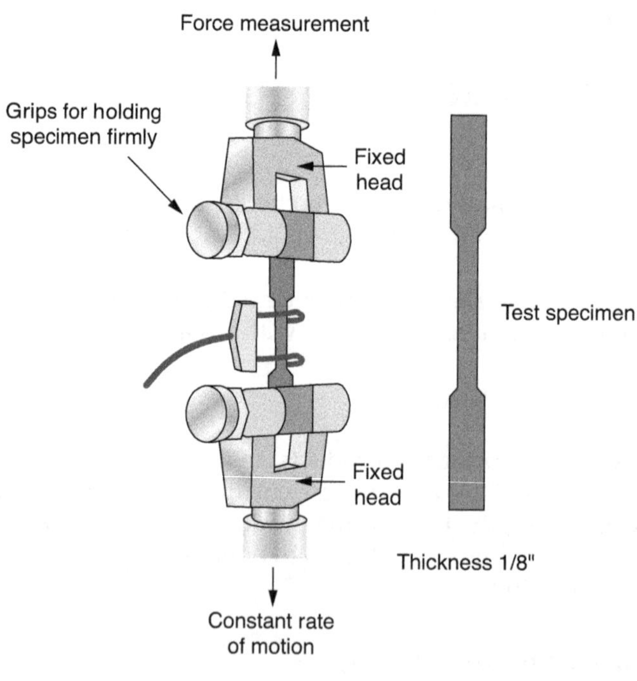

Figure 18.11 Tension test machine.
Source: ADMET Material Testing (2006).

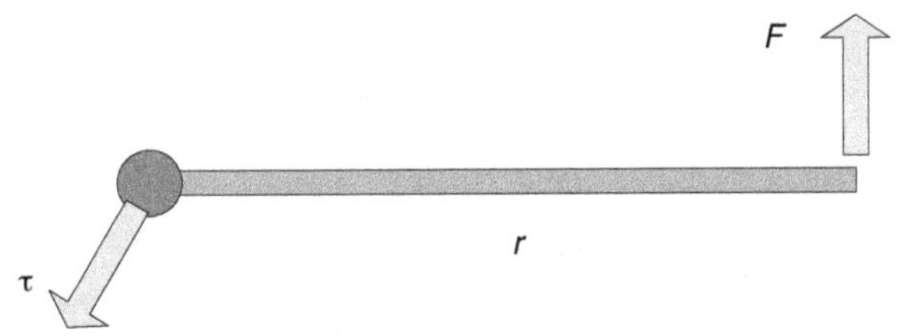

Figure 18.12 Torque can be calculated by multiplying the force applied to a lever by its distance from the lever's fulcrum.

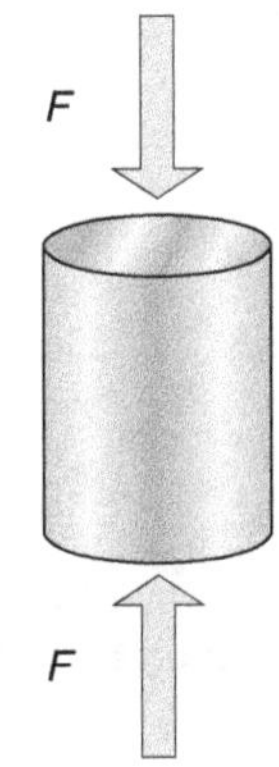

Figure 18.13 Compression test.

HARDNESS TESTING

Brinell

The *Brinell* hardness test is based on applying forces on an object using a steel or carbide ball that has a 10 mm diameter and is subjected to a load of 6614 lb, which can be reduced for softer material to avoid excessive indentation (see Figure 18.14). The diameter of the indentation is measured after a certain amount of time using a low-powered microscope, and then the Brinell hardness number is calculated by dividing the load applied by the surface area of the indentation (England 2012).

Compared to the other hardness test methods, the Brinell ball makes the deepest and widest indentation, so the test averages the hardness over a wider amount of material, which will more accurately account for multiple grain structures and any irregularities in the uniformity of the material (England 2012).

Rockwell

The *Rockwell* hardness test method (Figure 18.15) is also based on applying force on an object to create indentation, but it uses a diamond cone or hardened steel

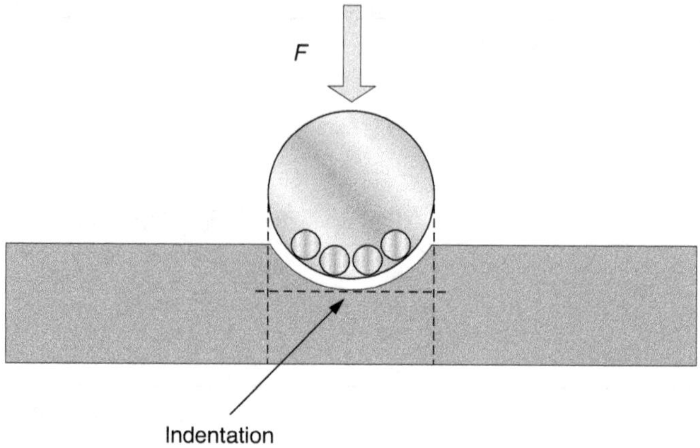

Figure 18.14 Brinell hardness test method.

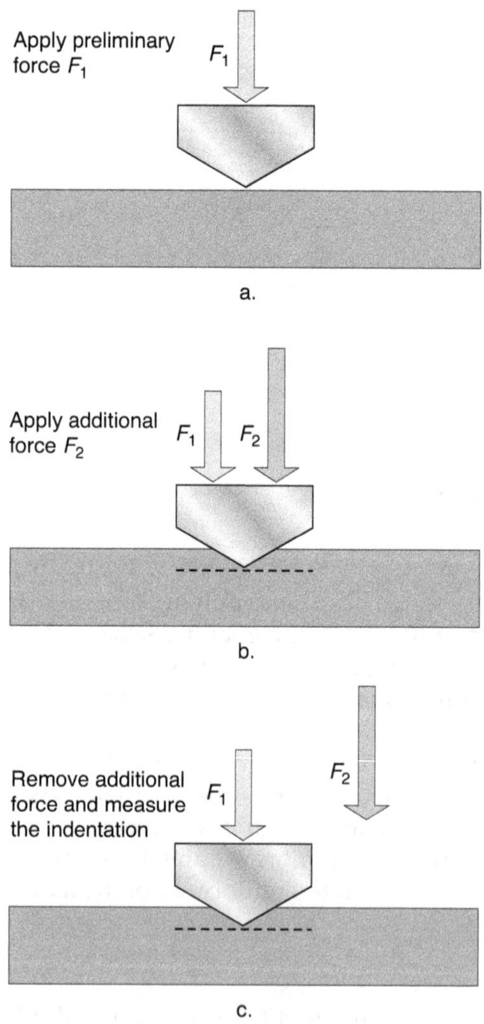

Figure 18.15 Rockwell hardness test method.

ball indenter. A preliminary force is applied on the indenter to be forced into the test material under minor load. When equilibrium has been reached, an additional major load is applied, with a resulting increase in penetration. When equilibrium has again been reached, the additional major load is removed, leaving the preliminary load as is. The removal of the additional major load allows a partial recovery. The indentation from that load is measured and is used to calculate the Rockwell hardness number (England 2012).

Advantages of the Rockwell hardness method include the direct Rockwell hardness number readout and rapid testing time. Disadvantages include many arbitrary nonrelated scales and possible effects from the iron sample object support block. The Vickers and Brinell methods don't suffer from this effect (England 2012).

Vickers

The *Vickers* hardness test (Figure 18.16) was developed as an alternative method to measure the hardness of materials. Unlike the Rockwell method, this method doesn't have the arbitrary nonrelated scales and is often easier to use than other hardness tests. The Vickers test can be used for all metals and can also be used on ceramic materials. It has one of the widest scales among hardness tests. The unit of hardness given by the test is known as the *Vickers pyramid number (HV)*. The hardness number is determined by the load over the surface area of the indentation, not the area normal to the force; therefore, it is not a pressure. Like the Brinell and Rockwell tests, this test uses an indenter to test the metal, but this indenter is in the shape of a diamond with 136° between opposite faces (England 2012; Meyers and Chawla 1999; CALCE and the University of Maryland 2001).

The two diagonals of the indentation left in the surface of the material after removal of the load are measured using a microscope, and their average is calculated.

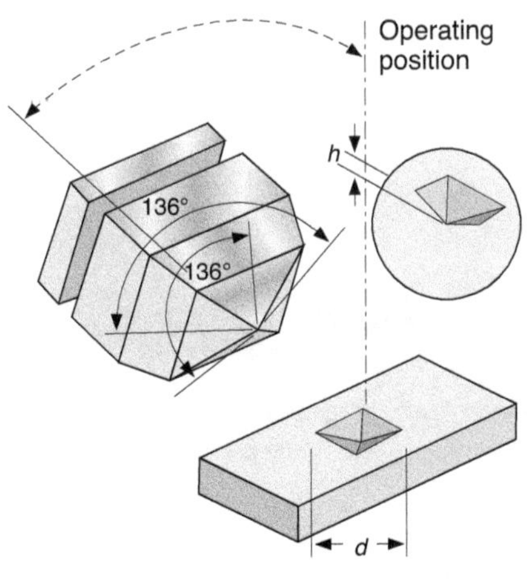

Figure 18.16 The Vickers hardness test.
Source: CALCE and the University of Maryland (2001).

The area of the sloping surface of the indentation is calculated. The Vickers hardness is the quotient obtained by dividing the kgf load by the square mm area of indentation (England 2012; Meyers and Chawla 1999; CALCE and the University of Maryland 2001).

The advantages of the Vickers hardness test are that extremely accurate readings can be taken, and only one type of indenter is used for all types of metals and surface treatments. Although thoroughly adaptable and very precise for testing the softest and hardest of materials, under varying loads, the Vickers machine is a floor-standing unit that is more expensive than the Brinell or Rockwell machines (England 2012; Meyers and Chawla 1999).

Microhardness Test

Microhardness testing of metals, ceramics, and composites is useful for a variety of applications for which other test methods cannot be used: testing very thin materials like foils, measuring individual microstructures within a larger matrix, or measuring the hardness gradients of a part along the cross section. Microhardness testing gives an allowable range of loads for testing with a diamond indenter; the resulting indentation is measured and converted to a hardness value (England 2012; CALCE and the University of Maryland 2001).

The term "microhardness test" usually refers to static indentations made with loads not exceeding 1 kgf. The indenter is either the Vickers diamond pyramid or the Knoop elongated diamond pyramid. The procedure for testing is very similar to that of the standard Vickers hardness test except that it is performed on a microscopic scale with higher-precision instruments. The surface being tested generally requires a metallographic finish; the smaller the load used, the higher the surface finish required. Precision microscopes are used to measure the indentations; these usually have a magnification of around ×500 and measure to an accuracy of ±0.5 micrometers (England 2012).

Chapter 19

Software for Test Equipment

> Identify and describe basic tools (e.g., safeguarding, functional checks, comparison of test results, and identification of attributes and parameters) used to ensure that the software for test equipment adequately and correctly performs its intended functions. Recognize various software validation methods. (Remember)
>
> **Body of Knowledge III.E**

Software that is integrated into measuring and testing devices allows for the automatic control, execution, and evaluation of the test process. It also facilitates documentation of results through automatic data acquisition and storage. Electrical components, medical devices, and automotive components can be effectively inspected using integrated test equipment software.

The use of test equipment software can reduce errors, assist with compliance with standards and regulations, improve quality assurance, and improve efficiency and safety. These outcomes stem from the automation of tasks throughout the test cycle. For example, software can automate (SPS Electronics):

- The test procedures themselves, which reduces test time and human error.
- Data collection and analysis in which measured values are captured and assessed in real time and displayed using graphical methods.
- Test reporting, which provides efficient and standardized quality control documentation.
- Safety functions which can automatically stop the test process in case of defects or unsafe conditions.
- Fault diagnosis and system monitoring that assures that the test stand operates properly.

BASIC TOOLS TO ENSURE SOFTWARE FUNCTIONALITY

To ensure that software for test equipment functions correctly and performs its intended tasks accurately, a range of basic tools and validation methods are used.

Here, we introduce functional checks, safeguarding, comparison of test results, and identification of attributes and parameters.

Functional Checks

Functional checks verify that the software behaves as expected in performing key functions. Automated or manual tests are designed to confirm that each software function operates correctly. These checks can involve unit testing, integration testing, and/or system testing.

In addition to functional checks, *functional testing* and *functional safety processes* can be performed. This type of testing focuses on verifying that each feature of a software application functions as intended, compared to the specified requirements. It helps ensure that the system behaves as expected, aligning with user expectations and the software's integrity.

Safeguarding

Safeguarding refers to the broader concept of protecting the software and its users from potential harm or failure. It encompasses not only functional testing but also security testing, vulnerability assessment, and other measures to ensure the integrity and reliability of the software. Safeguarding implements fail-safes and safety checks to prevent equipment damage or safety incidents. For example, a safeguard may be software that halts a test if voltage exceeds a predefined threshold.

Comparison of Test Results

To ensure accuracy and consistency, current test results can be compared against known results (gold standards), previous test data, or tolerance specifications.

Identification of Attributes and Parameters

To clearly define what the software is measuring or controlling, attributes and parameters must be defined. The software must correctly recognize and use all attributes (e.g., signal type) and parameters (e.g., voltage, time). This is critical for traceability and repeatability. For example, a check may be run to ensure the gauge software logs temperature in Celsius and identifies the correct sensor location.

Verification and Validation

Verification of software for the test equipment determines whether the software fulfills the requirements, purpose, and accuracy of the test intended. Software verification objectives include:

- Determining whether the software performs its intended functions correctly
- Ensuring that the software performs no unintended functions
- Measuring and assess the quality and reliability of the software

Validation, on the other hand, is the "confirmation, through the provision of objective evidence, that the requirements for a specific intended use or application have been fulfilled" (IEEE/ISO/IEC 2017). Validation is concerned with the process of evaluating the software to make certain that it meets its intended use and matches the stakeholders' needs (Westfall 2024).

SOFTWARE TESTING TERMINOLOGY[1]

Software testing includes the following activities:

1. Define the expected output or result, including the attributes and parameters.
2. Define how the software should function to get the results.
3. Compare test results with real results to ensure the consistency and accuracy of the software.
4. Include test cases for invalid or unexpected condition.
5. Test the code to see if it does what it is not supposed to do as well as what it is supposed to do.
6. Debug the software to minimize errors.
7. Retest for functionality until you reach high precision.

Definitions of a few software testing terms follow.

Automated Software Testing

Automated software testing involves the use of software to control the execution of tests, the comparison of actual outcomes to predicted outcomes, the setting up of test preconditions, and other test control and test reporting functions, without manual intervention.

Fault-Injection Testing

Fault-injection testing is a popular software engineering technique used to measure test coverage. Known bugs are randomly added to a program source code and the programmer is tasked to find them. The percentage of the known bugs not found gives an indication of the real bugs that remain.

Compatibility Testing

Compatibility testing is the process of testing to understand if software is compatible with other elements of a system with which it should operate, for example, browsers, operating systems, or hardware.

1. Reprinted from Original Software. 2007. *Glossary of Software Testing Terms*. Westmont, IL: Original Software. http://www.origsoft.com.

Functional Requirements

Functional requirements define the internal workings of the software, such as the calculations, technical details, data manipulation and processing, and other specific functionality that show how the use cases are to be satisfied. They are supported by nonfunctional requirements, which impose constraints on the design or implementation, such as performance requirements, security, quality standards, or design constraints.

Section IV

Quality Assurance and Improvement

Chapter 20 Quality Improvement

Chapter 21 Quality Audits

Chapter 22 Quality Tools and Techniques

Chapter 23 Statistical Process Control (SPC)

Chapter 24 Problem-Solving Tools and Continous Improvement Techniques

Chapter 25 Resources

Chapter 20

Quality Improvement

1. **Terms and concepts.** Define basic quality improvement concepts, such as defect detection and prevention, the cost of poor quality, total quality management (TQM), and the importance of customer satisfaction (Understand)
2. **Products and processes.** Define and distinguish between products and processes. Describe the interrelationships of product design, materials used, manufacturing processes, and final output, and how individual steps in a process can affect the final product or the system as a whole. (Understand)

Body of Knowledge IV.A

TERMS AND CONCEPTS

Prevention versus Detection

Prevention versus detection is a quality improvement concept wherein it is far more desirable to prevent errors and mistakes prior to their occurrence than to attempt to detect errors after they are committed. The logic of prevention versus detection is generally acknowledged to be based on the following:

1. The avoidance of escalating costs
2. The likelihood of missing errors as part of the normal inspection process

In the case of avoiding escalating costs, the further into production or service delivery that an error passes, the more expensive it is to rework or scrap the product or service delivery due to the increasing value-added effort and materials consumed. In fact, much effort has been expended by industrial research and development organizations as well as academic research to quantify exactly the financial resources consumed, and ultimately wasted, for each step or phase into production or service delivery that bad products or services cost. Further, of the many mathematical models that quantify costs of errors in specific production or service delivery operations, the most common model is based on a factor of 10, wherein it is assumed that the cost to correct an error increases 10 times for each stage or step in a production or service delivery operation that is past the point where the original error was made.

In the case of missing errors as part of the normal inspection process, we know that the sampling methods used to support the inspection process are based on some form of rational subgrouping and assumed or calculated probabilities that errors will be detected. Since errors are assumed to be randomly distributed within samples, there is the likelihood that errors will not be detected. While it is beyond the scope of this section of the book to provide a mathematical basis for determining the probability of missing errors in a sample, the point is nonetheless important. Sampling and inspection do not guarantee that all errors will be detected.

The logic of prevention versus detection should, at this point, be clear. As quality professionals, we prefer to prevent errors from happening rather than trying to detect errors that have already occurred.

The Importance of Customer Satisfaction

Quality has been defined by many individuals to include two broad components:

1. Product or process characteristics and/or performance consistent with some form of specification
2. The perceptions of customers

These two component parts of quality, taken together, mean that customers in a contemporary, globally based market economy ultimately judge the quality of products and services, and their judgments are based on quantifiable (specifications) and nonquantifiable (perceptions) components. Customers in a contemporary global economy expect all the products and services they purchase to meet or exceed their expectations—even expectations they are not yet aware they have. In the event that products and/or services do not meet customer expectations, customers are quite readily dissatisfied and react in a very predictable manner, including:

1. sharing their dissatisfaction with vendors along with requested changes,
2. not sharing their dissatisfaction with a selected vendor and changing to a new vendor (with and without telling others of the dissatisfaction), or
3. sharing their dissatisfaction with a selected vendor and changing to a new vendor (with and without telling others of the dissatisfaction).

A key point relating to the importance of customer satisfaction is that dissatisfied customers don't buy products or services, and they commonly share with several other organizations and people when and why they are dissatisfied. Further, since a primary component in how we define quality is based on perceptions, we cannot afford to have dissatisfied customers.

The Cost of Poor Quality

The cost of quality is a topic that addresses costs associated with specific activities and events intended to ensure the quality of products and services. Gryna (2001) identifies four categories of quality costs:

1. Prevention—costs associated with preventing errors (i.e., training)
2. Appraisal—costs associated with validating that errors have not occurred (i.e., measurement and inspection)
3. Internal failure—costs associated with correcting errors detected prior to customers' receipt of products or services (i.e., correction of errors found during inspection and testing)
4. External failure—costs associated with correcting errors detected after customers have received products or services (i.e., warranty costs)

The true cost of quality is recognized to be misunderstood and understated. A common analogy associated with the cost of quality is an iceberg—wherein its tip, which is roughly 10%–15% of the iceberg, is what we typically can identify. In accordance with the analogy, some 85%–90% of the true cost of quality lies beneath the surface of the water, thus is misunderstood and understated.

Juran and Godfrey (1999) have continued a common use of the "cost of quality" concept to refer to the topic as the "cost of poor quality." The cost of poor quality, Juran and Godfrey note, applies differently to different people, but it does refer to the costs associated with ensuring customer satisfaction. In addition to the costs associated with the cost of quality, as identified earlier, the cost of poor quality generally refers to costs associated with lost goodwill when customers are dissatisfied with products or services. This lost goodwill influences customer perceptions, which, as explained earlier, influences a customer's willingness to buy products and services from a given vendor. It also influences the potential recommendations given by a customer to other customers. The costs associated with this occurrence are difficult to quantify, but they are very real costs nonetheless.

Total Quality Management

Total quality management is a management approach centered on quality, based on organization-wide participation, and aimed at long-term success through customer satisfaction.

TQM focuses on customers, both internal (within the organization, the next party in the work process) and external (end users, stakeholders, regulatory agencies).

Given the fluctuating nature of customer satisfaction, continuous improvement is critical to an organization's survival. The concept applies to processes and the people who operate them as well as products. The plan-do-check-act (PDCA) cycle is a well-known model for continuous process improvement. The four-step process is also referred to as the Shewhart cycle (for Walter A. Shewhart), the Deming cycle (for W. Edwards Deming), and the PDSA cycle (with the S standing for "study"). First, a plan to effect improvement is developed. Next, the plan is carried out, preferably on a small scale. Then effects of the plan are observed. Last, results are studied to determine what was learned and what can be predicted.

TQM's emphasis on participation recognizes every activity contributes to or detracts from quality and productivity. Leadership from management and employee involvement are crucial for success. Management's role in TQM is to develop a quality strategy aligned with organizational business objectives and based on customer and stakeholder needs. After that strategy is defined, managers must participate in its deployment regularly and at every level.

Employee involvement can take several forms. Typically, quality improvement requires teams involving employees across functional boundaries. When employees are involved in quality, their organizations are more likely to make well-informed quality decisions and feel responsible for those decisions. Organizations empower employees by allowing them to make decisions that improve work processes within defined boundaries.

PRODUCTS AND PROCESSES

Product

For purposes of this book, and for the Certified Quality Inspector Body of Knowledge, a *product* is any tangible item intended for use by a customer. A product may be a physical item such as a unit of production, or the delivery of a service. Whether a physical item or the delivery of a service, some form of a process is required to produce a product.

Process

For purposes of this book, and for the Certified Quality Inspector Body of Knowledge, a *process* is a set of interrelated steps used to produce a product or deliver a service. A process may or may not require the use of tools and equipment.

The Interrelatedness of Products and Processes

Products, and the processes used to produce them, are integrally connected. Under a concept called *concurrent engineering*; customer expectations, product and/or service delivery designs, materials and procurement, process designs, packaging and distribution, and disposal are all considered simultaneously in acknowledgment of their interrelatedness.

Chapter 21

Quality Audits

1. **Types of audits.** Define and describe various types of audits, including internal, external, system, product, and process. (Understand)
2. **Audit process.** Explain various stages of the audit process (planning, performance, and closure), such as audit scope and purpose, resources needed, audit schedule, opening meeting, interviewing, data gathering, document and record review, analysis of results, closing meeting, audit documentation (reporting), recordkeeping, and verification of corrective actions, etc. (Understand)
3. **Audit tools.** Define and describe the purpose of checklists, log sheets, sampling plans, record and document reviews, and forward- and backward-tracing. (Understand)
4. **Communication tools and techniques.** Define and describe the use of graphs, charts, diagrams, and other aids for written and oral presentations, including interview techniques and listening skills. (Understand)
5. **Corrective action requests (CARs).** Describe how CARs from audits can support quality improvement. (Understand)

Body of Knowledge IV.B.

Quality auditing is an important topic for the ASQ Certified Quality Inspector (CQI). While quality auditing is a career field in and of itself, and there are three ASQ professional certifications pertaining to various aspects of quality auditing (ASQ Certified Quality Auditor, ASQ Certified Biomedical Auditor, and ASQ Certified HACCP Auditor), we will focus on the depth of content required for the ASQ CQI. Accordingly, in this section we will cover basic terms and definitions, types of audits, the audit process, audit tools, communication tools and techniques, and corrective action requests (CARs).

ASQ/ANSI/ISO 19011:2011 defines a quality audit as follows:

> A systematic, independent and documented process for obtaining audit evidence and evaluating it objectively to determine the extent to which audit criteria are fulfilled.

A ***quality audit*** is a fact-finding process that compares actual results with specified standards and plans. The audit provides feedback for improvement and differs from inspection, which emphasizes acceptance or rejection, and from surveillance, which is ongoing continuous monitoring. There are several key terms and definitions that apply specifically to the preparation, conduct, and reporting of quality audits.

TERMS AND DEFINITIONS

Auditee—The person or organization being audited

Auditor—The person or organization conducting the audit

Client—The person or organization requesting the audit

Interested party—A stakeholder to the audit process or results

Internal quality audit—Also known as a *first-party audit*. An audit conducted within a company or organization by one or more employees of that company or organization

External quality audit—Also known as a *second-party audit*. An audit conducted within a company or organization by one or more employees from a different company or organization

Third-party audit—An audit conducted by a regulatory agency or registrar

TYPES OF AUDITS

Quality audits may be classified according to the party conducting them or their scope.

When the auditor is an employee of the organization being audited (auditee), the audit is classified as an *internal quality audit*. For the purposes of maintaining objectivity and minimizing bias, internal auditors must be independent from the activity being audited. When the auditors are employees of the client or an independent organization, that is, a third party hired for the purpose, the audit is termed an *external quality audit*. In this case, the auditors are clearly independent of the auditee and are in a position to provide the client with an unbiased, objective assessment. This is the type of audit required to permit a listing in a register or to meet mandatory quality requirements. However, the time required and costs involved in an external audit are much higher compared to internal audits.

Another way to classify quality audits is by scope. An audit may be as comprehensive as needed or requested by a client. Arter (2003) identifies three types of audits, each with a different scope.

The most comprehensive type of audit is the quality *system audit*. System audits typically involve many areas and functions within a company or organization to examine the suitability and effectiveness of the quality system as a whole. System audits focus on the design and specification of specific controls and procedures, quality system documentation and interpretation by employees, and implementation aspects of the quality system. Reasons for initiating a system audit may range from evaluating a potential supplier to verifying an organization's own system. Audits of specific elements of a system, such as processes and products/services, are also possible, but they are more limited in scope than the system audit.

The *process audit*, while more limited in scope than a system audit, is a more detailed audit focused on one or more processes. Russell (2005) explains that a "process audit is verification by evaluation of an operation or method against predetermined instructions or standards, to measure conformance to these standards and the effectiveness of the instructions." Accordingly, a process audit ensures the effectiveness of process controls designed, documented, and implemented to ensure customer satisfaction with products and services.

The *product/service audit*, more limited in scope than the process audit, is conducted to ensure that finished products and services conform to customer expectations, specifications, and tolerances. A defining characteristic of a product or service audit is that such an audit is conducted after production or service delivery is complete. A product or service audit typically involves the use of measurement equipment to evaluate conformance to specifications and/or performance tests; in this regard, a product/service audit may be mistaken for final inspection. Final inspection is a normal step in production or service delivery operations by shop-floor personnel or inspectors to ensure that the product or service meets customer expectations. Product/service audits are not a normal step in day-to-day production or service delivery operations by shop-floor or inspection personnel; rather, they are audits conducted periodically by any interested party to ensure the product or service meets customer expectations.

It is important to note that these classifications are not mutually exclusive, and, in practice, cross-classifications of a quality audit are possible.

The following purposes of quality audits are listed in ASQ/ANSI/ISO 19011:2011:

- To contribute to the improvement of a management system and its performance
- To fulfill external requirements (e.g., certification to a management system standard)
- To verify conformity with contractual requirements
- To obtain and maintain confidence in the capability of a supplier
- To determine the effectiveness of the management system
- To evaluate the compatibility and alignment of the management system objectives with the management system policy and the overall organizational objectives.

ROLES AND RESPONSIBILITIES IN AUDITS

Each of the three parties involved in an audit—the client, the auditor, and the auditee—plays a role that contributes to the success of the audit. The *client*, the party that initiates the audit, selects the auditor and determines the reference standard to be used. The client, typically the end user of the audit results, determines the type of audit needed (system, process, or product) as well as its time and duration.

The selected *auditor*, whether an individual or a group, needs to adhere to the role of a third party, that is, the auditor must maintain objectivity and avoid bias in conducting the audit. The auditor must comply with any confidentiality requirements mandated by the auditee. An experienced individual is appointed the lead auditor to communicate audit requirements, manage the auditing activities, and report the results. For rules, qualifications, and evaluation criteria for an auditor, see ASQ/ANSI/ISO 19011:2011.

Finally, the *auditee* is responsible for accommodating the audit, which entails providing the auditors access to the facilities involved and providing copies of all relevant documentation. The auditee is also expected to provide the resources required and to select staff members to accompany the auditors.

AUDIT PROCESS

Proper planning is a key factor in achieving an efficient quality audit. Planning should be conducted in consideration of the client expectations. This includes the scope, depth, and time frame of the audit. The lead auditor is responsible for planning and conducting the audit and should be authorized to perform these activities.

Planning an audit, as with any activity, should address the questions of what, when, how, and who. What elements of the quality system are to be audited? Against what document or reference standard should the quality system be audited? The answers to both questions are determined by the client and should be clearly communicated to the auditee. When should the audit start, and when should it conclude? A schedule of the audit activities needs to be prepared and communicated to both the client and the auditee. The lead auditor is responsible for informing the client of any delays, reporting the reasons for the delays, and updating the completion date of the audit.

The method of conducting the audit should also be addressed. Working documents need to be prepared, including checklists of the elements to examine, questions to ask, and activities to monitor. A number of references provide generic checklists that can be used as templates. However, it is best to design a checklist to suit the audit at hand and its specific scope and objectives. Forms for collecting auditors' observations and the supporting evidence should also be included in the working document. Working documents are typically reviewed by an experienced auditor and approved by the lead auditor before implementation. It is recommended that the auditor explain the methods planned to the auditee. This should help the organization better prepare for the audit and ease the fear usually attached to the process.

The question of who will examine specific elements, processes, or products addresses the qualifications and experience of the individual auditors (assessors) to be appointed. With the client expectations in mind, the lead auditor should assign the various tasks among his or her team.

An audit is usually conducted in three steps. A pre-examination or opening meeting with the auditee marks the beginning of the process. During this meeting, the lead auditor introduces team members to the senior management of the auditee and explains the objectives of the audit and the methods used. The auditee selects representative members of the organization to facilitate and assist in the process, and submits a documented description of the quality system or element to be examined. Issues regarding proprietary information are typically addressed and resolved before the audit is started.

The next step involves a suitability audit of the documented procedures against the selected reference standard. Observed nonconformities at this stage of the audit should be reported to both the client and the auditee for immediate action. The auditing process should pause to allow for corrective measures.

In the third step, the auditor examines in depth the implementation of the quality system. The auditor maintains records of all nonconformities observed and the supporting data. Provisions should be made in the audit plan to allow additional investigation of clues suggesting nonconformities revealed by the data collected. The auditee management should be made aware of, and acknowledge, all the nonconformities observed during the audit. This step concludes with a closing meeting with the auditee's management for a representation of findings. In some cases, the auditor may be required to recommend corrective measures for improving the system. However, it is up to the auditee to plan and implement these measures in a way that best suits the organization.

AUDIT TOOLS

To successfully complete any type of audit, several tools are required. Since many of these tools have been discussed in detail in other sections of this book, detailed discussions of tools already covered will not be repeated here.

Checklists

Checklists are used during audits to record tabular data such as the number of observations or occurrences of specific phenomena such as defects, errors, or mistakes. Checklists also commonly provide a rough graphic or illustration of a product or flowchart of a service delivery operation to provide auditors future reference to where the phenomenon occurred. Data from checklists inform the audit process, particularly when making observations, reporting areas for improvement, and, later, verifying that corrective actions have been implemented and are effective.

Log Sheets

Log sheets are historical records of when specific actions were taken or functions were performed. Log sheets help identify who took specific actions, what actions were taken, and the timing and sequence of actions taken; often they refer to job instructions or procedures to be used. Log sheets are valuable sources of information and inform the audit process, ensuring necessary tasks or functions are performed. Since many of the types of functions recorded on log sheets are related to maintenance activities, maintenance interval requirements, cleaning requirements, replenishment of materials and supplies, use of consumables, process adjustments, and identification of sources of variability or errors; log sheets are important audit tools.

Sampling Plans

Sampling plans are statistics based tools that provide guidance in terms of what, when, how frequently, and how much to sample. As sampling plans are covered elsewhere in this book, they will not be discussed in detail here; however, as sampling plans are related to quality auditing, we will address that application.

Sampling plans are used in quality audits primarily to identify key characteristics of when, where, how, and how much to sample the supplies, materials, consumables, components, subassemblies, and so on, of the products or service delivery applications being audited. Sampling plans are also used in audits to define when, where, how, and how much data auditors are to investigate. Regardless of whether sampling plans are used to look at products, processes, or functions to be audited themselves, the sampling plans used are the same.

COMMUNICATION TOOLS AND TECHNIQUES

Tools and techniques for communication and use in presentations have been covered extensively throughout this book. We will not duplicate those discussions here. It is important to note that the many communication tools presented will be used by the ASQ CQI, and knowledge of these tools and techniques will be tested on the CQI exam. Of particular importance will be making sure the CQI understands the graphs, charts, and diagrams discussed throughout this book.

CORRECTIVE ACTION REQUESTS (CARs)

Corrective action requests (CARs) arising from audit processes are intended to address and correct deficiencies and improve processes. CARs create a historical record that (1) auditees can use to guide future work efforts, (2) individuals responsible for making repairs and process improvements can use as justification for resource requirements, and (3) auditors can use to verify that repairs and process improvements have been completed. CARs form an important part of the audit report for the auditee and the auditor.

AUDIT OUTCOMES

Working papers are a set of documents used to guide any form of audit. Working papers are used to ensure an efficient and effective review of selected systems, processes, and/or products/services. Included in the working papers are many examples and instances of discoveries, observations, and conversations, which become part of an audit. In particular, these discoveries, observations, and conversations are recorded during the information-gathering phase of an audit and are used later to provide the basis of comparisons with applicable standards, specifications, policies, procedures, and work instructions.

As a result of documenting the audit in the working papers, and in comparing actual operations with documentation of intended operations, two types of findings generally result—examples of effective or noteworthy accomplishments and examples of discrepancies. Examples of effective or noteworthy accomplishments, such as effective components of the quality system or innovative processes that prevent mistakes, are documented in the working papers as well as in the final audit report to encourage continued use of further implementation. Examples of discrepancies, such as inadequate instructions, incomplete training, and noncompliance with customer specifications, are documented in the working papers as well as in the final audit report to initiate corrective action.

When the need for corrective action is identified during an audit, a CAR is generated. A CAR is a document that identifies a discrepant condition, the source or standard giving rise to the discrepant condition, when the discrepant condition was discovered and by whom, who is responsible for addressing the discrepant condition, the allowable time frame for corrective action, and so on. The primary intent of the CAR is to trigger action that either eliminates or reduces the negative effects of the condition. Ultimately, the CAR is intended to be the vehicle that drives process improvement. From this perspective, the CAR directly impacts customer satisfaction. CARs are so important, they are normally cited in the audit report with a required action to verify the CAR was acted upon and actually resulted in one or more process improvements.

Audit Reporting and Follow-up

A final report is submitted to the client indicating the facts of the audit and conclusions regarding the ability of the subject system, element, process, or product to achieve quality objectives. Proper planning and execution of the audit facilitates the preparation of this report and provides data to support its conclusions. The lead auditor is responsible for the accuracy of the report and the validity of its conclusions. The report should be submitted to the client, who in turn is responsible for providing a copy to the auditee.

The audit final report should include, at a minimum, the following:

1. Type of audit conducted
2. Objectives of the audit
3. Identification of involved parties: auditor, auditee, and third party
4. Audit team members
5. Critical nonconformities and other observations (effective or noteworthy accomplishments)
6. Audit standards and reference documents used
7. Determination of proper corrective action(s)
8. Duration of audit
9. Audit report distribution and date
10. Audit results and recommendations
11. Audit-related records

Should the auditee initiate improvement efforts to correct nonconformities via CARs, the three parties should agree on a follow-up audit to verify the results. The plan, audit, report, and improve cycle may be repeated whenever systems and/or requirements change. The results attained provide a measure of the effectiveness of the audit. Improvement efforts should also be directed toward identifying and eliminating the root causes of reported nonconformities and identifying the corrective action(s) to be taken. Root causes represent the main reason behind the occurrence of a nonconformance or an undesirable condition or status. The corrective actions may then be validated by performing tests, inspections, or further audits.

Chapter 22

Quality Tools and Techniques

Apply the following quality tools and techniques. (Apply)

1. Pareto charts
2. Cause and effect diagrams
3. Flowcharts
4. Check sheets
5. Scatter diagrams
6. Histograms

Body of Knowledge IV.C

Facility in the use of basic quality tools and techniques is critical for a Certified Quality Inspector (CQI) to be effective and to function within the larger quality community. As many as 85% of the problems encountered in production or service delivery operations can be solved with these basic quality tools and techniques.

SEVEN BASIC QUALITY CONTROL TOOLS

The seven basic quality control tools are:

1. Pareto charts
2. Cause-and-effect diagrams (also known as fishbone or Ishikawa diagrams)
3. Flowcharts
4. Check Sheets
5. Scatter diagrams
6. Histograms
7. Control charts

A quality inspector may use any of these tools individually, in combination with other tools, and in any order that makes sense given the situation.

We will introduce the first six of these tools here. Note that the last tool, control charts, is addressed in detail in Chapter 23, "Statistical Process Control."

Pareto Charts

The *Pareto chart,* also introduced in Chapter 7, "Basic Statistics and Applications," helps the team home in on the major causes driving a problem. The Pareto chart is named after Vilfredo Pareto. Pareto was an Italian economist who used the chart to model wealth distribution in 19th-century Europe, in which 80% of the wealth was controlled by 20% of the population. Dr. Joseph Juran applied the "Pareto Principle" to the quality field and coined the term the "80/20 Rule," in which 80% of a problem is driven by 20% of the causes. Juran also described this phenomenon as the "vital few and the trivial many." At its core, a Pareto chart is a column chart presenting categories in descending order of frequency. In quality, these categories are often failure or defect types.

It is important for a team to identify which failures occur most frequently so it can focus its attention. Arranging and displaying the failure categories from highest to lowest frequency enables a team to prioritize the use of resources. Using the Pareto chart, the team can first eliminate those failures that have the greatest effect on quality of the product or delivery of service.

Example Pareto Chart

An inspection team collected data on failure types and their frequencies over the course of one month. The data were used to create a Pareto chart, shown in Figure 22.1.

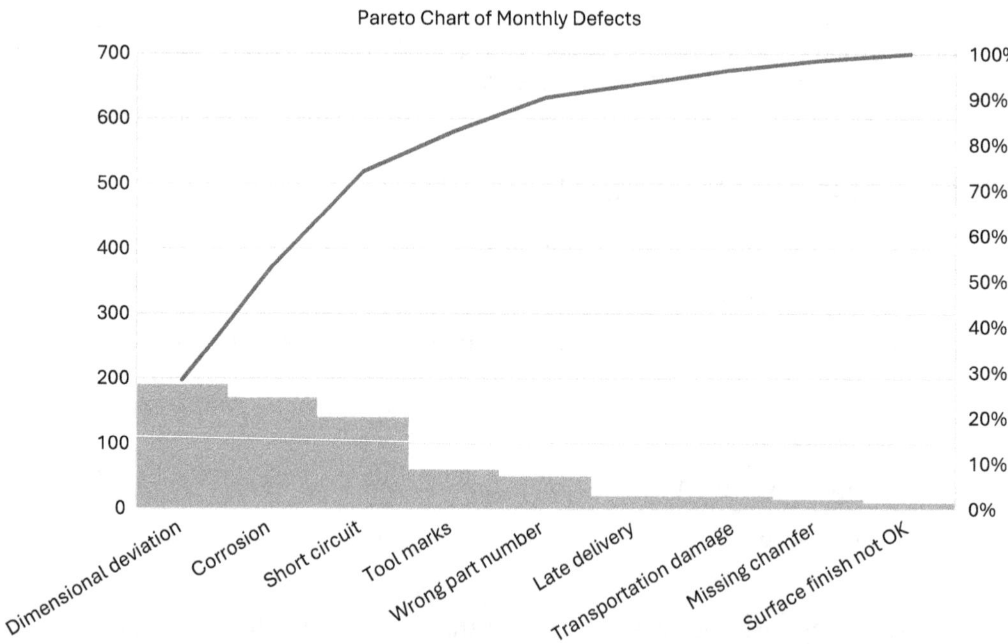

Figure 22.1 Pareto chart for failure types.

Figure 22.1 is a column chart with defect types along the horizontal axis, and their frequency along the vertical axis. The defects are arranged from highest to lowest frequency, which is a hallmark of the Pareto chart.

The chart shows that 80% of the defects stem from only three of the 10 categories of data: dimensional deviation, corrosion, and short circuit. This chart demonstrates the 80/20 phenomenon, in which roughly 80% of the defects occur because of approximately 20% of the defect categories.

Here, to make the biggest impact on reducing failures, the team should first attempt to eliminate or at least reduce failures because of dimensional deviation, then corrosion, then short circuits, and so on. By eliminating the top three defects, the number of failures would be reduced dramatically.

If it is possible to eliminate more than one defect simultaneously, a team should consider eliminating them even though some of the defects may occur less frequently. Furthermore, after one or more defects are eliminated or reduced, the team should collect the data again. From this new data, a new Pareto chart can be drawn to determine whether the priority has changed (e.g., another defect is now occurring more frequently). In that case, the team may refocus resources to eliminate the newly prioritized cause of failure first.

Note that even though the Pareto chart prioritizes the frequency of occurrence from highest to lowest, not all failures or defects may be equally costly or serious. In that situation, frequencies can be weighted by their relative cost or seriousness.

Cause-and-Effect Diagrams

A cause-and-effect diagram helps users identify the many *possible* causes of a problem (a problem is also referred to as the "effect") by sorting ideas into useful categories. Each of the categories of potential causes is referred to as a bone, and the diagram itself resembles the skeleton of a fish. Accordingly, the cause-and-effect diagram is also known as a *fishbone diagram* because of its shape. The tool is also known as an *Ishikawa diagram*, after Kaoru Ishikawa who developed the technique in the 1940s.

The cause-and-effect diagram helps to identify and isolate the potential causes of a particular problem. The diagram also creates a visual image of how potential causes may relate to the problem at hand.

To prepare a cause-and-effect diagram, it is quite common to first use a technique commonly known as *brainstorming*. The brainstorming technique is a form of creative and collaborative thinking used in a team setting. The team usually includes personnel from production, inspection, purchasing, design, and management, and any other stakeholders associated with the product or service delivery being discussed.

Asking the five Ws and one H (what, why, when, where, who, and how) can be effective in brainstorming the elements of the cause-and-effect diagram. In addition to using the five Ws and one H, consider starting with the six Ms which are used as the bones of the fish:

- Man (people/operators)
- Machine(equipment)
- Methods (operating procedures)

- Materials
- Measurement
- Mother Nature (environment)

Optional categories include:

- Money
- Management

Note that in service or office environments, categories such as people, processes, policies, technology, environment, and equipment are often used.

To construct the fishbone diagram, the problem being addressed is written at the head of the fish, and a spine is drawn with bones branching off it. Then, each idea is written as a bone branching off the category bone.

This tool is relatively easy to use and yet very powerful. The completed diagram shows at a glance the factors in the system or process that may be contributing to the problem at hand. A completed fishbone diagram is shown in the following example.

Example Cause-and-Effect Diagram

A manufacturing team convenes to brainstorm the source of periodic iron contamination in its product. The team uses the six Ms to prompt ideas during its brainstorming session. The resulting fishbone diagram is shown in Figure 22.2. Layers of branches show thorough thinking about the causes of the problem.

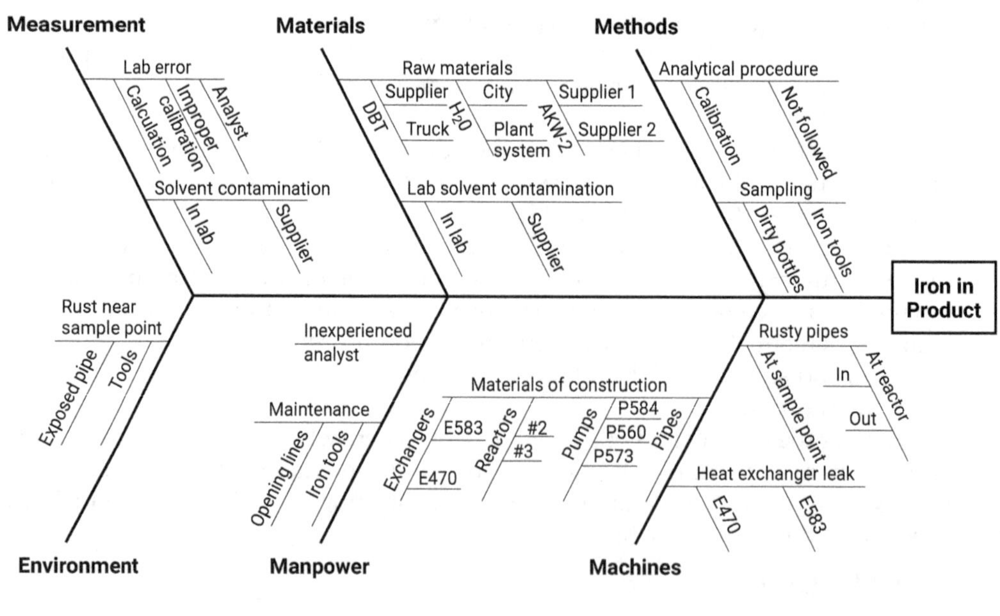

Figure 22.2 Fishbone diagram for iron contamination.
Source: N.R. Tague. *The Quality Toolbox*, 3rd ed. (Milwaukee: ASQ Quality Press, 2023): 248.

For example, under the heading Machines, the idea "materials of construction" shows four types of equipment and several specific machine numbers. Note that some ideas appear in two places. For example, calibration is listed under Methods as a factor in the analytical procedure, as well as under Measurement as a cause of lab error. The idea "iron tools" can be considered a methods problem when taking samples or a manpower problem with maintenance personnel.

Once the team lists the possible causes in each major category, the next step is to investigate each potential cause and, if the potential cause turns out to be an actual cause of the problem, eliminate it.

Flowcharts

Flowcharts are graphical representations of processes based on standard symbols and line work to depict logic flow and relationships.

The following symbols are commonly used in flowcharts:

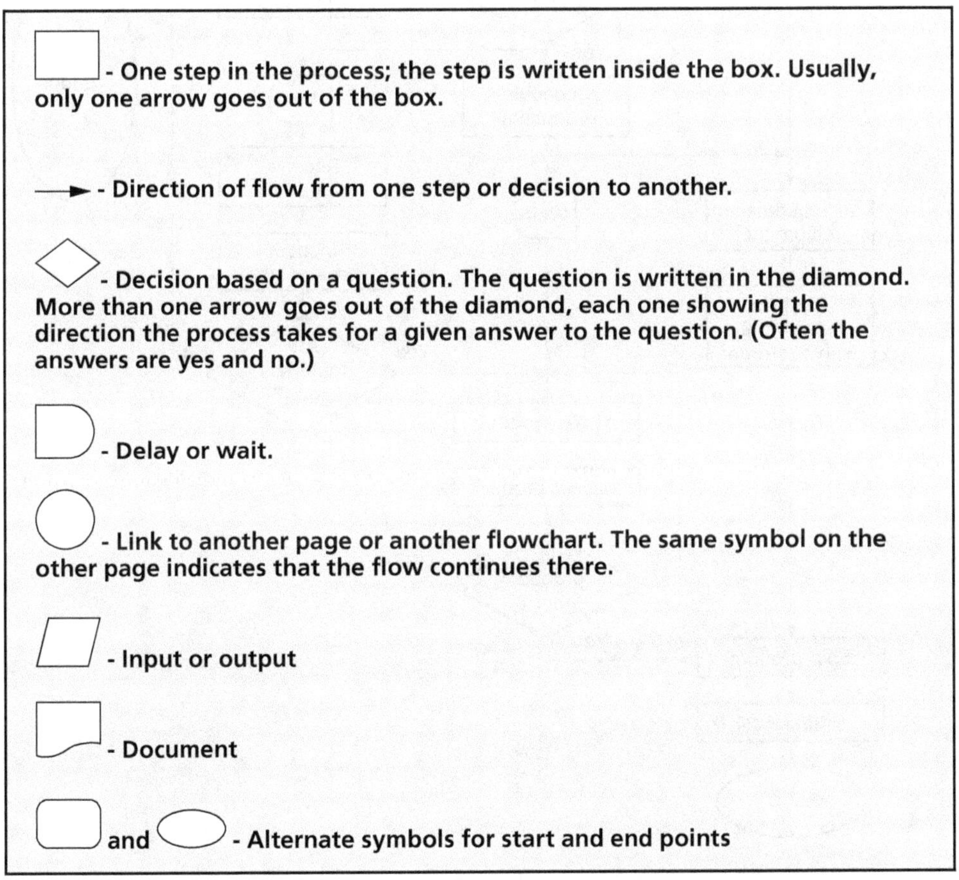

Figure 22.3 shows a flowchart for an order-filling process. The goal in using a flowchart is to document the current process with appropriate symbols and logic.

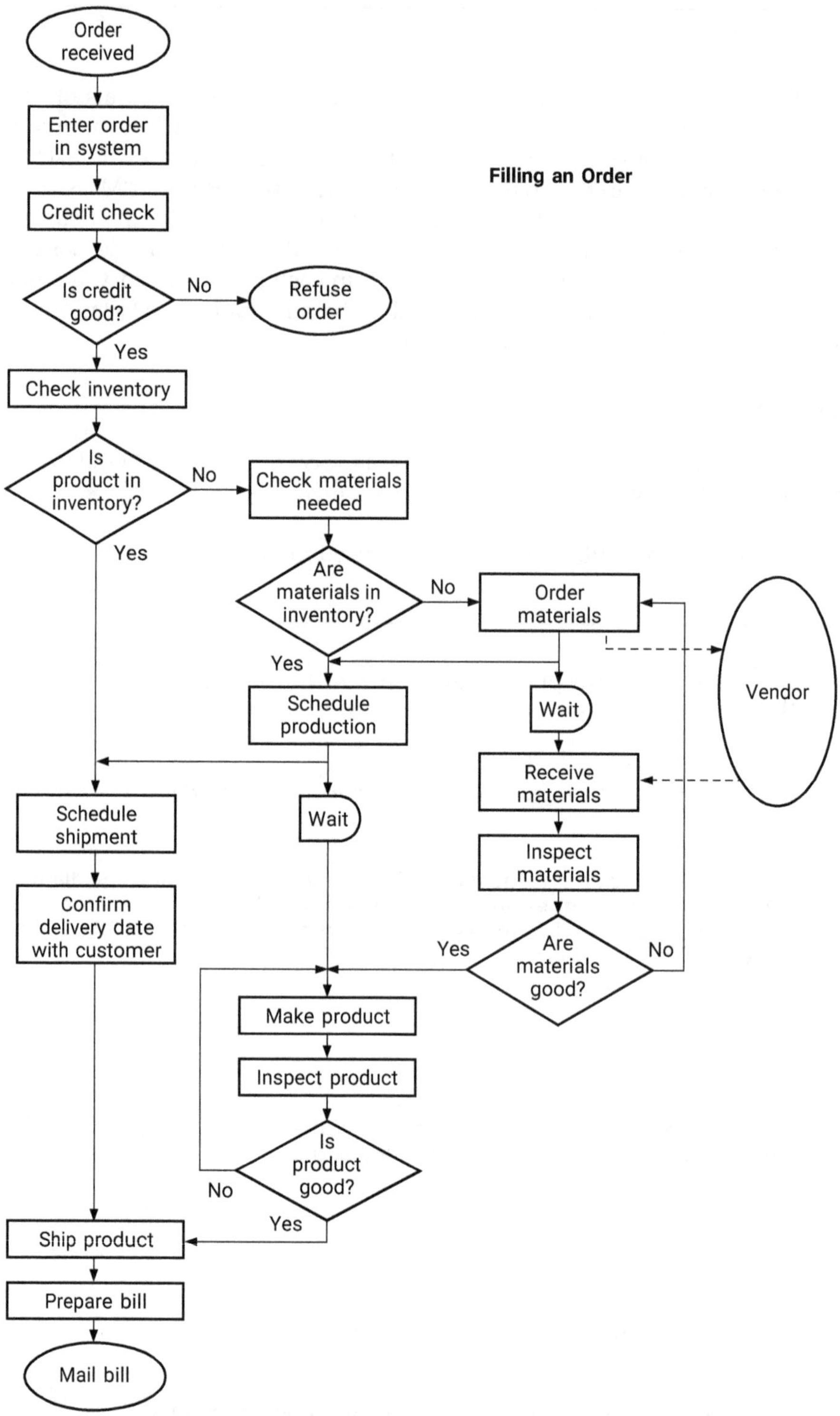

Figure 22.3 Flowchart of an order-filling process.
Source: N.R. Tague. *The Quality Toolbox*, Third Edition. (Milwaukee: ASQ Quality Press, 2023): 265.

One advantage of applying flowcharts is that they enable users to put boundaries on process start and stop points. In large and complex processes, understanding process boundaries enables users to drill down into the process steps where problems may be occurring. In this regard, flowcharts enable users to understand the context, timing, sequence of events, and logic that relate to (or may be causing) one or more problems. Accordingly, when flowcharts are applied, they commonly lead the team to use other quality tools and techniques.

Check Sheets

Check sheets, also known as tally sheets, are structured forms used to collect and then analyze data. This tool was also introduced in Chapter 7, "Basic Statistics and Applications."

Check sheets pre-categorize potential outcomes for data collection using sets of words, tally lists, or graphics. Figure 22.4 shows a completed check sheet used to collect data related to a paint mixing process. This tool provides a straightforward method for collecting data. When complete, the check sheet can point us toward other tools that can be used to better understand the unwanted process behavior. For example, a Pareto chart can highlight the most frequent defect, or a run chart can show trends in defect frequencies.

The basic steps in making a check sheet are:

1. Decide what event or problem will be observed. Develop operational definitions.
2. Decide when data will be collected and for how long.
3. Design the form. Set it up so data can be recorded simply by making check marks or an X or similar symbol, and so data do not have to be recopied for analysis.
4. Label all spaces on the form.
5. Test the check sheet for a short trial period to be sure it collects the appropriate data and is easy to use.
6. Each time the targeted event or problem occurs, record the data on the check sheet.

Paint color defect causes

Reason	Day of the week							Weekly total
	Mon	**Tue**	**Wed**	**Thu**	**Fri**	**Sat**	**Sun**	
Operator misread instructions	I	I	0	II	𝍸	𝍸 II	III	19
Wrong pigment used	III	I	𝍸	III	IIII	IIII	𝍸 𝍸	30
Wrong color code from customer	II	I	II	0	0	𝍸	0	10
Outdated base paint	0	0	I	0	III	0	IIII	8
Daily total	6	3	8	5	12	16	17	67

Figure 22.4 Example of a check sheet.
Source: R.A. Munro et. al. *The ASQ Certified Six Sigma Green Belt Handbook*, Third Edition. (Milwaukee: ASQExcellence, 2022): 120.

Scatter Diagrams

Scatter diagrams graph pairs of numerical data, with one variable of data corresponding to the x axis and the other to the y axis. In looking at the plot of the paired data, we are looking for a relationship, or correlation, between the two variables. Scatter diagrams are covered in more detail in Chapter 7, "Basic Statistics and Applications."

Correlations may be positive, negative, or the variables may not be correlated at all. For positive correlations, an increase in the x variable tends to correspond to an increase in the y variable. In the plot, positive correlations are relationships that result in points increasing from the lower left to the upper right.

Conversely, for a negative correlation, an increase in the x-variable tends to correspond to a decrease in the y-variable. In the plot, the data points decrease from the upper left to the lower right. For both positive and negative correlations, the variability or "scatter" of the data allows us to categorize the correlation as strong, moderate or weak. Less scatter results in a stronger correlation.

When there is no correlation (either positive or negative), a change in the x variable does not explain the behavior of the y variable.

We should note that if two variables (x, y) are correlated, it does not imply that a change in the x variable *causes* the y variable to change. Instead, a strong correlation tells us that x can be used to predict the behavior of the y variable. To determine a causal relationship, more sophisticated tools, such as design of experiments (DoE) can be used.

Figure 22.5 shows a scatter plot of percentage purity versus iron in parts per million (ppm). The plot demonstrates a moderate negative correlation. As the iron content measured in ppm, shown on the x axis, increases, the percentage purity tends to decrease. While not a strong negative relationship, the relationship is observable nonetheless.

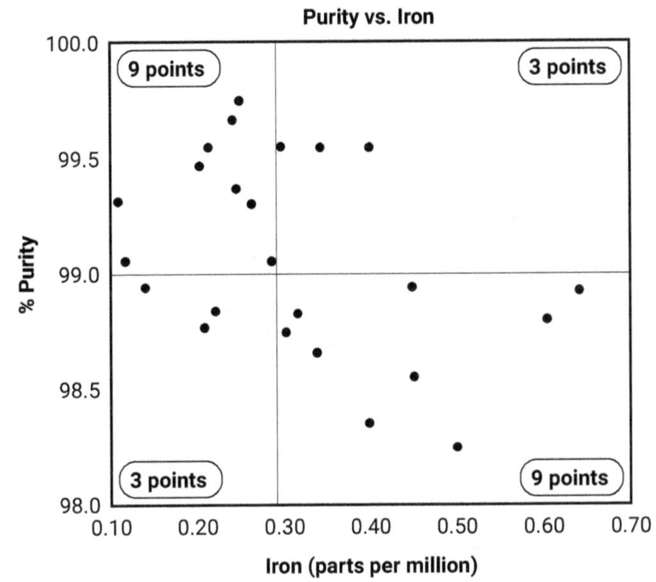

Figure 22.5 Scatter plot of purity versus iron in ppm.
Source: N.R. Tague. *The Quality Toolbox*, Third Edition (Milwaukee: ASQ Quality Press, 2023): 492.

By using a scatter diagram, the user can see graphically that one variable does, or does not, impact another variable. In the event the user determines there is a correlation between two variables, more sophisticated tools such as regression analysis or design of experiments (DOE) may be required to support further analysis and process improvement efforts.

Histograms

Histograms are bar graphs used to represent the shape, central tendency, and spread of the distribution of a quantitative variable. Such information is not usually apparent from the raw data. Generally, at least 50 data points are needed to create a histogram that can be interpreted reliably. The data are then sorted and divided into "classes" of equal width. Classes are chosen so that each data point can be assigned to one and only one class. Then the number of data points that fall in each class are recorded in a frequency table.

A histogram is then drawn as a column chart with the classes in numerical order on the x axis and the frequencies of each class on the y axis. Histograms can be used to gain insight into the shape of distribution. For example, the graph can show that a distribution is symmetric, bimodal, or skewed, as shown in Figure 22.6.

Figure 22.6 Histogram example.
Source: N.R. Tague. *The Quality Toolbox*, Third Edition. (Milwaukee: ASQ Quality Press, 2023): 299.

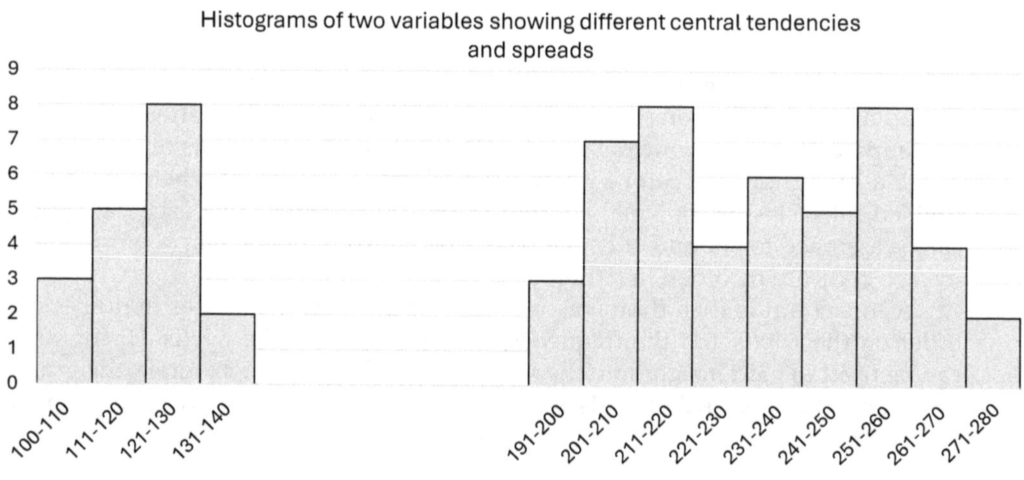

Figure 22.7 Histogram showing central tendency and spread of two variables.

The histogram can give a sense of the central tendency and spread of the data, as shown in Figure 22.7.

Example Histogram

The weights of fifty shipments of welding equipment, in pounds, were recorded. The data were sorted, as shown in Table 22.1.

Table 22.1 Shipment weights, in pounds.

300	620	710	920	1150
360	630	720	920	1160
440	640	760	950	1200
460	650	790	960	1230
480	660	800	960	1300
480	670	850	1000	1400
550	670	850	1000	1450
560	680	890	1030	1590
570	700	900	1070	1650
610	700	910	1100	1790

The weights were then used to create a frequency table in which the data were divided into intervals of 300 pounds. The number of data points in each class were recorded, as shown in Table 22.2.

Table 22.2 Frequency table for shipment weights.

Weight	Frequency
300-600	9
601-900	20
901-1200	14
1201-1500	4
1501-1800	3

A histogram of the data was then created, shown in Figure 22.8. We can see that the spread of the data spans from 300 to 1800 pounds, with the most frequent shipment weights between 601 to 900 pounds. The distribution itself is skewed to the right.

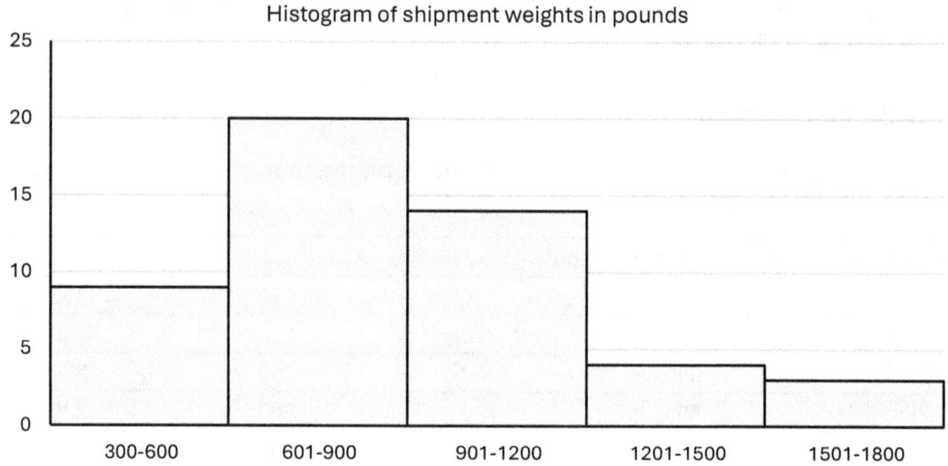

Figure 22.8 Histogram of shipping weights.

In summary, these basic quality tools form a simple but powerful structure for quality improvement. Once inspectors, technicians, and operators or service personnel become fully familiar with these tools, management must get involved to sustain their use for an ongoing quality-improvement process. Management must create an environment where these tools become part of the day-to-day production or service processes.

Control Charts

Control charts are an important process improvement tool for hunting special causes of variation and statistical process control programs. Control charts are discussed in detail in Chapter 23, "Statistical Process Control."

Chapter 23

Statistical Process Control (SPC)

1. **Common and special cause variation.** Explain the difference between these causes of variation. Determine whether a process is in statistical control by monitoring data patterns (e.g., runs, trends, and hugging), and identify what actions should be taken in response. (Apply)
2. **Control limits and specification limits.** Define, describe, and illustrate the differences between these limits as used in SPC. (Apply)
3. **Variables charts.** Identify characteristics of and apply $\bar{X}-R$ and $\bar{X}-S$ charts. (Apply)
4. **Attributes charts.** Identify characteristics of and apply *p*, *np*, *c*, and *u* charts. (Apply)
5. **Process capability analysis.** Define and describe the differences between C_p, C_{pk}, P_p, and P_{pk} studies, and identify their application to various types of data. (Understand)

Body of Knowledge IV.D

The concept of quality is centuries old. However, the concept of statistical quality control (SQC) is less than a century old. SQC is merely a set of interrelated tools such as statistical process control (SPC), design of experiments (DOE), acceptance sampling plans, and other tools that are used to monitor and improve quality. SPC consists of several useful tools for process monitoring. One of these tools is the quality control chart, which is the center of our discussion in this chapter. Walter A. Shewhart presented the first quality control chart in 1924.

CONTROL CHARTS

A *control chart* can be defined as:

1. A device for describing in concrete terms what a state of statistical control is.

2. A device for judging whether control has been attained, thus detecting whether assignable causes are present.
3. A device for attaining a stable process.

Suppose that we take a sample of size n from a process at approximately regular intervals and that for each sample we compute a sample statistic X. This statistic may be the sample mean, fraction of nonconforming product, or any other appropriate measure. Now, since X is a statistic, it is subject to some fluctuation or variation. If no special causes are present, the variation in X will have characteristics that can be described by some statistical distribution. By taking enough samples, we can estimate the desired characteristics of such a distribution. For instance, we now suppose that the statistic X is distributed as normal, and we divide the vertical scale of a graph in units of X and the horizontal scale in units of time or any other such characteristic. Then, we draw horizontal lines through the mean and the extreme values of X, called the centerline and the upper and lower control limits, respectively, which results in the tool known as a *control chart*. Figure 23.1 shows a typical control chart.

The main goal of using control charts is to reduce the variation in the process and bring the process target value to the desired level.

If we plot data pertaining to a process on a control chart, and if the data conform to a pattern of random variation that falls within the upper and lower limits, we say that the process is in statistical control. If, however, the data fall outside these control limits and/or do not conform to a pattern of random variation, the process is considered to be out of control. In the latter case, an investigation is launched to track down the special causes responsible for the process being out of control, and to correct them.

If any particular cause of variation is on the unfavorable side, an effort is made to eliminate it. If, however, the cause of variation is on the favorable side, an effort is made to detect such a cause and perpetuate it instead of eliminating it. In this manner, the process can eventually be brought into a state of statistical control.

Dr. Shewhart, the inventor of control charts, recommended very strongly that a process should not be judged to be in control unless the pattern of random variation has persisted for some time and for a sizable volume of output.

Note: If a control chart shows a process in statistical control, this does not mean that all special causes have been completely eliminated; rather, it means that for all practical purposes, it is reasonable to assume or adopt a hypothesis of common causes only.

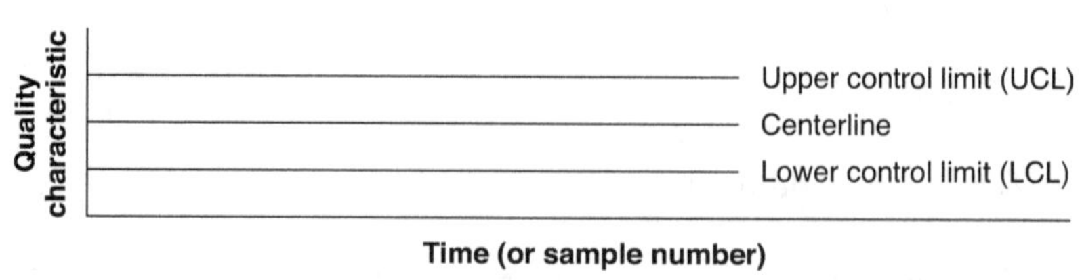

Figure 23.1 A pictorial representation of the components of a control chart.

Preparation for Use of Control Charts

For control charts to serve their intended purpose, it is important to take a few preparatory steps prior to their implementation:

1. *Establish an environment suitable for action:* Any statistical method will fail unless management has prepared a responsive environment.
2. *Define the process and determine the characteristics to be studied:* The process must be understood in terms of its relationship to the other operations/users and in terms of the process elements (e.g., people, equipment, materials, methods, and environment) that affect it at each stage. Some of the techniques discussed earlier, such as the Pareto analysis or the fishbone chart, can help make these relationships visible. Once the process is well understood, the next step is to determine which characteristics affect the process, which should be studied in depth, and which should be controlled.
3. *Look for a correlation between characteristics:* For an efficient and effective study, take advantage of the relationship between characteristics. If several characteristics of an item tend to vary together, it may be sufficient to chart only one of them. If some characteristics are negatively correlated, a deeper study is required before any corrective action on such characteristics can be taken.
4. *Define the measurement system:* The characteristic must be operationally defined so that the findings can be communicated to all concerned in ways that have the same meaning today as they did yesterday. This includes specifying what information is to be gathered, where, how, and under what conditions. The operational definition is very important for collecting data since it can impact the control charts in many ways. Moreover, the analysis of data depends on how the data are collected. Thus, it is extremely important that the data contain the pertinent information and are valid (e.g., appropriate sampling schemes are used) so that their analysis and the interpretation of the results is done appropriately. Moreover, one should always keep in mind that each measurement system has its own inherent variability. Thus, the accuracy of any measurement system is as important as the elimination of the special causes affecting the process.
5. *Minimize unnecessary variation:* Unnecessary external causes of variation should be reduced before the study begins. This includes overcontrolling the process or avoiding obvious problems that could and should be corrected even without the use of control charts.
6. *Meet the customer's needs:* This includes any subsequent processes that use the product or service as an input as well as the final end-item customer. For example, a computer manufacturing company is a customer of the semiconductor industry; a car manufacturing company is a customer of tire manufacturing companies; and in a paper mill, the paper-making unit is a customer of the pulp-making unit.

Note that in all cases, a *process log* should be kept. It should include all relevant events (big or small), such as procedural changes, the use of new raw materials, or a change of operators. This will aid in subsequent problem analysis.

Benefits of Control Charts

Properly used, control charts can:

1. Be used by operators for ongoing control of a process
2. Help the process perform consistently and predictably
3. Allow the process to achieve higher quality, higher effective capacity (since there will be either no or fewer rejections), and hence lower cost per unit
4. Provide a common language for discussing process performance
5. Help distinguish special causes from common causes for variability, hence serving as a guide for management to take local action or action on the system

Rational Subgroups for a Control Chart

It is very important to note that the *rational subgroup* or *sample* used to prepare a control chart should represent subgroups of output that are as homogeneous as possible. In other words, the subgroups should be such that if special causes are present, they will show up in differences between the subgroups rather than in differences between the members of a subgroup. A natural subgroup, for example, would be the output of a given shift. It is not correct to take the product for an arbitrarily selected period of time as a subgroup, especially if it overlapped two or more shifts. This is because if a sample comes from two or more shifts, any difference between the shifts will be averaged out, and, consequently, the plotted point won't indicate the presence of any special cause due to shifts. As another example, if the process used six machines, it would be better to take a separate sample from the output of each machine than to have samples that consist of items from all six machines. This is because the difference between machines may be the special cause of variation. It will be hard to detect this special cause if the samples are not taken from individual machines. Thus, it can be said that careful selection of a subgroup or sample is perhaps the most important item in setting up a control chart. The next question in the selection of a sample is to determine the sample size. Factors usually taken into consideration for determining the sample size and the frequency of the samples are the average run length and the operating characteristic curve, both of which we discuss here.

COMMON CAUSES VERSUS SPECIAL CAUSES

The quality of the final product depends on how well a process is designed and executed. However, no matter how perfectly a process is designed and how well it is executed, no two items produced by the process are identical. The difference between two items is called the *variation*. Such variation occurs due to two causes:

1. Common causes or random causes
2. Special causes or assignable causes

To better understand these causes, it is important to understand a process and to have a concrete plan for future actions that would be needed to improve the process.

Process Evaluation

Process evaluation includes the study of not only the final product but also all the *intermediate steps or outputs* that describe the actual operating state of the process. For example, in the paper-production process, wood chips and pulp may be considered as intermediate outputs. If the data on process evaluation are collected, analyzed, and interpreted correctly, they can show where and when a corrective action is necessary to make the whole process work more efficiently.

Action on Process

Action on process is very important for any process, since it prevents the production of an out-of-specification product. An action on process may include changes in raw material, operator training, change of some equipment, change of design, or other measures. Effects of such actions on a process should be monitored closely and further action should be taken if necessary.

Action on Output

If the process evaluation does not indicate that any action on the process is necessary, the first is to ship the final product to its destination. Note that some people believe action on output consists of sampling plans and discarding out-of-specification product that has already been produced. Obviously, such an action on the output is futile and expensive. We are interested in correcting the output before it is produced. This goal can be achieved by using control charts.

Variation

No process can produce two products that are alike or possess the same characteristics. Any process is bound to contain some sources of variation. A difference between two products may be large, moderate, small, or undetectable depending on the source of variation, but certainly there is always some difference. For example, the moisture content in any two rolls of paper, the opacity in any two spools of paper, and the brightness of any two lots of pulp will always vary. Our aim is to trace as far as possible the sources of such variation and eliminate them. The first step is to separate the common and special causes of such sources of variation.

Common Causes or Random Causes

Common causes or *random causes* refer to the various sources of variation within a process that is in statistical control. They behave like a constant system of chance causes. While individual measured values may all be different, as a group they tend to form a pattern that can be explained by a statistical distribution that is generally characterized by:

1. Location parameter
2. Dispersion parameter
3. Shape (the pattern of variation—symmetrical, right-skewed, or left-skewed)

Special Causes or Assignable Causes

Special causes or *assignable causes* refer to any source of variation that cannot be adequately explained by any single distribution of the process output, as would be the case if the process were in statistical control. Unless all the special causes of variation are identified and corrected, they will continue to affect the process output in an unpredictable way. Any process with assignable causes is considered unstable, hence not in statistical control. However, any process free of assignable causes is considered stable, therefore in statistical control. Assignable causes can be corrected by local actions, while common causes or random causes can be corrected only by actions on the system.

Local Actions and Actions on the System
The following are characteristics of local actions:

1. Are usually required to eliminate special causes of variation
2. Can usually be taken by people close to the process
3. Can correct about 15% of process problems
4. Deming (1982) believed that as much as 6% of all system variations are due to special or assignable causes, while no more than 94% of the variations are due to common causes

The following are characteristics of actions on the system:

1. Are usually required to reduce the variation due to common causes
2. Almost always require management's action for correction
3. Are needed to correct about 85% of process problems

Deming (1951) pointed out that there is an important relationship between the two types of variations and the two types of actions needed to reduce such variations. In what follows, we discuss this point in more detail.

Special causes of variation can be detected by simple statistical techniques. These causes of variation are not the same in all the operations involved. The detection of special causes of variation and their removal is usually the responsibility of someone who is directly connected with the operation, although the management is sometimes in a little better position to correct them. The resolution of a special cause of variation usually requires local action.

The extent of common causes of variation can be indicated by simple statistical techniques, but the causes themselves need more exploration to be isolated. It is usually the responsibility of management to correct the common causes of variation, although other personnel directly involved with the operation are sometimes in a better position to identify such causes and pass them on to the management for an appropriate action. Overall, the resolution of common causes of variation usually requires action on the system.

We should note that about 15% (or according to Deming, 6%) of industrial process troubles is correctable by the local action taken by people directly involved with the operation, while 85% is correctable only by the management's action on the system. Confusion about the type of action required is very costly to the organization in terms of wasted efforts, delayed resolution of trouble, and other aggravating problems. It would be totally wrong, for example, to take local action (e.g., changing an operator or calibrating a machine) when, in fact, management action on the system is required (e.g., selecting a supplier that can provide better and consistent raw material).

CONTROL LIMITS VERSUS SPECIFICATION LIMITS

It is common for floor workers, technicians, inspectors, engineers, and managers to be confused between specification limits and control limits. However, specification limits and control limits are two entirely different and unrelated concepts. Moreover, there is no analytical relationship between them.

Control limits are defined by the process variability, which is usually determined by the process standard deviation σ. Specification limits, however, are set by the manufacturing team, management, or the customer. The most important point worth noting is that specification limits are used to evaluate the individual values, whereas control limits are used to evaluate the sample values.

CONTROL CHARTS FOR VARIABLES

In this section, we study various control charts for variables.

Shewhart $\bar{X}$ and R Control Chart

Certain rules are widely used in preparing Shewhart $\bar{X}$ and R control charts:

1. Take a series of samples from the process under investigation. Samples consisting of four or five items taken frequently are usually good, for the following reasons:

   a. Rational subgroups or samples of size 4 or 5 are more cost effective.

   b. If samples are larger than 10, the estimate of process standard deviation obtained using the range is not efficient. Moreover, the R chart is not effective.

   c. With samples of size 4 or 5, there are fewer chances for occurrence of any special causes during the collection of a sample. If the type of variation is changing (special cause versus common cause variation), the sample size should be as small as possible so that the averages of samples do not mask the changes.

2. Enough samples should be collected so that the major source of variation has an opportunity to occur. Generally, at least 25 samples of size 4 or 5 is considered sufficient to give a good test for process stability.
3. During the collection of data, a complete log of any changes in the process, such as changes in raw materials, operators, or tools or any calibration of tools or machines, must be maintained. Keeping a log is important for finding the special causes in a process (Duncan 1986).

Calculation of Sample Statistics

The sample statistics that need to be determined initially to prepare the Shewhart $\bar{X}$ and R control chart are the sample mean ($\bar{x}$) and the sample range (R). Thus, for example, let $x_1, x_2, \ldots, x_n$ be a random sample from the process under investigation. Then, we have:

$$\bar{x} = \frac{x_1 + x_2 + \cdots + x_s}{n} \tag{23.1}$$

$$R = \text{Max}\ (x_i) - \text{Min}\ (x_i) \tag{23.2}$$

Example 23.1

Let 5, 6, 9, 7, 8, 6, 9, 7, 6, and 5 be a random sample from a process under investigation. Find the sample mean and the sample range.

Solution

Using the data, we have:

$$\bar{x} = \frac{5+6+9+7+8+6+9+7+6+5}{10} = 6.8$$

$$R = \text{Max}\ (x_i) - \text{Min}\ (x_i) = 9 - 5 = 4$$

Calculation of Control Limits

Step 1: First calculate $\bar{x}$ and R_i for the ith sample, for $i = 1, 2, 3, \ldots, m$, where m is the number of samples collected during the study period.

Step 2: Calculate:

$$\bar{R} = \frac{R_1 + R_2 + \cdots + R_m}{m} \tag{23.3}$$

$$\bar{\bar{x}} = \frac{\bar{x}_1 + \bar{x}_2 + \cdots + \bar{x}_m}{m} \tag{23.4}$$

Step 3: Calculate the 3-sigma control limits for the $\bar{X}$ control chart

$$\begin{aligned} UCL &= \bar{\bar{x}} + 3\hat{\sigma}_{\bar{x}} \\ &= \bar{\bar{x}} + 3\frac{\hat{\sigma}}{\sqrt{n}} \\ &= \bar{\bar{x}} + 3\frac{\bar{R}}{d_2\sqrt{n}} \\ &= \bar{\bar{x}} + A_2\bar{R} \end{aligned} \tag{23.5}$$

$$CL = \bar{\bar{x}} \tag{23.6}$$

$$\begin{aligned} LCL &= \bar{\bar{x}} - 3\hat{\sigma}_{\bar{x}} \\ &= \bar{\bar{x}} - 3\frac{\hat{\sigma}}{\sqrt{n}} \\ &= \bar{\bar{x}} - 3\frac{\bar{R}}{d_2\sqrt{n}} \\ &= \bar{\bar{x}} - A_2\bar{R} \end{aligned} \tag{23.7}$$

where the values of A_2 and d_2 for various sample sizes are given in Appendix E.

Note: Instead of calculating 3-sigma limits (common in the United States), we can calculate the probability limits (common in Europe) at the desired level of significance α by replacing 3 with $z_{a/2}$ in equations (23.5) and (23.7). Thus, the control limits will be:

$$\begin{aligned} UCL &= \bar{\bar{x}} + z_{\alpha/2}\frac{\hat{\sigma}}{\sqrt{n}} \\ &= \bar{\bar{x}} + z_{\alpha/2}\frac{\bar{R}}{d_2\sqrt{n}} \end{aligned} \tag{23.8}$$

$$\begin{aligned} LCL &= \bar{\bar{x}} - z_{\alpha/2}\frac{\hat{\sigma}}{\sqrt{n}} \\ &= \bar{\bar{x}} - z_{\alpha/2}\frac{\bar{R}}{d_2\sqrt{n}} \end{aligned} \tag{23.9}$$

Step 4: Calculate the control limits for the R control chart

$$\begin{aligned} UCL &= \bar{R} + 3\hat{\sigma}_R \\ &= \bar{R} + 3d_3\frac{\bar{R}}{d_2} \\ &= \left(1 + 3\frac{d_3}{d_2}\right)\bar{R} \\ &= D_4\bar{R} \end{aligned} \tag{23.10}$$

$$CL = \bar{R} \tag{23.11}$$

$$
\begin{aligned}
LCL &= \bar{R} - 3\sigma_R \\
&= \bar{R} - 3d_3\frac{\bar{R}}{d_2} \\
&= \left(1 - 3\frac{d_2}{d_2}\right)\bar{R} \\
&= D_3\bar{R}
\end{aligned}
\qquad (23.12)
$$

where the values of D_3 and D_4 for various sample sizes are given in Appendix E.

The first implementation of control charts is referred to as phase I. In phase I, it is important that we calculate the preliminary control limits. We do this to find the extent of variation in sample means and sample ranges if the process is stable. In other words, at this point, only common causes would affect the process. If all the plotted points fall within the control limits and there is no evidence of a pattern, the control limits are suitable for the current or a future process. However, if some points exceed the control limits, then such points are ignored and every effort is made to eliminate any evident special causes that may be present in the process. Then, fresh control limits are calculated by using the remaining data, and the whole process is repeated. Remember that ignoring the points that exceed the control limits without eliminating the special causes may result in unnecessarily narrow control limits, which may cause issues such as putting the points beyond the control limits when, in fact, they should not be. Further, it is highly recommended that for preliminary control limits, we use at least 25 samples of size 4 or 5. Otherwise, the control limits may not be suitable for the current or a future process.

Example 23.2

Table 23.1 provides data on the diameter measurements of ball bearings used in the wheels of heavy construction equipment. Twenty-five samples, each of size 4, are taken directly from the production line. Samples come from all three shifts, and no sample contains data from two or more shifts. Use these data to construct an $\bar{X}$ and R chart and to verify that the process is stable.

Solution

From Appendix E, for a sample of size $n=4$, we have $D_3=0$ and $D_4=2.282$. Thus, the control limits for the R chart are:

$$
\begin{aligned}
\text{LCL} &= D_3\bar{R} = 0 \times 0.03479 = 0 \\
\text{UCL} &= D_4\bar{R} = 2.282 \times 0.03479 = 0.07936
\end{aligned}
$$

It is customary to prepare the R chart first and verify that all the plotted points fall within the control limits, and then proceed to construct the $\bar{X}$ chart. In fact, the concept of first bringing the process variability under control and then controlling the average makes sense. This is because it is almost impossible to bring the process average under control without controlling the process variability.

Table 23.1 Diameter measurements (mm) of ball bearings used in the wheels of heavy construction equipment.

Sample	Observations				$\bar{x}_i$	R_i
1	15.155	15.195	15.145	15.125	15.155	0.070
2	15.095	15.162	15.168	15.163	15.147	0.073
3	15.115	15.126	15.176	15.183	15.150	0.068
4	15.122	15.135	15.148	15.155	15.140	0.033
5	15.148	15.152	15.192	15.148	15.160	0.044
6	15.169	15.159	15.173	15.175	15.169	0.016
7	15.163	15.147	15.137	15.145	15.148	0.026
8	15.150	15.164	15.156	15.170	15.160	0.020
9	15.148	15.162	15.163	15.147	15.155	0.016
10	15.152	15.138	15.167	15.155	15.153	0.029
11	15.147	15.158	15.175	15.160	15.160	0.028
12	15.158	15.172	15.142	15.120	15.148	0.052
13	15.133	15.177	15.145	15.165	15.155	0.044
14	15.148	15.174	15.155	15.175	15.155	0.027
15	15.143	15.137	15.164	15.156	15.150	0.027
16	15.142	15.150	15.168	15.152	15.153	0.026
17	15.132	15.168	15.154	15.146	15.150	0.036
18	15.172	15.188	15.178	15.194	15.183	0.022
19	15.174	15.166	15.186	15.194	15.180	0.028
20	15.166	15.178	15.192	15.184	15.180	0.026
21	15.172	15.187	15.193	15.180	15.183	0.021
22	15.182	15.198	15.185	15.195	15.190	0.016
23	15.170	15.150	15.192	15.180	15.173	0.042
24	15.186	15.194	15.175	15.185	15.185	0.019
25	15.178	15.192	15.168	15.182	15.180	0.024
					$\bar{\bar{X}}$ **=15.1628**	$\bar{R}$**=0.03479**

The R chart for the data is given in Figure 23.2 which shows that all the plotted points fall within the control limits and that there is no evidence of any special pattern. Thus, we may conclude that the only variation present in the process is due to common causes. In this case, we can proceed to calculate the control limits for the $\bar{X}$ chart. From Appendix E, for a sample of size $n=4$, we get $A_2=0.729$. Thus, we have:

$$\text{LCL} = \bar{\bar{x}} - A_2\bar{R} = 15.1628 - 0.729 \times 0.03479 = 15.13749$$
$$\text{UCL} = \bar{\bar{x}} + A_2\bar{R} = 15.1628 + 0.729 \times 0.03479 = 15.18814$$

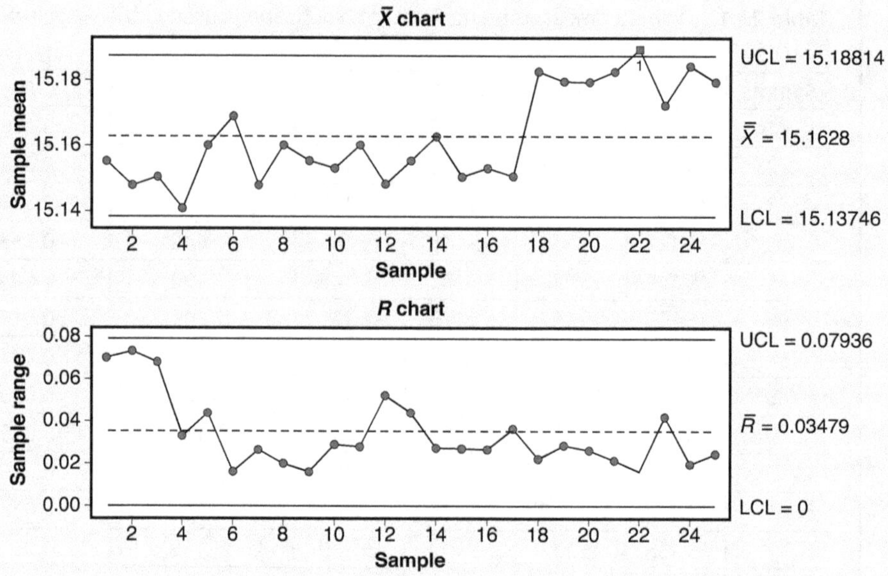

Figure 23.2 X-bar and *R* control chart for the ball-bearing data in Table 23.1.

The $\bar{X}$ chart for the data is given in Figure 23.2 which shows that point 22 exceeds the upper control limits. Moreover, too many consecutive points fall below the centerline. This indicates that the process is not under control and that some special causes are present that are affecting the process average. Thus, a thorough investigation should be launched to find the special causes, and appropriate action should be taken to eliminate them before we proceed to recalculate the control limits for an ongoing process.

A process is considered out of control not only when the points exceed the control limits but also when the points show some patterns of non-randomness. The *Statistical Quality Control Handbook* gives a set of decision rules for determining nonrandom patterns on control charts (Western Electric 1956). In particular, it suggests the patterns are nonrandom if either:

1. Two out of three successive points exceed the 2σ warning limits
2. Four out of five successive points fall at a distance of σ or beyond from the center line
3. Eight successive points fall on one side of the center line
4. Seven successive points run either upward or downward

Interpretation of Shewhart $\bar{X}$ and *R* Control Charts

We should investigate any out-of-control points—that is, points on or beyond 3σ control limits—or any patterns of nonrandomness on the *R* chart before interpreting the $\bar{X}$ chart. As discussed earlier, the reason for doing this is simple. It is not possible to bring the average under control without first bringing the variability under control. Normally the $\bar{X}$ chart is placed above the *R* chart, and they are aligned with each other such that the average and the range for any sample are plotted on the same vertical line. Examine whether one chart, both charts, or neither chart indicates the process is out of control for any given sample. If any point exceeds the control limits in one or both charts, the sample did not come from a stable process. In other words, there are some special or assignable causes present in the system. More precisely, if the plotted point exceeds the control limits in the *R* chart, then it is evident that the variability of the process has changed. In such cases, before a full-blown investigation is launched, some preliminary checks should be made. For example,

1. Check that all the calculations are correct and that the data are entered in the computer correctly.
2. Check whether there has been a change in workers, machines, or suppliers of raw material.

If the points exceed the control limits in the $\bar{X}$ chart, the process mean has changed. Again, follow the preliminary checks before a full-blown investigation is launched.

If points exceed the limits in both the $\bar{X}$ chart and the *R* chart, this usually indicates that a sudden shift occurred in the lot from which the samples were taken. In such cases, after the preliminary checks, there should be a full investigation concentrating on the period during which that lot was produced. Depending on the process, the possibility of stopping production until the special causes are detected should be considered.

In addition to points exceeding the control limits, nonrandom patterns such as a run of seven points moving upward or downward, or a run of eight successive points falling above or below the centerline should be checked.

An upward run or a run above the centerline in an *R* chart indicates:

1. A greater variability or a tendency of perpetuating a greater variability in the output of the process is occurring. This may be due to new material of undesirable low quality or a difference between the shifts. Immediate attention to detect special causes is warranted.
2. The measurement system has changed.

A downward run or a run below the centerline in an *R* chart indicates:

1. A smaller variability or a tendency of perpetuating a smaller variability in the output of the process is occurring. This is usually a good sign for the process. A thorough investigation should be made so that similar conditions are maintained as long as possible. Similar conditions should be implemented in the process elsewhere.
2. The measurement system has changed.

A run relative to the $\bar{X}$ chart indicates:

1. The process average has changed or is still changing.
2. The measurement system has changed.

Shewhart $\bar{X}$ and S Control Chart

The $\bar{X}$ and S control charts, like the $\bar{X}$ and R control charts, are developed from a measured process output data, and both the $\bar{X}$ chart and the S chart are used together. The standard deviation s is usually a more efficient indicator of process variability than the range, particularly when the sample sizes are large (10 or greater). The sample standard deviation s for the S chart is calculated using all the data points rather than just the maximum and minimum values in a data set, as is done for the R chart. S charts are usually preferred over R charts in the following situations:

- Samples are of size 10 or larger
- Sample size is variable
- The process is automated, so the s for each sample can be calculated easily

The sample standard deviation s is determined using the formula:

$$s = \sqrt{\frac{1}{n-1}\left(\sum_{i=1}^{n} x_i^2 - \frac{1}{n}\left(\sum_{i=1}^{n} x_i\right)^2\right)} \tag{23.13}$$

Then the control limits for the $\bar{X}$ and S control charts are determined, as shown in the following section.

CALCULATION OF CONTROL LIMITS

Step 1: First, calculate $\bar{x}_i$ and s_i for the ith sample, for $i = 1, 2, 3, \ldots, m$, where m is the number of samples collected during the study period.

Step 2: Calculate

$$\bar{s} = \frac{s_1 + s_2 + \cdots + s_m}{m} \tag{23.14}$$

$$\bar{\bar{x}} = \frac{\bar{x}_1 + \bar{x}_2 + \cdots \bar{x}_m}{m} \tag{23.15}$$

Step 3: Calculate the control limits for the $\bar{X}$ chart

$$\begin{aligned} \text{UCL} &= \bar{\bar{x}} + 3\frac{\hat{\sigma}}{\sqrt{n}} \\ &= \bar{\bar{x}} + 3\frac{\bar{s}}{c_4\sqrt{n}} \\ &= \bar{\bar{x}} + A_3\bar{s} \end{aligned} \tag{23.16}$$

$$\text{CL} = \bar{\bar{x}} \tag{23.17}$$

$$\begin{aligned} \text{UCL} &= \bar{\bar{x}} - 3\frac{\hat{\sigma}}{\sqrt{n}} \\ &= \bar{\bar{x}} - 3\frac{\bar{s}}{c_4\sqrt{n}} \\ &= \bar{\bar{x}} - A_3\bar{s} \end{aligned} \tag{23.18}$$

where the values of A_3 and c_4 for various sample sizes are given in Appendix E.

Step 4: Calculate the control limits for the S chart

$$\begin{aligned} \text{UCL} &= \bar{s} + 3\hat{\sigma}_s \\ &= \bar{s} + 3\frac{\bar{s}}{c_4}\sqrt{1-c_4^2} \\ &= \left(1 + 3\frac{1}{c_4}\sqrt{1-c_4^2}\right)\bar{s} \\ &= B_4\bar{s} \end{aligned} \tag{23.19}$$

$$\text{CL} = \bar{s} \tag{23.20}$$

$$\begin{aligned} \text{UCL} &= \bar{s} - 3\hat{\sigma}_s \\ &= \bar{s} - 3\frac{\bar{s}}{c_4}\sqrt{1-c_4^2} \\ &= \left(1 - 3\frac{1}{c_4}\sqrt{1-c_4^2}\right)\bar{s} \\ &= B_3\bar{s} \end{aligned} \tag{23.21}$$

We illustrate the development of the $\bar{X}$ and S control charts with Example 23.3.

Example 23.3

Use the ball-bearing data in Table 23.1 to construct the $\bar{X}$ and S control charts.

Solution

From Appendix E, for a sample of size $n=4$, we have $B_3=0$ and $B_4=2.266$. Thus, the control limits for the S control chart are:

$$\text{LCL} = B_3\bar{s} = 0 \times 0.01557 = 0$$
$$\text{UCL} = B_4\bar{s} = 2.266 \times 0.01557 = 0.03527$$

It is customary to prepare the S control chart first and verify that all the plotted points fall within the control limits, and then proceed to construct the $\bar{X}$ control chart. As described earlier, the concept of first bringing the process variability under control and then controlling the average makes sense, since without controlling the process variability, it is practically impossible to bring the process average under control.

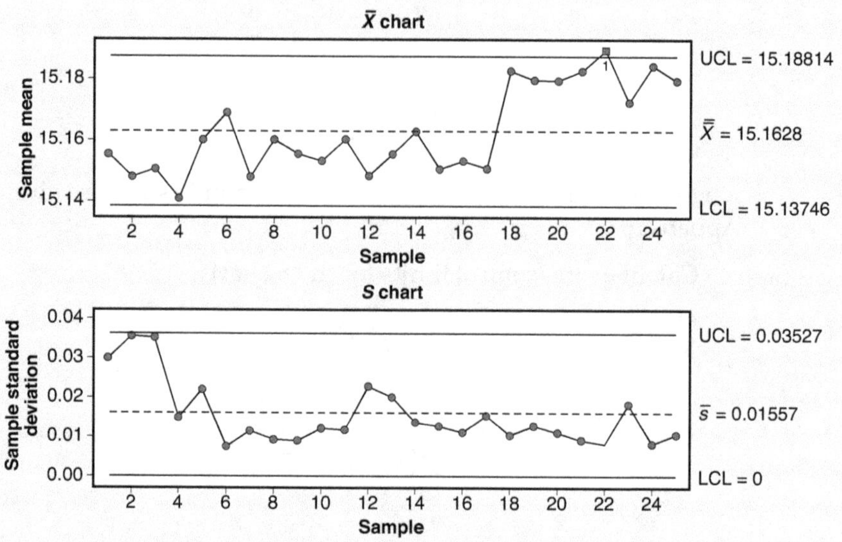

Figure 23.3 *X*-bar and *S* control chart for the ball-bearing data in Table 23.1.

The *S* chart for the data is given in Figure 23.3 which shows that points 2 and 3 almost coincide with the upper control limit. Moreover, point 17 is almost on the centerline. If this point were clearly below the centerline, we would have had a run of nine points below the centerline. These observations indicate that the process variability is marginally under control, therefore, the process should be carefully monitored. Since the process variability is under control, though marginally, we can proceed to calculate the control limits for the $\bar{X}$ chart. From Appendix E, for a sample of size $n=4$, we get $A_3 = 1.628$. Thus, we have:

$$LCL = \bar{\bar{x}} - A_3\bar{s} = 15.1628 - 1.628 \times 0.01557 = 15.13746$$
$$UCL = \bar{\bar{x}} + A_3\bar{s} = 15.1628 + 1.628 \times 0.01557 = 15.18814$$

The $\bar{X}$ chart for the data in Table 23.3 is given in Figure 23.3, which shows that point 22 exceeds the upper control limits. Moreover, too many consecutive points fall below the centerline. This indicates that the process is not under control and that some special causes are present that are affecting the process average. Thus, a thorough investigation should be launched to find the special causes, and appropriate action should be taken to eliminate them before we proceed to recalculate the control limits for an ongoing process.

CONTROL CHARTS FOR ATTRIBUTES

As noted earlier in this chapter, quality characteristics are usually of two types, called variables and attributes. In the previous section, we studied control charts for variables for detecting large shifts. However, not all quality characteristics can be measured numerically. For example, we may be interested in finding whether the new car paint meets the specifications in terms of shine, uniformity, and scratches. Clearly, in this example we cannot quantify the shine, blemishes, and scratches. In other words, we cannot measure the shine, uniformity, or the scratches numerically; consequently, to study the quality of paint on new cars, we cannot use control charts for variables. This type of situation can arise when a quality characteristic is measurable, but because of cost, time, or both, we do not want to take the measurements. We may prefer to use a more economical method such as a go/no-go gauge. Therefore, it is important that we study control charts that are appropriate for quality characteristics that cannot be measured numerically. Such control charts are called control charts for attributes. In this section, we study various control charts for attributes for detecting large shifts, which usually occur in phase I implementation of SPC.

When a quality characteristic cannot be measured numerically, we classify such a product as defective or non-defective; in SPC it has become more common to use the terminology *conforming* or *nonconforming*. Thus, in this section, we shall continue using the terminology *conforming* or *nonconforming*. A quality characteristic that classifies any product as conforming or nonconforming is called an *attribute*.

Quality characteristics such as the determination that a soft drink can is not leaking, a stud has regular edges, a rod fits into a slot, a 100-watt lightbulb meets the desired standard, or a steel rivet meets the manufacturer's quality specifications are examples of attributes. Note that the data collected on a quality characteristic, which is an attribute, are count data. Moreover, the sample sizes when using control charts for attributes are normally much larger (usually in the hundreds) than a sample of size 4 or 5, which we usually use in control charts for variables.

In general, variable control charts are more informative and effective in detecting a defect even before it occurs, whereas attribute charts are used only after the defects have occurred. There are cases, however, when variable control charts have limitations. For example, consider a product that is nonconforming due to any one of 10 quality characteristics that do not conform to specifications. Clearly, we cannot control all 10 quality characteristics by using one variable control chart, since one variable control chart can control only one quality characteristic at a time. Thus, in this case, to control all 10 quality characteristics, we will have to use 10 different variable control charts. On the other hand, one attribute control chart can study all the quality characteristics because a nonconforming unit is nonconforming irrespective of the number of quality characteristics that do not conform to specifications. Thus, we can conclude that both variable and attribute control charts have pros and cons.

In some cases, the quality characteristic is such that instead of classifying a unit as conforming or nonconforming, it records the number of nonconformities per manufactured unit. For example, the number of holes in a roll of paper, the number of irregularities per unit area of a spool of cloth, the number of blemishes on a painted surface, the number of loose ends in a circuit board, the number of nonconformities per unit length of a cable, the number of nonconformities of all types in an assembled unit, and so on. In such cases, we use control charts that fall under the category of control charts for attributes. These control charts are used to reduce the number of nonconformities per unit length, area, or volume of a single manufactured unit, or to reduce the number of nonconformities per manufactured or assembled unit.

The control charts for attributes are quite similar to the control charts for variables; that is, the centerline and the control limits of both charts are set in the same manner. However, it is important to note that the reasons for using control charts for variables and control charts for attributes are quite distinct. As noted earlier, the purpose of using control charts for variables in any process is to reduce the variability due to special or assignable causes, whereas the control charts for attributes are used to reduce the number of nonconforming units; the number of nonconformities per manufactured or assembled unit; or the number of nonconformities per unit length, area, or volume of a single manufactured unit.

In this section, we will study four types of control charts for attributes: the *p* chart, *np* chart, *c* chart, and *u* chart. In Table 23.2 we give a brief description of these charts, which can help determine the appropriate type of control chart that should be used for the quality characteristic under investigation.

The *p* Chart: Control Chart for Fraction of Nonconforming Units

The most frequently used attribute control chart is the *p* chart. It is used to find the fraction or percentage of units that do not conform to the specifications in a situation where the observed quality characteristic is an attribute or a variable measured by a go/no-go gauge. A *p* chart can be used to study one or more quality characteristics simultaneously. Since each inspected unit is classified as conforming or nonconforming and it is assumed that the conformity or nonconformity of each unit is defined independently, which is true only if the process is stable, the probability of occurrence of a nonconforming unit at any given time is the same. Then the basic rules of the *p* chart are governed by the binomial probability distribution with parameters n and p, where n is the sample size and p is the fraction of nonconforming units produced by the process under investigation.

The binomial probability distribution function of a random variable X with parameters n and p is defined by:

$$P(X=x)=\binom{n}{x}p^x(1-p)^{n-x} \quad x=0,1,\ldots,n \tag{23.22}$$

The mean and the standard deviation of the random variable X are given by np and $\sqrt{np(1-p)}$ respectively. For more details on binomial distribution, please refer to Gupta and Walker (2007a).

Table 23.2 Control charts for attributes.

Control chart	Quality characteristic under investigation
p chart	Percent or fraction of nonconforming units in a subgroup or a sample, where sample size can be variable
np chart	Number of nonconforming units in a sample
c chart	Number of nonconformities in a sample or in one or more inspection units
u chart	Number of nonconformities per unit, where sample size can be variable

Control Limits for the *p* Chart

To develop a p control chart, proceed as follows:

1. From the process under investigation, select m ($m \geq 25$) samples of size n ($n \geq 50$) units. Note, however, if we have some prior information or any clue that the process is producing a very small fraction of nonconforming units, the sample size should be large enough that the probability that it contains some nonconforming units is relatively high.
2. Find the number of nonconforming units in each sample.
3. Find for each sample the fraction p_i of nonconforming units, that is

$$p_i = \frac{x_i}{n} \tag{23.23}$$

   where x is the number of nonconforming units in the ith ($i = 1, 2, \ldots, m$) sample.

4. Find the average nonconforming $\bar{p}$ over the m samples, that is,

$$\bar{p} = \frac{p_1 + p_2 + \cdots + p_m}{m} \tag{23.24}$$

   The value of $\bar{p}$ determines the centerline for the p chart and is an estimate of p, the process fraction of nonconforming units.

5. Using the well-known result that the binomial distribution with parameters n and p for large n can be approximated by the normal distribution with mean np and variance $np(1-p)$, it can easily be seen that $\bar{p}$ will be approximately normally distributed with mean p and standard deviation $\sqrt{\frac{p(1-p)}{n}}$.

Hence, the upper and lower 3σ control limits and the centerline for the p chart are as shown in the following equations:

$$\text{UCL} = \bar{p} + 3\sqrt{\frac{\bar{p}(1-\bar{p})}{n}} \tag{23.25}$$

$$\text{CL} = \bar{p} \tag{23.26}$$

$$\text{LCL} = \bar{p} - 3\sqrt{\frac{\bar{p}(1-\bar{p})}{n}} \tag{23.27}$$

Note that in equations (23.25) and (23.27) $\sqrt{\frac{\bar{p}(1-\bar{p})}{n}}$ is an estimator of $\sqrt{\frac{p(1-p)}{n}}$, the standard deviation of $\bar{p}$. Further, if the control charts are being implemented for the first time, the control limits given by equations (23.25)–(23.27) should be treated as the trial limits. In other words, before using these control limits any further, the points corresponding to all the samples used to determine these limits should be plotted and verified. Additionally, it should be established that all the points fall within these control limits and that no pattern is evident. If any sample points exceed the control limits, or if there is any pattern, the possible special causes should be detected and eliminated before recalculating the control limits for future use. When recalculating the control limits, points that exceeded the trial control limits should be ignored, provided any special causes related to such points have been detected and eliminated.

Note: Sometimes for small values of $\overline{p}$, n, or both, the value of the lower control limit may turn out to be negative. In such cases, we always set the lower control limit at zero. This is because the fraction of nonconforming units can never go below zero.

Interpreting the Control Chart for Fraction Nonconforming

1. If any point or points exceed the upper or lower control limit, we conclude that the process is not stable and that some special causes are present in the process.
2. The presence of special causes, which may be favorable or unfavorable, must be investigated, and appropriate action(s) should be taken.
3. A point above the upper control limit is generally an indication that:
   - The control limit or the plotted point is in error.
   - The process performance has deteriorated or is deteriorating.
   - The measurement system has changed.
4. A point below the lower control limit is generally an indication that:
   - The control limit or the plotted point is in error.
   - The process performance has improved or is improving. This condition of the process should be investigated very carefully so that such conditions of improvement are implemented on a permanent basis at this location and elsewhere in the industry.
   - The measurement system has changed.
5. As in the case of the $\overline{X}$ and R or the $\overline{X}$ and S charts, the presence of any unusual patterns or trends is either an indication of an unstable process or an advance warning of conditions that, if left unattended or without any appropriate action, could make the process unstable.
6. If $\overline{p}$ is moderately high ($n\overline{p} \geq 5$), an approximately equal number of points should fall on either side of the centerline. Therefore, either of the following conditions could indicate that the process has shifted or a trend of shift has started:
   - A run of seven or more points going up or down.
   - A run of seven or more points falls either below or above the centerline.
7. A run above the centerline or a run going up generally indicates:

- The process performance has deteriorated and may still be deteriorating.
- The measurement system has changed.

8. A run below the centerline or a run going down generally indicates:
   - The process performance has improved and may still be improving.
   - The measurement system has changed.

To illustrate the construction of the p chart, we consider the data in Table 23.3 of Example 23.4.

Example 23.4

A semiconductor industry tracks the number of nonconforming computer chips produced each day. A team of Six Sigma Green Belts wants to improve the overall quality by reducing the fraction of nonconforming computer chips. To achieve this goal, the team decided to set up a p chart based on daily inspections of 1000 chips over a period of 30 days. Table 23.3 gives the number of nonconforming chips out of 1000 inspected chips each day during the study period of 30 days.

Solution

Using the data in Table 23.3 we develop the trial control limits of the p chart as follows.

First we calculate the sample fraction nonconforming values (p_i), which are listed in columns 3 and 6. Substituting the sample fraction nonconforming values in equation (23.24), we get:

Table 23.3 Number of nonconforming computer chips out of 1000 inspected each day during the study period of 30 days.

	Number of nonconforming	Sample fraction nonconforming		Number of nonconforming	Sample fraction nonconforming
Day	x	pi	Day	x	pi
1	9	0.009	16	12	0.012
2	5	0.005	17	5	0.005
3	6	0.006	18	6	0.006
4	11	0.011	19	12	0.012
5	11	0.011	20	10	0.010
6	12	0.012	21	6	0.006
7	7	0.007	22	7	0.007
8	11	0.011	23	11	0.011
9	6	0.006	24	11	0.011
10	6	0.006	25	9	0.009
11	8	0.008	26	5	0.005
12	5	0.005	27	12	0.012
13	8	0.008	28	11	0.011
14	5	0.005	29	7	0.007
15	8	0.008	30	9	0.009

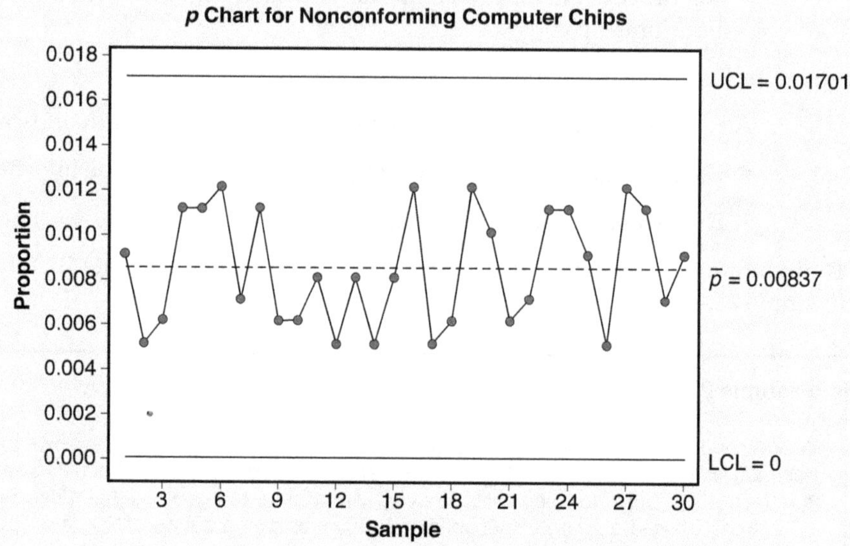

Figure 23.4 *p* chart for nonconforming computer chips, using trial control limits for the data in Table 23.3.

$$\bar{p} = 0.00837$$

Plugging the value of $\bar{p} = 0.00837$ and $n = 1000$ into Equations (23.25) through (23.27), we get the control limits for the *p* chart, that is:

$$UCL = 0.01701$$

$$CL = 0.00837$$

$$LCL = 0.0$$

Figure 23.4 shows the *p* control chart for the data in Table 23.3.

From the control chart in Figure 23.4, we observe that all the points are well within the control limits. However, starting with point 9, seven successive points fall below the centerline. This indicates that from day 9 to day 15, the number of nonconforming chips was relatively low. The process conditions on these days should be investigated so that similar conditions could be implemented for future use. Otherwise, since all the points of the current data fall within the control limits, the trial control limits can be extended for use over the next 30 days, when the control chart should again be reevaluated.

The *p* Chart: Control Chart for Fraction Nonconforming with Variable Samples

There are times when, for various reasons, it is not possible to select samples of equal sizes. This is particularly true when the samples consist of 100% inspection during a fixed period of time on each day of the study. The procedure to develop a p chart with variable sample sizes is similar to the procedure for a p chart with constant sample size.

Thus, for example, suppose we have m samples of sizes $n_1, n_2, n_3, \ldots, n_m$. To develop a p chart for variable sample sizes, we proceed as follows:

1. From the process under investigation, select m ($m \geq 25$) samples of sizes $n_1, n_2, n_3, \ldots, n_m$ ($n_i \geq 50$) units.
2. Find the number of nonconforming units in each sample.
3. Find for each sample the fraction p_i of nonconforming units

$$p_i = \frac{x_i}{n_i} \tag{23.28}$$

where x is the number of nonconforming units in the ith ($i=1, 2, \ldots, m$) sample.

4. Find the average fraction $\bar{p}$ of nonconforming units over the m samples, that is

$$\bar{p} = \frac{n_1p_1 + n_2p_2 + \cdots + n_mp_m}{n_1 + n_2 + n_2 + \cdots + n_m} \tag{23.29}$$

The value of $\bar{p}$ determines the centerline for the p chart and is an estimate of p, the process fraction of nonconforming units.

5. The control limits for p with variable sample sizes are determined for each sample separately. Thus, for example, the upper and lower 3-sigma control limits for the ith sample are

$$\text{UCL} = \bar{p} + 3\sqrt{\frac{\bar{p}(1-\bar{p})}{n_i}} \tag{23.30}$$

$$\text{CL} = \bar{p} \tag{23.31}$$

$$\text{LCL} = \bar{p} - 3\sqrt{\frac{\bar{p}(1-\bar{p})}{n_i}} \tag{23.32}$$

Note that the centerline is the same for all samples, whereas the control limits will be different for different samples.

To illustrate the construction of the p chart with variable sample size, we consider the data in Table 23.4 in Example 23.5.

Example 23.5

Suppose in Example 23.4 that all the chips manufactured during a certain fixed period of time are inspected each day. However, the number of computer chips manufactured varies during that fixed period on each day. The data collected for the study period of 30 days is shown in Table 23.4. Construct the p chart for these data and determine whether the process is stable.

Table 23.4 Number of nonconforming computer chips with different size samples inspected each day during the study period of 30 days.

Number of nonconforming			Number of nonconforming		
Day	x	**Sample size**	**Day**	x	**Sample size**
1	7	908	16	7	962
2	11	986	17	11	926
3	8	976	18	7	917
4	7	991	19	9	978
5	7	944	20	7	961
6	5	906	21	6	970
7	11	928	22	9	905
8	5	948	23	9	962
9	10	994	24	8	900
10	8	960	25	11	998
11	7	982	26	5	935
12	6	921	27	6	970
13	7	938	28	6	967
14	10	1000	29	9	983
15	6	982	30	8	976

Solution

Using the data in Table 23.4 and Equations (23.30) through (23.32), we get the trial control limits as UCL=0.01675, CL=0.00813, and LCL=0.0. Figure 23.5 shows the p chart for the data in Table 23.4.

From Figure 23.5, we observe that all the points are well within the control limits and that there is no apparent pattern or trend in the chart. Thus, the process is stable. Further, in Figure 23.4 we had a run of seven points that fell below the centerline, whereas in Figure 23.5 there is no such run even though we are dealing with the same process. Such differences are normal when samples are taken at different times.

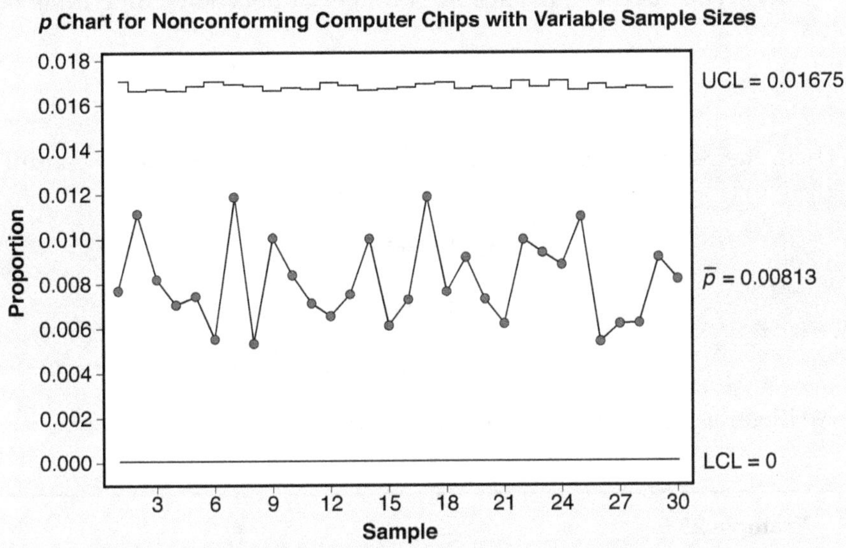

Figure 23.5 *p* chart for nonconforming chips with variable sample sizes, using trial control limits for the data in Table 23.4.

The *np* Chart: Control Chart for Number of Nonconforming Units

In the *np* chart, we plot the number of nonconforming units in an inspected sample instead of the fraction of nonconforming units in the inspected sample. In *p* charts sample sizes can be equal or unequal, whereas in *np* charts sample sizes are equal. In all other terms, the *np* chart is similar to the *p* chart and can be implemented under the same circumstances. We summarize here some points pertinent for the *np* chart:

- The inspection sample sizes should be equal.
- The sample size should be large enough to include some nonconforming units.
- Record the sample size and the number of observed nonconforming in each sample, and plot the number of nonconforming on the control chart.

Control Limits for *np* Control Chart

Select m samples each of size n from the process under investigation and determine the number of nonconforming units in each sample. Let the number of nonconforming units found be denoted by $x_1, x_2, x_3, \ldots, x_m$. Then the control limits are found as follows.

First, calculate $n\bar{p}$, the average number of nonconforming units per sample, that is:

$$n\bar{p} = \frac{x_1 + x_2 + x_3 + \cdots + x_m}{m} \tag{23.33}$$

Then, the 3-sigma control limits and the centerline for the *np* control chart are given by:

$$UCL = n\bar{p} + 3\sqrt{n\bar{p}(1-\bar{p})} \tag{23.34}$$

$$CL = n\bar{p} \tag{23.35}$$

$$LCL = n\bar{p} - 3\sqrt{n\bar{p}(1-\bar{p})} \tag{23.36}$$

We illustrate construction of the *np* control chart with Example 23.6.

Example 23.6

Consider the data on computer chips in Table 23.3. Construct an *np* chart for these data and verify whether the process is stable.

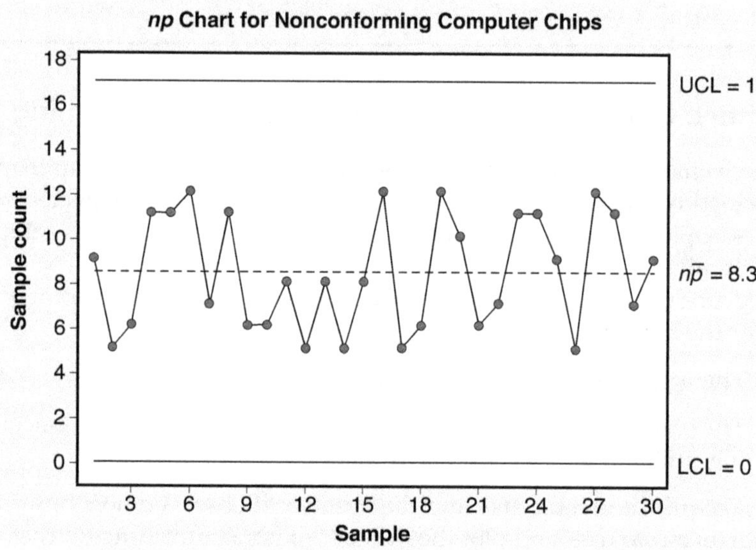

Figure 23.6 *np* chart for nonconforming computer chips, using trial control limits for the data in Table 23.3.

From the control chart in Figure 23.6, we observe that all the points are well within the control limits. However, as in Figure 23.4, starting from point 9, seven successive points fall below the centerline. This indicates that from day 9 to day 15, the number of nonconforming units was relatively low. The process conditions on these days should be investigated so that similar conditions could be extended for future use. Otherwise, since all the points of the current data fall within the control limits, the trial control limits can be extended for use over the next 30 days, when the control chart should be reevaluated.

The *c* Chart (Nonconformities versus Nonconforming Units)

In many situations we are interested in studying the number of nonconformities in a sample, which is also known as an inspection unit, rather than studying the fraction or total number of nonconforming units. This is particularly true when a unit is nonconforming due to various types of nonconformities. For example, we may be interested in studying the quality of electric motors, which could be nonconforming due to defective bearings, a defective gear, a defective seal, or a defective terminal connection of the winding.

As another example, suppose we are interested in studying the quality of printed circuit boards for laptops, which could be nonconforming due to shorted trace, cold solder joint, or a solder joint. In these cases, we may be more likely to study the nonconformities rather than the nonconforming units. The control chart that is most commonly used to study nonconformities is the c control chart. Whereas the p control chart studies the fraction of nonconforming units and the np control chart studies the total number of nonconforming units in each sample, the c chart studies the total number of nonconformities in each sample, or what is known as an inspection unit. The letter c in c control chart denotes the total number of nonconformities, which may be of one or various kinds, in an inspection unit.

To illustrate, suppose that to develop trial control limits for a c control chart for electric motors we select samples of size 100 motors each, where each sample would be considered as one inspection unit or several inspection units depending on how the inspection units are defined. Note that the size of an inspection unit is purely a matter of convenience. But the c control charts are constructed with sample sizes or number of inspection units in a sample being equal. It can be shown that under certain conditions (such as those of a Poisson process, which are briefly discussed in Gupta and Walker [2007a]), the number of nonconformities, c, is distributed according to a Poisson probability distribution with parameter λ, where λ is the average number of nonconformities per inspection unit. The Poisson probability distribution is defined as:

$$p(x) = \frac{e^{-\lambda}\lambda^{x}}{x!}, x = 0,1,2,3,\ldots \tag{23.37}$$

where the mean and the variance of the Poisson distribution are given by λ. Now suppose that we select m samples with each sample being one inspection unit and let the number of nonconformities in these samples be $c_1, c_2, c_3, \ldots, c_m$. Then the parameter λ, which is usually unknown, is estimated by:

$$\hat{\lambda} = \bar{c} = \frac{c_1 + c_2 + c_3 + \cdots + c_m}{m} \tag{23.38}$$

Then the 3-sigma control limits for the c control chart are defined as follows:

$$\text{UCL} = \bar{c} + 3\sqrt{\bar{c}} \tag{23.39}$$

$$\text{CL} = \bar{c} \tag{23.40}$$

$$\text{LCL} = \bar{c} - 3\sqrt{\bar{c}} \tag{23.41}$$

Note that for small values of $\bar{c}(\le 5)$ the Poisson distribution is asymmetric; the value of a type I error (α) above the upper control limit and below the lower control limit is usually not the same. Thus, for small values of $\overline{c}$ it may be more prudent to use probability control limits rather than the 3-sigma control limits. The probability control limits can be found using Poisson distribution tables.

To illustrate the construction of a *c* control chart using 3-sigma control limits, we consider the data in Table 23.5 of Example 23.7.

Example 23.7

A paper mill has detected that nonconformities of two types—holes and wrinkles in the paper—are responsible for almost 90% of rejected paper rolls. The Six Sigma Green Belt team in the mill decided to set up control charts to reduce or eliminate the number of these nonconformities. To set up the control charts, the team decided to collect some data by taking random samples of 5 rolls each day for 30 days and count the number of nonconformities (holes and wrinkles) in each sample. The data are shown in Table 23.5. Set up a *c* control chart using these data.

Solution

From the data in Table 23.5, the estimate of the population parameter is given by:

$$\hat{\lambda} = \overline{c} = \frac{\sum_{i=1}^{30} c_i}{30} = \frac{222}{30} = 7.4$$

Table 23.5 Total number of nonconformities in samples of five rolls of paper.

Day	Total number of nonconformities	Day	Total number of nonconformities	Day	Total number of nonconformities
1	8	11	7	21	9
2	6	12	6	22	6
3	7	13	6	23	8
4	7	14	8	24	7
5	8	15	6	25	6
6	7	16	6	26	9
7	8	17	8	27	9
8	7	18	9	28	7
9	6	19	8	29	7
10	9	20	9	30	8

Therefore, using equations (23.39)-(23.41), the 3-sigma control limits of the phase I *c* control chart are given by:

$$UCL = 7.4 + 3\sqrt{7.4} = 15.56$$

$$CL = 7.4$$

$$LCL = 7.4 - 3\sqrt{7.4} = -0.76 = 0$$

(continued)

Note that if the lower control limit turns out to be negative, as in this example, we set it at zero since the number of nonconformities cannot be negative. Figure 23.7 shows the c control chart for the data in Table 23.5.

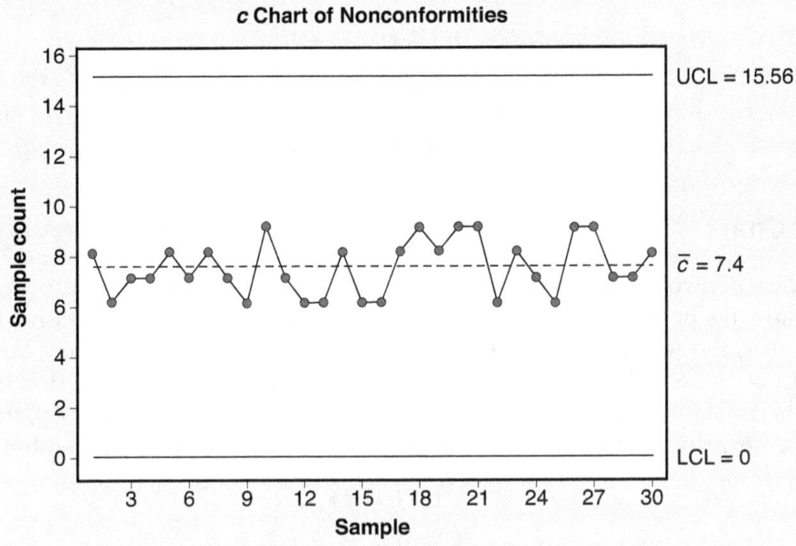

Figure 23.7 c control chart of nonconformities for the data in Table 23.5.

From Figure 23.7, it is clear that the process is stable. In other words, no special causes are present, and the only causes affecting the process are the common causes. Thus, to eliminate the imperfections in the paper, the management must take action on the system, such as examining the quality of wood chips and pulp, changing old equipment, or providing more training for the workers. Also, to further enhance the process and eliminate the nonconformities, the quality engineers should use the techniques available in the design of experiment.

If the economic factors and time allow, one should take samples or inspection units large enough so that the lower control limit is positive. The lower control limit can be positive only if $\bar{c} > 9$. This means that the sample size should be such that it can catch 9 or more nonconformities with high probability. An advantage of a positive lower control limit is that it will allow us to see the conditions under which the nonconformities are very low and consequently give us the opportunity to perpetuate these conditions on-site and implement them elsewhere in the industry.

As noted earlier, the size of the inspection unit is usually determined by what is convenient. However, to determine the actual inspection unit size, one should also take into consideration the statistical characteristics of the process, such as the average run length, the state of the process (i.e., whether the process has deteriorated or improved), and other factors that may require us to increase or decrease the sample size. Thus, while using control charts for nonconformities, particularly in phase I, situations may arise when the sample size may vary. In such cases, we use the u control chart instead of the c control chart. We discuss the u control chart in the following section.

Note that if the samples consist of n inspection units, the control limits for the c control chart are given by:

$$\text{UCL} = n\bar{c} + 3\sqrt{n\bar{c}} \tag{23.42}$$

$$\text{CL} = n\bar{c} \tag{23.43}$$

$$\text{LCL} = n\bar{c} - 3\sqrt{n\bar{c}} \tag{23.44}$$

The *u* Chart

The u control chart is essentially the c control chart except that the u control chart is always based on the number of nonconformities per inspection unit. In other words, the actual sample size may not be equal to one, or may vary, but the control limits of the u chart are always determined based on one inspection unit. Thus, if n is constant, one can use either a c chart or a u chart. For a u chart, the centerline is determined by $\bar{u} = \bar{c}/n$ and the 3-sigma control limits are given by:

$$\text{UCL} = \bar{u} + 3\sqrt{\bar{u}/n} \tag{23.45}$$

$$\text{LCL} = \bar{u} - 3\sqrt{\bar{u}/n} \tag{23.46}$$

If the sample size varies, we define $\bar{u}$ as

$$\bar{u} = \frac{c_1 + c_2 + \cdots + c_m}{n_1 + n_2 + \cdots + n_m} = \frac{\bar{c}}{\bar{n}} \tag{23.47}$$

where m is the number of samples selected during the study period; $c_1, c_2, c_3, \ldots, c_m$ are the number of nonconformities in the m samples; and $\bar{n}$ is the average sample size, given by

$$\bar{n} = \frac{n_1 + n_2 + \cdots + n_m}{m} \tag{23.48}$$

Thus, in this case, the centerline is fixed, but the control limits are different; that is, the centerline and the 3-sigma control limits for the ith sample are given by:

$$\text{UCL} = \bar{u} + 3\sqrt{\bar{u}/n_i} \tag{23.49}$$

$$\text{CL} = \bar{u} \tag{23.50}$$

$$\text{LCL} = \bar{u} - 3\sqrt{\bar{u}/n_i} \tag{23.51}$$

Sometimes, if the samples do not vary much, the n_i's in equations (23.49) and (23.51) are replaced by $\bar{n}$ so that the control limits are:

$$\text{UCL} = \bar{u} + 3\sqrt{\bar{u}/\bar{n}} \tag{23.52}$$

$$\text{CL} = \bar{u} \tag{23.53}$$

$$\text{LCL} = \bar{u} - 3\sqrt{\bar{u}/\bar{n}} \tag{23.54}$$

To illustrate the construction of the u chart, we consider the data in Table 23.6 of Example 23.8.

Example 23.8

A Six Sigma Green Belt team in a semiconductor industry found that the printed boards for laptops have nonconformities of several types, such as shorted trace, cold solder joint, and solder short, and the number of nonconformities is unacceptable. Thus, to reduce the number of nonconformities in the printed boards for laptops, the Six Sigma Green Belt team wants to set up a *u* chart. They collect some data by selecting samples of 5 inspection units, where each inspection unit consists of 30 boards. The data, shown in Table 23.6, were collected over a period of 30 days.

Solution

Using the data in Table 23.6, we have $\bar{c} = 41.333$

Table 23.6 Number of nonconformities on printed boards for laptops per sample; each sample consists of five inspection units.

Day	Number of nonconformities per sample	Day	Number of nonconformities per sample
1	48	16	42
2	49	17	34
3	38	18	30
4	49	19	49
5	43	20	44
6	37	21	47
7	45	22	33
8	48	23	37
9	39	24	33
10	46	25	34
11	40	26	49
12	44	27	50
13	43	28	49
14	35	29	35
15	31	30	39

Therefore,

$$\bar{u} = \frac{\bar{c}}{5} = \frac{41.333}{5} = 8.2667$$

Hence, the control limits are given by:

$$\begin{aligned} UCL &= \bar{u} + 3\sqrt{\bar{u}/n} = 8.2666 + 3\sqrt{8.2666/5} = 12.124 \\ CL &= \bar{u} = 8.2667 \\ UCL &= \bar{u} - 3\sqrt{\bar{u}/n} = 8.2666 - 3\sqrt{8.2666/5} = 4.409 \end{aligned}$$

Figure 23.8 shows the *u* chart for the data in Table 23.6.

The *u* chart in Figure 23.8 shows that there are no assignable causes present in the process. In other words, only common causes are affecting the process. Therefore, management needs to take action on the system.

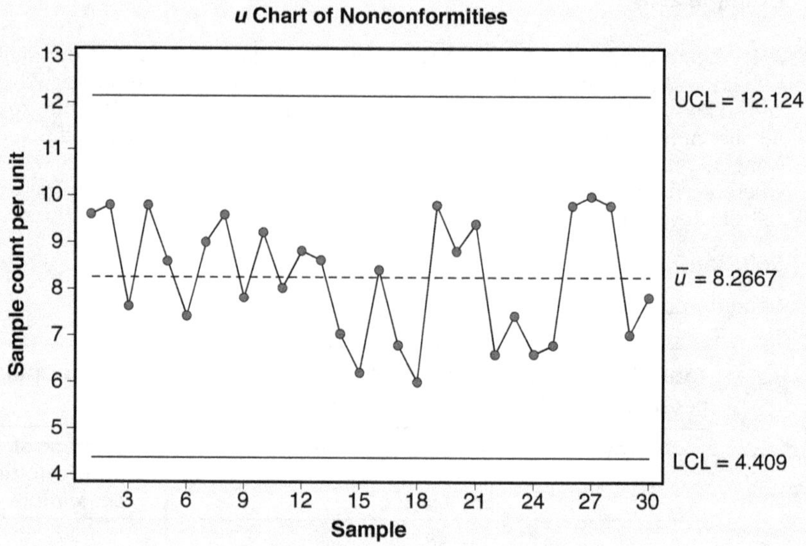

Figure 23.8 *u* chart of nonconformities for the data in Table 23.6, constructed using Minitab.

To illustrate the construction of the *u* chart when the sample sizes vary, we consider Example 23.9.

Example 23.9

Suppose in Example 23.8, due to some administrative reasons, it was not possible to examine 5 inspection units every day. In other words, the sample size varied and the data obtained are shown in Table 23.7. Construct and interpret a *u* chart for the data in Table 23.7.

Solution

Using the data in Table 23.7, we have

$$\bar{u} = \frac{33+40+38+\cdots+40}{3+5+3+\cdots+3} = \frac{1162}{126} = 9.22$$

(continued)

Table 23.7 Number of nonconformities on printed boards for laptops per sample, with varying sample size.

Day	Sample size	Number of nonconformities per sample	Day	Sample size	Number of nonconformities per sample
1	3	33	16	5	40
2	5	40	17	4	40
3	3	38	18	5	37
4	5	43	19	4	40
5	5	45	20	5	37
6	5	35	21	4	39
7	5	35	22	5	48
8	3	41	23	5	39
9	3	40	24	5	36
10	5	30	25	3	39
11	5	36	26	3	36
12	5	40	27	5	43
13	3	33	28	3	36
14	4	36	29	4	49
15	4	38	30	3	40

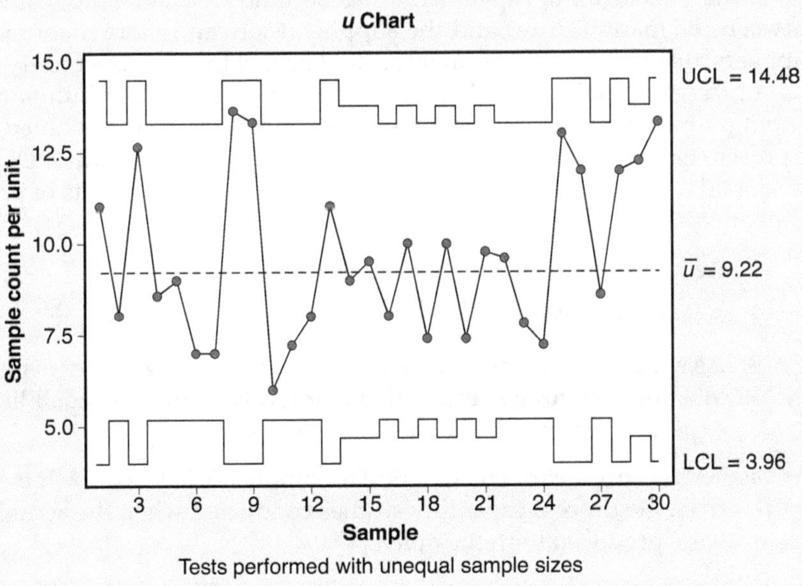

Figure 23.9 *u* chart of nonconformities for the data in Table 23.7, constructed using Minitab.

The control limits are calculated for each individual sample. Thus, for example, the control limits for sample 1 are given by:

$$UCL = \bar{u} + 3\sqrt{\bar{u}/n_1} = 9.22 + 3\sqrt{9.22/3} = 14.48$$
$$CL = \bar{u} = 9.22$$
$$UCL = \bar{u} - 3\sqrt{\bar{u}/n_1} = 9.22 - 3\sqrt{9.22/3} = 3.96$$

The control limits for the rest of the samples are calculated in the same manner. The centerline, however, remains the same. Figure 23.9 shows the u chart for the data in Table 23.7.

The *u* chart in Figure 23.9 shows that all the points are within the control limits. However, several points fall beyond the warning control limits. Moreover, points 8 and 9, in particular, fall very close to the upper control limits. All these observations about the process indicate that there may be some special causes present in the process, that is, the process may be on the verge of being unstable. Therefore, precautions should be taken to avoid the process becoming unstable, but without overcontrolling it.

PROCESS CAPABILITY ANALYSIS

Earlier in this chapter we defined what a process is. Once the process is defined and brought under statistical control, the manufacturer's next concern is to deliver product that the consumer wants. In other words, the manufacturer must determine whether the process is capable of delivering the desired product. One way to address such a concern is to quantify the capability. A *process capability index (PCI)* is a unitless measure that quantifies the process capability, specifically the relationship between the output and the designed tolerances. This measure has become an important tool in process capability analysis. Moreover, since the PCI is a unitless measure of capability, it has become an easy tool of communication between the manufacturer and the supplier. More and more manufacturers and suppliers are using the PCI as an important part of their contract to ensure quality.

A *process capability analysis* is the comparison of the distribution of a process output with the product tolerances. As Kotz and Lovelace (1998) noted, the results of process capability analysis have proved valuable in many ways. Deleryd (1996) developed a list of the 13 most common ways of using the results of process capability analysis:

1. As a basis in the improvement process.
2. As an alarm clock.
3. As specifications for investments. The purchasing process is facilitated by assigning specifications for levels of process capability indices expected to be reached by new machines.
4. As a certificate for customers. The supplier is able to attach the results from the process capability studies conducted when the actual products were produced with the delivery.
5. As a basis for new constructions. The capability of the production processes guides the designer on how to set reasonable specifications to make the product manufacturable.

6. For control of maintenance efforts. By continuously conducting process capability studies, it is possible to see if some machines are gradually deteriorating.
7. As specifications for introducing new products.
8. For assessing reasonableness of customer demands.
9. For motivation of coworkers.
10. For deciding priorities in the improvement process.
11. As a base for inspection activities.
12. As a receipt for improvement.
13. For formulating quality improvement programs.

To implement a process capability analysis, one needs to consider the following:

1. The target value specification, which is usually defined by the customer.
2. The specification limits, which should be defined by the customer, or by his or her technical staff, and should be agreed upon by the manufacturer. Further, the specification limits should be such that they allow manufacturing variability without jeopardizing proper function of the product.
3. An analysis of the process that allows the manufacturer to determine whether the product produced meets the customer's specifications.

Once production starts, the manufacturer conducts capability studies to compare the measures of the quality characteristic of the manufactured product with the specification limits. This is the point where PCIs are used.

The first-generation PCIs were established by the Japanese in the 1970s. They used the following indices:

C_p	Inherent capability of a process
k	Position of the process in relation to the target value
C_{pk}	Position of the 6σ process in relation to the target value
C_{pl}	Position of the 6σ process in relation to the lower specification limit
C_{pu}	Position of the 6σ process in relation to the upper specification limit

In this section, we will study these and some other capability indices that are frequently used in the process capability analysis. We will not address every aspect of these indices; however, an excellent reference for a more detailed study of these indices is Kotz and Lovelace (1998).

Throughout the study of these indices, we will assume that the process producing the desired quality characteristic is under control, thus predictable.

Process Capability Index: C_p

Let X be the process quality characteristic that we want to monitor. Let USL and LSL be the *upper specification limit* and the *lower specification limit*, respectively. The performance of the process with respect to these limits is defined as follows:

$$\text{Percentage of nonconforming produced by the process at the upper end} = P(X > \text{USL})$$
$$\text{Percentage of nonconforming produced by the process at the lower end} = P(X < \text{LSL})$$

Thus, the total percentage of nonconforming produced by the process is defined as:

$$P(X < \text{LSL or } X > \text{USL}) = 1 - P(\text{LSL} < X < \text{USL})$$

In the previous paragraph, we saw the performance of the process with respect to the specification limits. Now we look at the performance of the process with respect to the *natural tolerance limits*, that is, the upper natural tolerance limit (UNTL) and the lower natural tolerance limit (LNTL).

The performance of the process with respect to the natural tolerance limits is the percentage of the product produced by the process with its quality characteristic falling within the interval $(\mu-3\sigma, \mu+3\sigma)$, where μ and σ are the mean and the standard deviation of the process quality characteristic, respectively. Assuming that the process quality characteristic is normally distributed and that the process is under control, the percentage of the product produced by the process with its quality characteristic falling within the interval $(\mu-3\sigma, \mu+3\sigma)$ is approximately 99.73%. As noted earlier, a PCI is the comparison between what a process is expected to produce and what it is actually producing. Thus, we now define the PCI C_p, one of the first five indices used by the Japanese and proposed by Juran, Gryna, and Bingham (1974), as follows:

$$\begin{aligned} C_p &= \frac{\text{USL} - \text{LSL}}{\text{UNTL} - \text{LNTL}} \\ &= \frac{\text{USL} - \text{LSL}}{(\mu+3\sigma)-(\mu-3\sigma)} \\ &= \frac{\text{USL} - \text{LSL}}{6\sigma} \end{aligned} \tag{23.55}$$

Note that the numerator in equation (23.55) is the desired range of the process quality characteristic, whereas the denominator is the actual range of the process quality characteristic. From this, we can see that a process can produce product of desired quality and that it is capable only if the range in the numerator is at least as large as that in the denominator. In other words, the process is capable only if $C_p \geq 1$, and larger values of C_p are indicative of a process being more capable. For a 6σ process, $C_p = 2$. A predictable process, which is normally distributed with $C_p = 1$ and its mean located at the center of the specification limits, usually known as the *target value* of the process characteristic, is expected to produce 0.27% nonconforming units. Montgomery (2005) has given a comprehensive list of values of the PCI C_p and associated process nonconforming for one-sided specifications and two-sided specifications.

Since C_p is very easy to calculate, it is used widely in industry. However, the fact that it does not take into consideration the position of the process mean is a major drawback. A process could be incapable even if the value of C_p is large (>1). That is, a process could produce 100% defectives, for example, if the process mean falls outside the specification limits and is far from the target value. Further, note that the value of C_p will become even larger if the value of the process standard deviation σ decreases, while the process mean moves away from the target value.

Note that the numerator in equation (23.55) is always known, but the denominator is usually unknown. This is because in almost all practical applications, the process standard deviation σ is unknown. Thus, to calculate C_p we must replace σ in equation (23.55) with its estimator $\hat{\sigma}$. Now we know that σ can be estimated either by the sample standard deviation s or by $\bar{R}/d_2$. However, remember that the estimator $\bar{R}/d_2$ is used only when the process is under control and the sample size is fewer than 10. Thus, an estimated value of C_p is given by:

$$\hat{C}_p = \frac{USL - LSL}{6\hat{\sigma}} \tag{23.56}$$

We illustrate the computation of $\hat{C}_p$ with Example 23.10.

Example 23.10

Table 23.8 gives the summary statistics on $\bar{X}$ and R for 25 samples of size $n=5$ collected from a process producing tie rods for a certain type of cars. The measurement data are the lengths of tie rods, and the measurement scale is in millimeters.

Table 23.8 Data showing the lengths of tie rods for cars.

Sample number	$\bar{X}$	R	Sample number	$\bar{X}$	R
1	274	6	14	268	8
2	265	8	15	271	6
3	269	6	16	275	5
4	273	5	17	274	7
5	270	8	18	272	4
6	275	7	19	270	6
7	271	5	20	274	7
8	275	4	21	273	5
9	272	6	22	270	4
10	273	8	23	274	6
11	269	6	24	273	5
12	273	5	25	273	6
13	274	7			

The target value and the specification limits for the length of the rods are 272 and 272±8, respectively. Calculate the value of $\hat{C}_p$ assuming that the tie rod lengths are normally distributed, and find the percentage of nonconforming tie rods produced by the process.

Solution

To find the value of $\hat{C}_p$ and the percentage of nonconforming tie rods produced by the process, we first need to estimate the process mean μ and the process standard deviation σ. These estimates may be found by using $\overline{X}$ and $\overline{R}/d_2$, respectively. Thus, we get:

$$\hat{\mu} = \overline{\overline{X}} = \frac{1}{m}\sum_{i=1}^{m}\overline{X}_i = \frac{1}{25}(274+265+269+\cdots+273) = 272$$

$$\hat{\sigma} = \frac{\overline{R}}{d_2} = \frac{\frac{1}{m}\sum_{i=1}^{m}R_i}{d_2} = \frac{\frac{1}{25}(6+8+\cdots+5+6)}{2.326} = \frac{6}{2.326} = 2.58$$

where the value of d_2 for different sample sizes is found in Appendix E. Substituting the values of USL, LSL, and $\hat{\sigma}$ in equation (23.56), we get:

$$\hat{C}_p = \frac{280-284}{6(2.58)} = 1.03$$

which indicates that the process is capable. To find the percentage of nonconforming tie rods produced by the process, we proceed as follows:

$$\begin{aligned}\text{Percentage of nonconforming} &= (1 - P(264 \le X \le 280))\ 100\% \\ &= (1 - P(-3.1 \le Z \le 3.1))\ 100\% \\ &= (1-0.9980)\ 100\% = 0.2\%\end{aligned}$$

Thus, in this example, the percentage of nonconforming tie rods is as expected when we consider the value of $\hat{C}_p$. But, as noted earlier, this may not always be the case since C_p does not take into consideration where the process mean is located. To better explain this, we use Example 23.11.

Example 23.11

Suppose that the process in Example 23.10 had a setback, and as a result, the process had an upward shift. Further, suppose that after the process experienced this shift, we took another set of 25 random samples of size $n=5$ and these samples produced $\overline{\overline{X}} = 276$ and $\overline{R} = 6$. Clearly, in this example the value of $\overline{\overline{X}}$ changed from 272 to 276, while $\overline{R}$ remained the same.

Solution

From the given values of $\overline{\overline{X}}$ and $\overline{R}$, we obtain

$$\hat{\mu} = 276 \text{ and } \hat{\sigma} = 2.58$$

Since the process standard deviation did not change, the value of $\hat{C}_p$ remained the same, that is, $\hat{C}_p = 1.03$. However, the percentage of nonconforming tie rods produced by the process will be:

$$\begin{aligned}\text{Percentage of nonconforming} &= (1 - P(264 \le X \le 280))100\% \\ &= \left(1 - P\left(\frac{264-276}{2.58} \le \frac{X-276}{2.58} \le \frac{280-276}{2.58}\right)\right)100\% \\ &= (1 - P(-4.65 \le Z \le 1.55))100\% = 6.06\%\end{aligned}$$

Thus, even though the value of C_p did not change after the process mean experienced a shift, the process is producing nonconforming units at a rate that is 30 times higher than in the previous example. This implies that C_p did not measure the effect that the upward shift had on the ability of the process to produce products within the specification limits. This major drawback of C_p makes it less reliable than many other PCIs available in the literature. We will study some of them here. However, before we study other PCIs, let us see another alternative but equivalent interpretation of C_p, which is given by finding the percentage of specification band used, that is

$$\text{Percentage of specification band used} = \frac{1}{C_p} \times 100$$

A smaller percentage of specification band used indicates a better process. Again, for reasons discussed earlier, this interpretation sometimes can also be misleading.

The other two PCIs first used by the Japanese are C_{pl} and C_{pu}. These indices are related to the lower specification limit and the upper specification limit, respectively, and are defined as follows:

$$C_{pl} = \frac{\mu - \text{LSL}}{3\sigma} \tag{23.57}$$

$$C_{pu} = \frac{\text{USL} - \mu}{3\sigma} \tag{23.58}$$

The estimates of C_{pl} and C_{pu} are given by:

$$\hat{C}_{pl} = \frac{\overline{\overline{X}} - \text{LSL}}{3\hat{\sigma}} \tag{23.59}$$

$$\hat{C}_{pu} = \frac{\text{USL} - \overline{\overline{X}}}{3\hat{\sigma}} \tag{23.60}$$

To illustrate the computation of $\hat{C}_{pl}$ and $\hat{C}_{pu}$ we use the information in Examples 23.12 and 23.13.

Example 23.12

Using the information given in Example 23.10, compute $\hat{C}_{pl}$ and $\hat{c}_{pu}$.

Solution

$$\hat{C}_{pl} = \frac{\overline{X} - \text{LSL}}{3\hat{\sigma}} = \frac{272 - 264}{3(2.58)} = 1.03$$

$$\hat{C}_{pu} = \frac{\text{USL} - \overline{X}}{3\hat{\sigma}} = \frac{280 - 272}{3(2.58)} = 1.03$$

Note that both $\hat{C}_{pl}$ and $\hat{C}_{pu}$ are equal to $\hat{C}_p$, which will always be the case when the process mean is centered between the specification limits. Moreover, when both $\hat{C}_{pl}$ and $\hat{C}_{pu}$ are equal, the percentage of nonconforming units below the lower specification limit is the same as the percentage of nonconforming units above the upper specification limit.

Example 23.13

Using the information given in Example 23.11, compute $\hat{C}_{pl}$ and $\hat{C}_{pu}$.

Solution

$$\hat{C}_{pl} = \frac{\bar{X} - \text{LSL}}{3\hat{\sigma}} = \frac{276 - 264}{3(2.58)} = 1.55$$

$$\hat{C}_{pu} = \frac{\text{USL} - \bar{X}}{3\hat{\sigma}} = \frac{280 - 276}{3(2.58)} = 0.52$$

In this case, the value of $\hat{C}_{pl}$ is much larger than 1, whereas the value of $\hat{C}_{pu}$ is much smaller than 1, which indicates that most of the nonconforming units produced by the process are falling above the upper specification limit. Finally, note that both $\hat{C}_{pl}$ and $\hat{C}_{pu}$ are sensitive to where the process mean is located.

Process Capability Index: C_{pk}

To overcome the centering problem in C_p discussed in the preceding section, PCI C_{pk} was introduced. The PCI C_{pk}, which is again one of the first five used in Japan, is defined as:

$$C_{pk} = \min\,(C_{pl}, C_{pu}) \tag{23.61}$$

The index C_{pk} is related to the index C_p as:

$$C_{pk} = (1 - k)C_p \tag{23.62}$$

where

$$k = \frac{|(\text{USL} + \text{LSL}/2 - \mu)|}{(\text{USL} - \text{LSL})/2} \tag{23.63}$$

Further, it can easily be seen that $0 \le k \le 1$, so that C_{pk} is always less than or equal to C_p. Also, note that when the process mean μ coincides with the midpoint between the specification limits, $k = 0$, therefore C_{pk} equals C_p.

Note that C_{pk} takes care of the centering problem only if the process standard deviation remains the same. If it changes, the value of C_{pk} may not change even when the process mean moves away from the center. It can be seen that this will always be the scenario, provided the distance of the process mean from the nearest specification limit in terms of σ remains the same. For example, in Table 23.9 we consider four processes with the same specification limits but with different process means (μ) and different standard deviations (σ) such that the value of C_{pk} in each case remains the same.

The value of C_{pk} for each of the four processes remains the same though the process mean has been moving away from the center, because in each case the distance between the process mean and the nearest specification limit (in this example, the LSL) is three times the process standard deviation. Thus, we can say that in some ways, C_{pk} is also not an adequate measure of centering. Assuming the process characteristic is normally distributed, Table 23.10 gives the parts per million (PPM) of nonconforming units for different values of C_{pk}.

Table 23.9 Different processes with same value of C_{pk}.

Process	LSL	USL	Center	μ	σ	C_{pk}
1	12	36	24	24	4	1.00
2	12	36	24	22	3.33	1.00
3	12	36	24	20	2.67	1.00
4	12	36	24	18	2.00	1.00

Table 23.10 Parts per million of nonconforming units for different values of C_{pk}.

C_{pk}	1.00	1.33	1.67	2.00
PPM	**1350**	**30**	**1**	**0.001**

From Table 23.10, we can see that each of the processes in Table 23.9 will produce 1350 nonconforming PPM. This is only possible if the process standard deviation is shrinking while the process mean is shifting and the natural tolerance limits remain within the specification limits. In fact, in Table 23.9, the process with standard deviation $\sigma=2$ would be of Six Sigma quality if the process mean were at the center of the specification target. In this case, however, the process is producing 1350 nonconforming PPM because the process mean is off-center by 3σ, a shift larger than 1.5σ, which would be tolerable only in a Six Sigma quality process. For discussion on some other indices, see Gupta and Walker (2007b).

Process Capability Indices: P_p and P_{pk}

The capability indices we have studied so far have one characteristic in common: each is used when the process is stable. The American Automotive Industry Group (AAIG), consisting representatives from Ford Motor Company, General Motors Corporation, American Daimler/Chrysler Corporation, and the American Society for Quality, was founded in 1991. This group standardized the supplier quality reporting procedure for the industry, and advocated the use of two sets of PCIs—one consisting C_p and C_{pk}, and the other consisting P_p and P_{pk}. Further, it advised using C_p and C_{pk} when the process is stable and P_p and P_{pk} when the process is not stable, where the two sets of indices P_p, P_{pk} and C_p, C_{pk} are defined in exactly the same manner and only differ in their estimates. Thus, we have:

$$\hat{P}_p = \frac{USL - LSL}{6\hat{\sigma}_s} \tag{23.64}$$

$$\hat{C}_p = \frac{USL - LSL}{6\hat{\sigma}_{R/d_2}} \tag{23.65}$$

and

$$\hat{P}_{pk} = \min\left(\frac{\text{USL} - \overline{\overline{\text{X}}}}{3\hat{\sigma}_s}, \frac{\overline{\overline{\text{X}}} - \text{LSL}}{3\hat{\sigma}_s}\right) \tag{23.66}$$

$$\hat{C}_{pk} = \min\left(\frac{\text{USL} - \overline{\overline{\text{X}}}}{3\hat{\sigma}_{\bar{R}/d_2}}, \frac{\overline{\overline{\text{X}}} - \text{LSL}}{3\hat{\sigma}_{\bar{R}/d_2}}\right) \tag{23.67}$$

In other words, they differ only in the method in which one estimates the standard deviation. The estimate $\hat{\sigma}$ is the usual standard deviation:

$$s = \sqrt{\sum_{i=1}^{n}\sum_{j=1}^{m}\left(x_{ij} - \overline{\overline{x}}\right)^2 / (mn - 1)}$$

whereas $\hat{\sigma}_{\bar{R}/d_2} = \bar{R} / d_2$ is an estimate obtained using the subgroup ranges R_i, $i = 1, 2, \ldots, m$ and the corresponding value of d_2. Note that the estimate $\hat{\sigma}_s$ measures the process variability by using the variability within and between the subgroups, whereas $\hat{\sigma}_{\bar{R}/d_2}$ uses only the variability within the subgroups. When the process is stable, there is little variability between the subgroups, and the two sets of indices are essentially the same because the estimates $\hat{\sigma}_s$ and $\hat{\sigma}_{\bar{R}/d_2}$ are approximately equal. But if the process is not stable, there is potentially large variability between the subgroups. This would mean that $\hat{\sigma}_{\bar{R}/d_2}$ underestimates the process standard deviation, and, consequently, C_{pk} would overestimate the process capability. Fortunately, this situation will not arise, because the use of C_{pk} is recommended only when the process is stable.

Kotz and Lovelace (1998, 253) strongly argued against the use of P_p and P_{pk}. They recommend against using these indices when the process is not in statistical control. Under these conditions, the P-numbers are meaningless with regard to process capability, have no tractable statistical properties, and infer nothing about long-term capability of the process. Worse still, they provide no motivation to the user-companies to get their process in control. The P-numbers are a step backward in the efforts to properly quantify process capability, and a step toward statistical terrorism in its undiluted form.

Montgomery (2005, 349) agrees with Kotz and Lovelace. He writes, "The process performance indices P_p and P_{pk} are more than a step backwards. They are a waste of engineering and management effort—they tell you nothing." The authors wholeheartedly agree with Kotz and Lovelace and Montgomery. No one can judge a process when its future behavior is so unpredictable.

Chapter 24

Problem-Solving Tools and Continuous Improvement Techniques

> Define and describe the following tools and techniques in various situations.
>
> 1. Plan-do-check-act (PDCA) or plan-do-study-act (PDSA) cycle (Understand)
> 2. Lean tools for eliminating waste (e.g., error-proofing and value-stream mapping) and lean concepts (e.g., kaizen, flow, and pull) (Understand)
> 3. 5S/6S (Apply)
> 4. Six sigma phases: define, measure, analyze, improve, control (DMAIC) (Understand)
> 5. Failure mode and effects analysis (FMEA) (Understand)
> 6. 8D Methodology (Understand)
> 7. 5 Why's (Understand)
> 8. Fault tree analysis (Understand)
> 9. Corrective and preventive action (CAPA) (Understand)
>
> Body of Knowledge IV.E

Numerous tools and techniques are available to support problem solving and continuous improvement. These tools and techniques range from the most basic, such as the Plan-Do-Check-Act (PDCA) cycle, to the advanced, such as Six Sigma. It should be noted that a careful review of the current Certified Quality Inspector Body of Knowledge (CQI BoK) is required to assess the level of knowledge required for the CQI to master relative to Bloom's Taxonomy. For example, the CQI will be expected to know more about the PDCA cycle than about Six Sigma and failure mode and effects analysis (FMEA).

PLAN-DO-CHECK-ACT (PDCA) OR PLAN-DO-STUDY-ACT (PDSA) CYCLES

The Plan-Do-Check-Act (PDCA) and the Plan-Do-Study-Act (PDSA) cycles are common in use and in the language of the quality profession. Accordingly, we begin our discussion of these two cycles with a clarification: These two cycles were *not* created by the same person, are *not* structurally the same, and have different philosophical intents. As McKinley (2006) has documented, the original creators of the cycles are as follows:

- Walter Shewhart—Shewhart Cycle—PDCA—development date late 1930s. It should be noted that though Shewhart is credited with the "Shewhart Cycle," he did not label the steps of his cycle as PDCA, but used the terms "specification," "production," and "inspection."
- W. Edwards Deming—Deming Cycle—PDSA—development date late 1950s

Structurally, these cycles differ on the "Check" step advocated by Shewhart as compared to the "Study" step advocated by Deming. The primary emphasis of the Check step is to verify that the desired results have been obtained. The primary emphasis of the Study step is to gain quantitative knowledge of the process and its performance to learn how it may be further improved. This difference in emphasis between the Check step and the Study step explains the different philosophical intents of each cycle. It should also be noted that the points of difference explained here have been commonly confused, and PDCA and PDSA continue to be used interchangeably. Not only is this practice misleading and incorrect, but it obscures the points of origin and original intent of the two cycles.

Plan

Once a problem has been clearly defined, the first steps in solving it are to collect and analyze data, consider and analyze alternative solutions, and choose the best solution. These steps, although easy to state, can be extremely difficult and time-consuming to execute. Jointly, these steps constitute the Plan phase in the PDCA cycle. One approach to this phase is to list the goals, the barriers to reaching those goals, and a strategy for coping with the barriers. This approach provides guidance for the next steps.

In most situations, a team representing everyone affected by the problem and its solution should be formed and assigned the problem-solving task.

There is a tendency to jump to the Do phase of the cycle rather than taking the time to adequately execute the Plan phase. Before moving to the Do phase, thorough plans should be made regarding the collection of information during that phase.

In some situations it may be useful to apply a quick-and-dirty or Band-Aid solution to allow time to focus on a permanent solution. Of course, this approach creates the tendency to move on to the next problem because this one is "solved."

Do

Once a solution to the problem has been decided on, and a data collection scheme has been determined, the next phase is to try it. If possible, this should be done on a small scale and/or offline. Sometimes the proposed solution can be tried in a laboratory setting or outside the regular production process. During the Do phase, as much data as possible should be collected. In some situations, videotaping a process permits further data collection upon replay.

Check

The Check phase is used to analyze the information collected during the Do phase. The data must be validated using valid mathematical and statistical techniques.

Act

In the Act phase, action is taken based on the conclusions reached in the Check phase. If the data show that the proposed solution is a good solution for the problem, the Act phase consists of integrating the solution into the standard way of doing things. If the data show that another proposed solution is needed, the Act phase consists of initiating the search for another solution.

As the word "cycle" implies, the Act phase is followed by the Plan phase because quality improvement is a continuous journey, as shown in Figure 24.1.

The PDCA cycle is intended to be iterative; that is, the cycle is repeated until the desired improvement results have been achieved.

Gordon (2002) observes that many organizations are good at the Do and Check phases but fall short at the Plan and Act phases. Perhaps this is partly due to the "Don't just stand there, do something" impulse. There is a tendency, for instance, to provide a service, process, or product to customers without adequate care in the design (Plan) phase. The strategy is that customers will provide feedback, and much will be learned from the design mistakes. Automotive industries have attempted to combat this with such programs as advanced product quality

Figure 24.1 The Plan-Do-Check-Act cycle.

planning (APQP), potential failure mode and effects analysis (PFMEA), and production part approval process (PPAP).

LEAN TOOLS FOR ELIMINATING WASTE

Lean production refers to a set of tools and techniques intended to improve efficiency and reduce or eliminate waste in manufacturing and service delivery. Lean production has often been characterized as "doing more with less." This concept of doing more with less is related to the fact that there is generally substantial waste in manufacturing and service delivery operations. As the concept of lean came from Japan in the late 1980s and early 1990s, the Japanese refer to this waste as "muda." Taiichi Ohno originally identified seven types of waste:

1. Overproduction
2. Waiting
3. Unnecessary transportation or materials or products
4. Overprocessing
5. Inventory
6. Unnecessary employee movement
7. Production of defective parts

Ohno later identified an eighth waste, the underutilization of employees.

There are many tools that are considered applicable in lean operations; however, the most common lean tools are as follows.

Error-Proofing

Error-proofing is commonly known as "fail-safing" and is derived from the Japanese term "poka-yoke." *Poka-yoke* refers to efforts intended to eliminate human-based errors or mistakes before they occur. Poka-yoke concepts can be applied to the design, assembly or delivery, material selection, and handling and movement of products or service delivery.

ASQ provides guidance on when and how to implement error-proofing.

When to Use Error-Proofing

1. When a process step has been identified where human error can cause mistakes or defects to occur, especially in processes that rely on the worker's attention, skill, or experience.
2. In a service process, where the customer can make an error which affects the output.
3. At a hand-off step in a process, when output or (for service processes) the customer is transferred to another worker.
4. When a minor error early in the process causes major problems later in the process.
5. When the consequences of an error are expensive or dangerous.

Error-Proofing Procedure

1. Obtain or create a flowchart of the process. Review each step, thinking about where and when human errors are likely to occur.
2. For each potential error, work back through the process to find its source.
3. For each error, think of potential ways to make it impossible for the error to occur. Consider:
   - Elimination—eliminating the step that causes the error.
   - Replacement—replacing the step with an error-proof one.
   - Facilitation—making the correct action far easier than the error.
4. If you cannot make it impossible for the error to occur, think of ways to detect the error and minimize its effects. Consider inspection method, setting function, and regulatory function.
5. Choose the best error-proofing method or device for each error. Test it, then implement it. Three kinds of inspection methods provide rapid feedback:
   - Successive inspection is done at the next step of the process by the next worker.
   - Self-inspection means workers check their own work immediately after doing it.
   - Source inspection checks, before the process step takes place, that conditions are correct. Often it is automatic and keeps the process from proceeding until conditions are right.

Value Stream Mapping

Value stream mapping combines detailed flowcharting with key information to produce a thorough picture of manufacturing or service delivery operations. A key function of the value map is eliminating waste by identifying and categorizing operations that add value and those that do not add value (i.e., waste or muda). Operations that add value are those that change the form, fit, or function of a product or service; customers are thus willing to pay for operations that add value. Operations that do not change the form, fit, or function of a product or service do not add value for the customer; customers are thus not willing to pay for non-value-added operations. Non-value-added operations must be eliminated.

The process of developing a value stream map is much like that of developing a flowchart. While there are many approaches to value stream mapping, one approach is as follows:

1. Identify a product or service delivery operation for mapping
2. Perform a first-pass walk-through or inspection of the operation(s) to gain initial familiarity
3. Construct a map or diagram of the operation(s) in its current state using standard value-mapping symbology

4. Study the current-state map to identify areas of improvement, waste, or non-value-added activity
5. Validate the areas of improvement, waste, or non-value-added activity
6. Construct a map or diagram of the operation(s) in a desired future state using standardized value-mapping symbology
7. Gain consensus and/or approval on the desired future state of the operation(s)
8. Implement the changes needed for the desired future state of the operation(s)
9. Validate that the changes implemented for the desired future state of the operation(s) were effective and did not create new difficulties or waste
10. Monitor the desired state of the operation(s) to ensure the improvements are sustained

Lean Concepts

Kaizen

ASQ defines *kaizen* as "a Japanese term that means gradual unending improvement by doing little things better and setting and achieving increasingly higher standards."

In operational terms, kaizen generally refers to major projects or events intended to dramatically improve processes. Kaizen events generally involve teams of people who, working from current- and future-state maps (mentioned in the material on value stream mapping), reconfigure workspaces to improve productivity and eliminate waste.

Push

Push is a concept that refers to the production of products before they are needed or are ordered by a customer. Products produced under a push system are placed in an inventory with the assumption that customers will buy the product in a reasonable period of time as it is configured—this is not always a reasonable assumption and can lead to a great deal of waste. The waste associated with push production is unnecessary material storage space, risk of damage or theft, and the inability to recover in a timely manner any compensation for the value added to the product(s).

Pull

Pull is a concept that refers to the production of products only when they are needed or are ordered by a customer. Products produced with a pull system are generally not placed in an inventory, which means the customer may define his or her own specifications or configurations. The pull system eliminates a great deal of waste.

Flow

ASQ defines *flow* as "the progressive achievement of tasks along the value stream so a product proceeds from design to launch, order to delivery, and raw to finished materials in the hands of the customer with no stoppages, scrap, or backflows."

5S/6S

5S is an acronym and tool that focuses on cleanliness and order in the workplace. The concept of 5S is that if all elements in the workplace, such as tools, equipment, supplies, work surfaces, aisles, storage areas, and so on, are clean and organized, work will flow more rapidly through the workspace and production or service delivery will be faster with less waste, thus more efficient. The 5S's are sort, set in order, shine, standardize, and sustain.

Sort

Sorting refers to making sure that, as much as possible, only the things needed to perform specific and necessary work tasks—tools, equipment, supplies, instructions, paperwork, references, and so on—are in the area where the work is to be completed. By ensuring only those things needed for the work are in the workspace, clutter is removed. Clutter in the workplace slows production, adds to confusion and the potential for mistakes, and generally requires the unnecessary movement of material.

Set in Order

Setting in order refers to organization in the workplace that facilitates production. Setting in order applies to the physical arrangement of tools and equipment such that they are easily identifiable, located in predefined and easily accessible positions, and grouped in a way that improves the efficiency of their use when needed. Setting in order also refers to supplies wherein parts, pieces, or other items such as forms needed in service delivery operations are readily positioned for access and use with minimal travel or movement within the workspace.

Shine

Shine refers to keeping the workspace clean. A clean workspace improves productivity in at least three ways: (1) it ensures that when it is time to work, the work can begin promptly, (2) it contributes to safety by mitigating the potential for injuries, and (3) it improves employee morale.

Standardize

Standardize refers to making sure the basic principles of 5S are used consistently throughout the workplace. In many companies, standardize means preparing and implementing policies, procedures, and instructions intended to ensure all areas and work spaces use the same approaches to sorting, setting in order, and shining so that workspaces have certain elements of commonality and familiarity for all employees.

Sustain

Sustain refers to making sure each of the 5S elements remains in the workplace after initial implementation. Sustain is important because without a commitment to be diligent in implementing 5S, the workplace would likely revert to the way things were done prior to the 5S implementation. Companies use a variety of other lean tools to help ensure 5S is sustained. Many of these other lean tools focus on keeping key information and data visible and prominently displayed for employees in what are referred to as visual controls.

6S is an acronym that adds safety to the 5S model. Safety in the workplace has become increasingly important and is becoming embedded in several quality tools and techniques. Relative to quality tools and techniques, an emphasis on safety becomes evident when process controls and procedures are put in place to keep employees and professionals safe. Many safety-related program, such as SafeStart, seek to develop in employees and professional safe working practices that are applicable both on and off the job.

SIX SIGMA PHASES

Quality is, in part, defined by consistency. By extension, consistency means an absence of variability. The CQI should take this to mean that the approach and techniques used to solve problems are essentially processes, wherein as a process, problem solving is no different from production or service delivery operations. To ensure consistent results in problem-solving efforts, a structured approach is required. Countless problem-solving approaches have been developed and reported in the press over many years—some specifically related to quality, some not. While it is not the intent of the authors to advocate for or against any particular problem-solving approach, it should be emphasized that for reasons of consistency and thoroughness, some standardized approach to problem solving should be adopted and used consistently to guide problem-solving efforts. Two problem-solving approaches are of interest to the CQI:

- The scientific method (a method not specifically related to quality)
- Six Sigma (a method specifically related to quality)

The Scientific Method

Perhaps the most noted and recognized approach to structured problem solving is the scientific method. The *scientific method* is used in many fields of inquiry to guide problem-solving efforts in accordance with standardized steps, which include:

- Asking a question
- Collecting data
- Posing hypotheses
- Testing hypotheses
- Analyzing data
- Drawing conclusions
- Communicating results

It is true that the scientific method can be used to guide problem-solving efforts in quality, and if used properly, it could be quite effective. Many of the tools and techniques used in continuous improvement can readily be applied to support and enhance the scientific method. What differentiates the scientific method from other problem-solving methods used specifically to support continuous improvement is an absence of common language, tools, and techniques most commonly used in the quality community.

Six Sigma

Perhaps the most common problem-solving approach used within the quality community today is Six Sigma. *Six Sigma* is a five-step problem-solving approach with the steps as follows:

- Define
- Measure
- Analyze
- Improve
- Control

The five steps of Six Sigma are most commonly referred to as the DMAIC methodology. Upon further analysis, we can see that each of the outcomes of the scientific method is found within the DMAIC methodology. Two primary points differentiate the scientific method from Six Sigma: (1) Six Sigma's use of quality-specific language and tools, and (2) Six Sigma's use of the DMAIC methodology in quality-related problem solving when certain financial constraints or expectations have been met. The financial constraints or expectations are related to the amount of cost savings expected as a result of problem-solving activities to justify use of the comprehensive and systematic DMAIC methodology. It is beyond the scope of this book, and the CQI BoK, to provide further content on Six Sigma and the DMAIC methodology. It should be emphasized, however, that Six Sigma and the DMAIC methodology are, and will remain, a critically important problem-solving approach within the quality community. The CQI is strongly encouraged to continue his or her professional development by learning more about this topic.

The goal of all approaches to problem solving is to identify the reason(s) why a problem occurred. The importance of identifying the true underlying reason(s) a problem occurred is known as root cause analysis.

ROOT CAUSE ANALYSIS

Root cause analysis (RCA) is an approach to problem solving that is based on the identification of the true underlying reason(s) why a problem occurred. RCA is critical because any corrective action taken on a non-root-cause issue or item will result in the problem remaining active within a selected process. In other words, following some misapplied corrective action on a non-root-cause issue or item may change the appearance or manifestation of the original problem, but the original problem still exists.

RCA is similar to the scientific method and Six Sigma in many ways: RCA makes use of many of the quality-related tools and techniques discussed thus far in this book; it is intended to be iteratively applied, wherein by iteration we contribute to continuous improvement; and it has a structured approach, which includes the following:

- Define the problem(s)
- Identify manifestations of the problem(s)

- Collect data
- Identify root causes
- Propose solutions
- Test solutions
- Verify effectiveness of solutions

FAILURE MODE AND EFFECTS ANALYSIS (FMEA)

Failure mode and effects analysis (FMEA) is a team-based problem-solving tool intended to help users identify and eliminate, or reduce the negative effects of, potential failures before they occur in systems, subsystems, product or process design, or the delivery of a service. FMEA can be used as a stand-alone tool or as part of comprehensive quality programs such as ISO/QS 9000, APQP, or Six Sigma. Accordingly, our goal is to familiarize readers with terminology, theory, mechanics, and applications of FMEA consistent with the CQI BoK requirements and as it applies to product designs, process designs, and systems.

Selecting a Standard for FMEA

There are two primary standards for FMEA, the military standard (Mil-Std 1629A) and the Society of Automotive Engineers standard (SAE J1739). Both standards are limited in scope to address only design and process FMEAs. These standards provide general FMEA forms and documents, identify criteria for the quantification of risk associated with potential failures, and provide general guidelines on the mechanics of completing FMEAs. Mil-Std 1629A and SAE J1739 may be obtained by contacting the Department of the Navy and the Society of Automotive Engineers at the following addresses, respectively:

Department of the Navy

Navy Publishing and Printing Services Office

709 Robbins Avenue

Philadelphia, PA 19111–5094

Society of Automotive Engineers

400 Commonwealth Drive

Warrendale, PA 10596–0001

Planning for an FMEA

Planning for an FMEA involves a series of considerations that include, as a minimum, the following:

- Selecting appropriate applications for the analysis—An FMEA may be authorized by individuals at various levels within an organization, or it may be required by ISO/QS 9000, APQP, Six Sigma methodologies, internal quality programs, or customer requirements. However authorized or required, an FMEA is expensive to complete and should be completed only in those instances where the benefits of completing the analysis outweigh the costs.

- Identifying and allocating resources—These resources include FMEA team members and a reporting structure; physical space to conduct the analysis; and store documentation, time, and clerical/communications support.
- Defining scope—An FMEA can be conducted at a high level (i.e., the system level) or at a detailed level (i.e., the component level or service delivery level). Further, a high-level FMEA may lead to additional FMEAs at more detailed levels. Therefore, it is very important to set the scope of the analysis before beginning.
- Expectations and deliverables—The team-based nature of completing an FMEA means FMEA team members will have dual or multiple responsibilities and reporting structures in addition to the FMEA team. It is critical, therefore, to clearly define performance expectations for all FMEA team members and to communicate those expectations directly to appropriate supervisory or managerial personnel in reporting structures outside the FMEA team. It is equally important that all FMEA team members understand what deliverables will result from the analysis and their respective roles in developing those deliverables.
- Establishing milestones, due dates, and deadlines—Key milestones for an FMEA include authorizing the analysis, establishing a reporting structure, allocating resources (particularly FMEA team members), gathering input for the analysis, completing the analysis, taking and monitoring corrective action, and preparing documentation, report-outs, and debriefings. To ensure effectiveness, an FMEA should be conducted as a project from the perspective of establishing a schedule, specifying due dates and deadlines for each of the major milestones.

Establishing a Single Point of Responsibility

As mentioned earlier, FMEA is a team-based analysis. However, there is sufficient practical experience to support the idea that assigning responsibility to a cross-functional team rather than a single individual is not the most effective policy. For a variety of reasons, a single person should be assigned the responsibility of FMEA team leader with the authority to make decisions and allocate resources to complete the FMEA as planned.

FMEA Team Members

The belief that only the one or two people closest to a system, subsystem, product or process design, or service delivery should be assigned to an FMEA violates the very intent of the analysis. FMEA is intended to be completed by team members representing a broad cross-section of expertise—technical and nontechnical. At a minimum, an FMEA should have representation from the following functional groups:

- Design engineering
- Manufacturing engineering
- Production
- Quality/Reliability
- Purchasing/Material control
- Sales and marketing
- Customers

It cannot be overemphasized that for an FMEA to be truly effective, the viewpoints and perspectives of every functional group listed here must be included—particularly customers. Excluding the stakeholder input from the FMEA typically results in missed opportunities to detect and correct failure modes.

Inputs to an FMEA

To prepare for an FMEA, it is necessary to gather information from several sources; these inputs should be gathered prior to the initial FMEA team meeting so as to maximize the effectiveness of team members' time. Inputs to an FMEA include, as a minimum, the following:

- Process flowchart or functional block diagram
- Design specifications
- Customer requirements/specifications
- Testing data/results
- Data on similar process/design technology
- Warranty data
- Failure/Rework data
- Design/Configuration change data
- Prior FMEAs
- Results from quantitative analysis (design of experiments, statistical process control, process capability assessments, reliability assessments, etc.)

FMEA and Other Quality Tools

In addition to the inputs previously described, other quality tools are frequently used during the completion of an FMEA. These tools include, but are not limited to, the following:

- Cause-and-effect diagrams
- Process decision program charts
- Histograms
- Pareto diagrams
- Run charts
- Force field analysis
- Fault tree diagrams
- Root cause analysis

Outputs from an FMEA

Outputs or deliverables from an FMEA include the following:

- FMEA documentation
- System, subsystem, design, process, and/or service delivery documentation
- Recommendation reports
- Corrective action reports
- Design changes
- Compliance reports
- Debriefings and presentations

Basic Steps in an FMEA

Complexity in an FMEA is directly related to the number of levels of analysis dictated by the situation or team members. At the most fundamental level, however, every FMEA consists of the same basic steps:

- Identify a starting point for the analysis—A starting point will be a system, subsystem, product or process design, or service delivery system of interest.
- Gather all relevant inputs to support the analysis—Gathering inputs for an FMEA is a milestone to be completed prior to an initial FMEA meeting. It is far more effective, both from cost and efficiency perspectives, to have all team members at meetings participating in the analysis rather than leaving meetings to gather input.
- Identify potential failure modes as related to the following:
  - Who would be impacted by a failure?
  - What would happen in the event of a failure?

  - When would the failure occur?
  - Where would the failure occur?
  - Why would the failure occur?
  - How would the failure occur?
- Quantify the risk associated with each potential failure—Risk assessment is based on severity, occurrence, and detection of a potential failure.
- Develop a corrective action plan for the most significant risks.
- Iterate or repeat the analysis until all potential failures pose an "acceptable" level of risk—What constitutes an "acceptable" risk must be clearly defined by the individual or agent authorizing the FMEA.
- Document results.
- Report-out and/or present results.

Quantifying the Risk Associated with Each Potential Failure

To avoid confusion, the quantification of risk associated with potential failures will now be introduced—prior to detailed explanation of FMEA mechanics. While FMEA may appropriately be applied to systems, subsystems, product or process designs, and/or service delivery, risk assessment methods and metrics remain consistent regardless of analysis level. Where the actual criteria for risk assessment change (specifically between design and process FMEAs), the criteria will be provided as needed for the discussion. Readers are encouraged to familiarize themselves with this section on risk assessment and be prepared to review this section once specific analysis levels are discussed.

Risk Components

For purposes of FMEA, risk has three components, which are multiplied to produce a risk priority number (RPN). The three components of risk are as follows:

- Severity (S)—An indicator of the severity of a failure should a failure occur. Severity is described on a 10-point scale.
- Occurrence (O)—An indicator of the likelihood of a failure occurring. Occurrence is described on a 10-point scale.
- Detection (D)—An indicator of the likelihood of detecting a failure once it has occurred. Detection is described on a 10-point scale.

$$RPN_{min} = 1, \text{ while } RPN_{max} = 1000$$

Taking Action Based on an RPN

A common mistake in assessing FMEA risk is prioritizing corrective action based on the descending order of RPNs. Logic would suggest that the largest RPNs represent the highest risk—which is true, but only to a point. When multiplying the three risk components together, their importance relative to each other becomes obscured. Consider the following example:

	(S)	×	(O)	×	(D)	=	RPN
Potential failure 1	2		10		5		100
Potential failure 2	10		2		5		100
Potential failure 3	2		5		10		100
Potential failure 4	10		5		2		100

In each case the resulting RPN is 100. Therefore, it is unclear which potential failure to take corrective action on first. There is, however, a generally accepted strategy when taking action on an RPN:

1. Eliminate the occurrence
2. Reduce the severity
3. Reduce the occurrence
4. Improve detection

Getting back to the example, with the additional information provided by the generally accepted strategy when taking action on an RPN, it becomes clear how to proceed.

1. Eliminating occurrences would, mathematically, reorder the RPNs.
2. Reducing severity next would focus our attention on potential failures 2 and 4. But then what? We still have two potential failures with the same level of risk.
3. Reducing occurrence as the next step in this process focuses our attention on potential failure 4, which has a higher occurrence rating than potential failure 2.

Now our attention can turn to evaluating the remaining potential failures since potential failures 2 and 4 have been ranked as the two most important. Of the two remaining potential failures (1 and 3, respectively), potential failure 1 has the higher occurrence rating and is therefore ranked as the third most important potential failure; potential failure 3 drops to the least important position by default. The rank order by which the potential failures should be investigated for corrective action is as follows:

First priority: Potential failure 4

Second priority: Potential failure 2

Third priority: Potential failure 1

Fourth priority: Potential failure 3

Do We Rate the Failure Mode or the Cause?

A common point of confusion arises when considering what is actually rated as part of the risk assessment—the actual failure itself or the cause of a given failure. It is perfectly acceptable to rate either the failure or the cause—as long as the assumption is well documented (on actual FMEA charts, in written correspondence, and in all reports/presentations) and everyone on the FMEA team and in the reporting structure is aware of the assumption. Whether rating a failure itself or a cause of that failure, an FMEA should provide consistent results and corrective actions.

FMEAs Encountered by Quality Professionals

It was mentioned earlier in this chapter that FMEA can be applied to the system/subsystem, design or process, or service delivery levels. A brief synopsis of each FMEA application follows.

System FMEA
A *system*, or *subsystem*, is a collection of elements or components working together to accomplish a desired task or function. FMEA is applied at the system or subsystem level to identify potential failure modes and effects that could negatively impact system or subsystem performance. At the system or subsystem level, FMEA is focused at system or subsystem boundaries where potential failures are most likely to occur. The boundaries of interest for a system or subsystem FMEA include functional (i.e., expected outcomes assuming normal operation) or operational (i.e., specific outputs expected as compared with tolerances, specifications, and timing).

- *Design FMEA*. A *design*—or more accurately, a *product design*—is a set of specifications that describes all aspects of a product (i.e., major functions, operating parameters and tolerances, materials, dimensions, etc.). FMEA is applied to product designs as early in the product design process as feasible to identify potential failure modes that could result from a design flaw. Design FMEAs are a normal part of key milestones in the product development process, such as concept reviews, concept approvals, preliminary design reviews, and final design reviews.
- *Process FMEA*. A *process design* is a set of specifications that describes all aspects of a process (i.e., functional components, flow rates, process steps, equipment to be used, steps to be performed, operators or employees to be involved, etc.). Process design FMEA is applied to process designs at the earliest possible point to identify potential failure modes that could result from a design flaw. Process FMEAs, too, are a normal part of key milestones in the process development process.
- *Service delivery FMEA*. A *service delivery* is the completion of a set of tasks designed to meet one or more customer expectations. Service delivery FMEA is applied to service delivery designs to identify potential failure modes that, if experienced, would result in some level of dissatisfaction from customers. Service delivery FMEAs are also completed as early as possible in the design process and are a normal part of key milestones in the service delivery design process.

In most instances, the practicing quality professional can be expected to work primarily on design and/or process FMEAs. A quality inspector would expect fewer opportunities, if any, to work on system/subsystem and/or service delivery FMEAs. Accordingly, this chapter focuses on design and process FMEAs and omits system/subsystem and service delivery FMEAs.

Design and Process FMEA

Following the steps outlined earlier describing the planning functions preceding an FMEA, the analysis proceeds as the FMEA team completes appropriate documentation such as the FMEA form. For purposes of this discussion, one form applicable to either a design or a process FMEA will be described. Where the criteria change between a design and a process FMEA, both criteria will be provided. Figures 24.2 and 24.3 are blank FMEA forms applicable to design and process FMEA—each component of the forms is identified and described in the following sections.

Heading Information and Documentation

Product or process name	Provide the formal and/or commonly used (if different) name for the product or process.
Product or process description	Provide a brief description of the product or process that is meaningful to the FMEA team members.
FMEA number	Assign an FMEA number to each FMEA for tracking and documentation purposes. There are no standards for numbering FMEAs; however, a numbering system that links the FMEA to a specific period of time and product/process family is preferred.
Design/Process owner	Identify the individual or team assigned primary responsibility for the design or process for tracking and documentation purposes. This individual or team is also identified for reference, if needed, during the FMEA.
FMEA team leader	Identify the individual assigned primary responsibility for completion of the FMEA for documentation purposes. This individual is also identified so as to establish a point of contact should any stakeholder need information during or after the FMEA.
FMEA team	List each member of the FMEA team along with any key responsibilities relative to the FMEA.
FMEA date	Provide the date(s) during which the FMEA is completed to help establish a chronology of events. Revision dates should be noted here as well.
FMEA risk assessment based on	Indicate the basis of the risk assessment. The FMEA risk assessment may be based on either actual failures or failure causes. It is important to document the team's decision to assess risk based on failures or causes to ensure everyone evaluating the FMEA understands exactly how risk was assessed.

Potential
Failure Mode and Effects Analysis
(Design FMEA)

____ System
____ Subsystem
____ Component ________________

Design responsibility ________________

FMEA number ________

Page ______ of ______

Prepared by ________________

Model year(s)/vehicle(s) ____________

Key date ________________

FMEA date (orig.) ______ (rev.) ______

Core team ________________________________

Item / Function	Potential failure mode	Potential effect(s) of failure	Sev	Class	Potential cause(s)/ mechanism(s) of failure	Occur	Current design controls	Detec	RPN	Recommended action(s)	Responsibility and target completion date	Action results					
												Actions taken	Sev	Occ	Det	RPN

Figure 24.2 Blank design FMEA form.

Potential
Failure Mode and Effects Analysis
(Process FMEA)

Item ____________ Process responsibility ____________ FMEA number ____________

Model year(s)/vehicle(s) ____________ Key date ____________ Page ______ of ______

Core team ____________ Prepared by ____________

FMEA date (orig.) ______ (rev.) ______

Process function / Requirements	Potential failure mode	Potential effect(s) of failure	Sev	Class	Potential cause(s)/ mechanism(s) of failure	Occur	Current process controls	Detec	RPN	Recommended action(s)	Responsibility and target completion date	Action results				
												Actions taken	Sev	Occ	Det	RPN

Figure 24.3 Blank process FMEA form.

Analysis Content and Documentation

DFMEA part name, #, function, or PFMEA process function	Identify the product (i.e., part name, part number, and function) or process (i.e., functions to be completed as part of the process).
Potential failure mode	List each of the potential failure modes associated with the design or process. Design failure modes may include dented, deformed, fractured, loosened, leaking, warped, and so on. Process failure modes may include overheating, inoperable, visual defect, and so on.
Potential effect of failure mode	For each potential failure mode, indicate the potential effect on customers or production/process personnel—it is entirely possible to have multiple effects for each potential failure mode.
Severity	Indicate the seriousness of the effect of the potential failure using the severity criteria defined in Tables 24.1 and 24.2. Note: The severity rating applies only to the effect of the potential failure.
Classification	Classify any special characteristics that may require additional process controls. SAE J1739 identifies classifications that include critical, key, major, and significant.
Potential cause of failure mode	For each potential effect of each failure mode, identify all possible causes—it is entirely possible to have more than one cause for each potential effect.
Occurrence	Indicate how frequently each failure is expected to occur using the occurrence criteria defined in Tables 24.3 and 24.4.
DFMEA design verifications, or PFMEA current process controls	For a design FMEA, identify the actions completed that ensure or verify the adequacy of the design. For a process FMEA, identify the control currently in place that prevents a failure mode from occurring.
Detection	Indicate the ability of design verification or current process controls to detect a potential failure mode in the event that failure actually occurs. Use the detection criteria defined in Tables 24.5 and 24.6.
Risk priority number (RPN)	For each potential failure mode, multiply the severity (S), occurrence (O), and detection (D) assessments together. Since each scale (S, O, and D) ranges from 1 to 10, $RPN_{min} = 1$ and $RPN_{max} = 1000$.
Recommended actions	For each potential failure mode, list one or more recommended corrective actions. For further direction and guidance on prioritizing recommended corrective actions, readers are encouraged to refer to the "Taking Action Based on an RPN" section of this chapter.
Individual/Team responsible and completion date	For each recommended action, assign an appropriate individual or team and an expected completion date.
Actions taken	Provide a brief description of the actual actions taken and their respective action dates.
Resulting RPN analysis	Following each action taken, reiterate the severity, occurrence, and detection assessments and calculate a new resulting RPN. Actions taken based on RPNs and resulting RPNs continue until the risk assessment for each potential failure is "acceptable" to the customer and/or authorizing agent for the FMEA.

Table 24.1 Design FMEA severity criteria.

Effect	Severity criteria	Ranking
Hazardous without warning	Very high ranking when potential failure mode affects safe operation and/or regulation noncompliance. Failure occurs without warning.	10
Hazardous with warning	Very high ranking when potential failure mode affects safe operation and/or regulation noncompliance. Failure occurs with warning.	9
Very high	Item or product is inoperable, with loss of function. Customer very dissatisfied.	8
High	Item or product is operable, with loss of performance. Customer dissatisfied.	7
Moderate	Item or product is operable, but comfort/convenience items inoperable. Customer experiences discomfort.	6
Low	Item or product is operable, but with loss of performance of comfort/convenience items. Customer has some dissatisfaction.	5
Very low	Certain characteristics do not conform. Noticed by most customers.	4
Minor	Certain characteristics do not conform. Noticed by average customers.	3
Very minor	Certain characteristics do not conform. Noticed by discriminating customers.	2
None	No effect.	1

S × O × D = risk priority number (RPN)

Derived from Technical Standard SAE J1739.

Reprinted by permission of The Society of Automotive Engineers (SAE).

Table 24.2 Process FMEA severity criteria.

Effect	Severity criteria	Ranking
Hazardous without warning	May endanger machine or assembly operator. Very high severity ranking when a potential failure mode affects safe operation and/or involves noncompliance with regulation. Failure will occur without warning.	10
Hazardous with warning	May endanger machine or assembly operator. Very high severity ranking when a potential failure mode affects safe operation and/or involves noncompliance with regulation. Failure will occur with warning.	9
Very high	Major disruption to production line. 100% of product may have to be scrapped. Item inoperable, loss of primary function. Customer very dissatisfied.	8
High	Minor disruption to production line. A portion of product may have to be sorted and scrapped. Item operable, but at reduced level. Customer dissatisfied.	7

(*continued*)

Table 24.2 Process FMEA severity criteria. (continued)

Effect	Severity criteria	Ranking
Moderate	Minor disruption to production line. A portion of product may have to be scrapped (no sorting). Item operable, but some comfort items inoperable. Customer experiences discomfort.	6
Low	Minor disruption to production line. 100% of product may have to be reworked. Item operable, but some comfort items operable at reduced level of performance. Customer experiences some dissatisfaction.	5
Very low	Minor disruption to production line. Product may have to be sorted and a portion reworked. Minor adjustments do not conform. Defect noticed by customer.	4
Minor	Minor disruption to production line. Product may have to be reworked online, but out of station. Minor adjustments do not conform. Defect noticed by average customer.	3
Very minor	Minor disruption to production line. Product may have to be reworked online, but out of station. Minor adjustments do not conform. Defect noticed by discriminating customer.	2
None	No effect.	1

Derived from Technical Standard SAE J1739.

Table 24.3 Design FMEA occurrence criteria.

Probability of failure	Possible failure rates	Ranking
Very high: Failure almost inevitable	> 1 in 2	10
	1 in 3	9
High: Repeated failures	1 in 8	8
	1 in 20	7
Moderate: Occasional failures	1 in 80	6
	1 in 400	5
	1 in 2000	4
Low: Relatively few failures	1 in 15,000	3
	1 in 150,000	2
Remote: Failure is unlikely	< 1 in 1,500,000	1

Derived from Technical Standard SAE J1739.

Table 24.4 Process FMEA occurrence criteria.

Probability of failure	Possible failure rates	Ranking
Very high: Failure almost inevitable.	> 1 in 2 1 in 3	10 9
High: Generally associated with processes similar to previous processes that have often failed.	1 in 8 1 in 20	8 7
Moderate: Generally associated with processes similar to previous processes that have experienced occasional failures.	1 in 80 1 in 400 1 in 2000	6 5 4
Low: Isolated failures associated with similar processes.	1 in 15,000	3
Very low: Only isolated failures associated with almost identical processes.	1 in 150,000	2
Remote: Failure is unlikely. No failures associated with almost identical processes.	< 1 in 1,500,000	1

Derived from Technical Standard SAE J1739.

Reprinted by permission of The Society of Automotive Engineers (SAE).

Table 24.5 Design FMEA detection criteria.

Effect	Detection criteria	Ranking
Absolute uncertainty	Design control will not and/or cannot detect a potential cause/ mechanism and subsequent failure mode or there is no design control.	10
Very remote	Very remote chance the design control will detect a potential cause/mechanism and subsequent failure mode.	9
Remote	Remote chance the design control will detect a potential cause/ mechanism and subsequent failure mode.	8
Very low	Very low chance the design control will detect a potential cause/ mechanism and subsequent failure mode.	7
Low	Low chance the design control will detect a potential cause/ mechanism and subsequent failure mode.	6
Moderate	Moderate chance the design control will detect a potential cause/mechanism and subsequent failure mode.	5
Moderately high	Moderately high chance the design control will detect a potential cause/mechanism and subsequent failure mode.	4

(continued)

Table 24.5 Design FMEA detection criteria. (continued)

Effect	Detection criteria	Ranking
High	High chance the design control will detect a potential cause/ mechanism and subsequent failure mode.	3
Very high	Very high chance the design control will detect a potential cause/mechanism and subsequent failure mode.	2
Almost Certain	Design control will almost certainly detect a potential cause/ mechanism and subsequent failure mode.	1

Derived from Technical Standard SAE J1739.

Reprinted by permission of The Society of Automotive Engineers (SAE).

Table 24.6 Process FMEA detection criteria.

Effect	Detection criteria	Ranking
Absolutely impossible	No known controls to detect failure mode.	10
Very remote	Very remote likelihood current controls will detect failure mode.	9
Remote	Remote likelihood current controls will detect failure mode.	8
Very low	Very low likelihood current controls will detect failure mode.	7
Low	Low likelihood current controls will detect failure mode.	6
Moderate	Moderate likelihood current controls will detect failure mode.	5
Moderately high	Moderately high likelihood current controls will detect failure mode.	4
High	High likelihood current controls will detect failure mode.	3
Very high	Very high likelihood current controls will detect failure mode.	2
Almost certain	Current controls will almost certainly detect a failure mode. Reliable detection controls are known with similar processes.	1

Derived from Technical Standard SAE J1739.

Reprinted by permission of The Society of Automotive Engineers (SAE).

A Final Word on Taking Corrective Action

An FMEA represents an in-depth, objective, quantitative analysis of the risk associated with potential failures that results in the calculation of one or more RPNs. Once RPNs have been calculated and the FMEA team prepares to take corrective action, the analysis necessarily takes on a subjective element as FMEA team members use the risk assessment to guide prioritization of corrective actions.

As was mentioned earlier in this chapter, the most common practice used to prioritize corrective action is based on RPNs. Prioritization of corrective action based solely on RPNs works effectively, however, only as long as there is a "comfortable" difference among the RPN values. When there are clusters of RPN values that are the same, or very close (say within 25 to 50 points), taking action based on RPNs alone is not straightforward. In such cases, additional guidance in prioritizing corrective action is needed as follows:

1. Rank the RPNs in descending order
2. For those RPNs that cluster within a predefined range of, say, 25 to 50 points, use the strategy suggested earlier in this chapter (i.e., first eliminate occurrence, then reduce severity, then reduce occurrence, and finally improve detection)
3. Plan, take, and monitor corrective action on the largest nonclustered RPNs
4. Plan, take, and monitor corrective action on RPN clusters as defined in step two
5. Repeat steps three and four as needed to address all potential failures identified in the analysis

As another means of eliminating the subjectivity in prioritizing corrective actions based on RPNs, the Criticality Analysis was developed as part of Mil-Std 1629A.

Assessing Criticality

Mil-Std 1629A defines two very important terms and concepts with respect to risk assessment:

Criticality—"A relative measure of the consequences of a failure mode and its frequency of its occurrences."

Criticality Analysis—"A procedure by which each potential failure mode is ranked according to the combined influence of severity and probability of occurrence."

When criticality is considered in an FMEA, the name is changed to failure mode effects and criticality analysis (FMECA). FMECA can be a qualitative or quantitative assessment of risk that leads to a prioritization of corrective action based on severity (S) and occurrence (O) assessments. In the qualitative approach to risk assessment in FMECA, risk is categorized as frequent, reasonably probable, occasional, remote, or extremely unlikely. In the quantitative approach to risk assessment in FMECA, failure rate data, failure effect probability data, individual part failure data, and operating time data are required as input to one or more protocols.

The key result of an FMECA is a criticality matrix that ranks potential failures with respect to severity. The matrix then identifies a prioritization scheme for corrective actions based on the severity of potential failure modes. As displayed on an FMECA criticality matrix, potential failures plotted farther away from the matrix origin on a diagonal line represent higher potential risk of failure, thus warrant increased need for corrective action.

For additional discussion of FMECA, readers are encouraged to consult Mil-Std 1629A for guidance in completing a criticality assessment.

A Caution about Using FMEA

FMEA can be a powerful and effective tool for system/subsystem, design or process, or service delivery improvement. It should be remembered, however, that completing an FMEA involves significant costs. One may be tempted to follow the results of an FMEA to further levels of refinement and specificity—readers are cautioned to remember that there is a cost/benefit relationship associated with the use of FMEA.

Design and Process FMEA Examples

Figures 24.4 and 24.5 are, respectively, examples of design and process FMEAs to help guide the reader through an actual analysis.

FMEA Summary

FMEA is a tool to help cross-functional teams identify, eliminate, and/or reduce the negative effects of potential failures—before they happen. FMEA is widely used as a stand-alone tool or as part of comprehensive quality systems/programs, which is why ASQ has incorporated FMEA into the CQI BoK.

Two standards were discussed in this chapter that relate to FMEA, namely, Mil-Std 1629A and SAE J1739. It is important to note that both standards have been designed such that they will lead to the same or similar results when they are correctly used. Much of the preparation for, inputs to, and prioritization of corrective action guidance are not covered in the standards to support an FMEA.

8D METHODOLOGY

The *8D methodology* is a problem solving and analysis tool used to ". . . identify, correct, and eliminate recurring problems . . ." (ASQ Quality Press). The term "8D" refers to eight stages in the methodology which initially were referred to as "disciplines." The term "disciplines" as intended with the 8D methodology refers to process steps to be consistently and deliberately followed versus organizational functions such as engineering, finance, marketing/sales, etc.

The 8D methodology is useful in process improvement efforts in production operations as well as in service delivery applications. As a methodology, 8D is a step-by-step process to be followed while investigating and/or improving processes. It brings together several other statistical (quantitative) and qualitative

Potential
Failure Mode and Effects Analysis
(Design FMEA)

___ System
_X_ Subsystem
___ Component 01.03/Body closures (2)

Design responsibility Body engineering (3)

Model year(s)/vehicle(s) 199X/Lion 4dr/wagon (5)

Key date 9X 03 01 ER (6)

Core team T. Fender—Car product dev., Childers—Manufacturing, J. Ford—Assy ops (Dalton, Fraser, Henley assembly plants) (8)

FMEA number 1234 (1)

Page 1 of 1

Prepared by A. Tate—X6412—Body engr (4)

FMEA date (orig.) 8X 03 22 (rev.) 8X 07 14 (7)

Item (9) Function	Potential failure mode (10)	Potential effect(s) of failure (11) (12)►	Sev	Class	◄(13) Potential cause(s)/ mechanism(s) of failure (14)	Occur	◄(15) Current design controls (16) (17)►	Detec	RPN	◄(18) Recommended action(s) (19)	Responsibility and target completion date (20)	Action results (22): Actions taken (21)	Sev	Occ	Det	RPN
Front door L.H. H8HX-0000-A • Ingress to and egress from vehicle • Occupant protection from weather, noise, and side impact • Support anchorage for door hardware including mirror, hinges, latch, and window regulator • Provide proper surface for appearance items • Paint and soft trim	Corroded interior lower door panels	Deteriorated life of door leading to: • Unsatisfactory appearance due to rust through paint over time • Impaired function of interior door hardware	7		Upper edge of protective wax application specified for inner door panels is too low	6	Vehicle general durability test vah. T-118 T-109 T-301	7	294	Add laboratory accelerated corrosion testing	A Tate-Body Engrg 8X 09 30	Based on test results (Test No. 1481) upper edge spec raised 125mm	7	2	2	28
					Insufficient wax thickness specified	4	Vehicle general durability testing—as above	7	196	Add laboratory accelerated corrosion testing Conduct Design of Experiments (DOE) on wax thickness	Combine w/test for wax upper edge verification A Tate Body Engrg 9X 01 15	Test results (Test No. 1481) show specified thickness is adequate. DOE shows 25% variation in specified thickness is acceptable	7	2	2	28
					Inappropriate wax formulation specified	2	Physical and Chem Lab test—Report No. 1265	2	28	None						
					Entrapped air prevents wax from entering corner/edge access	5	Design aid investigation with non-functioning spray head	8	280	Add team evaluation using production spray equipment and specified wax	Body Engrg & Assy Ops 8X 11 15		7	1	3	21
					Wax application plugs door drain holes	3	Laboratory test using "worst case" wax application and hole size	1	21	None		Based on test, 3 additional vent holes provided in affected areas				
					Insufficient room between panels for spray head access	4	Drawing evaluation of spray head access	4	112	Add team evaluation using design aid buck and spray head	Body Engrg & Assy Ops	Evaluation showed adequate access	7	1	1	7

SAMPLE

Figure 24.4 Design FMEA example.

Potential
Failure Mode and Effects Analysis
(Process FMEA)

Item Front door L.H./H8HX-000-A ②
Model year(s)/vehicle(s) 199X/Lion 4dr/wagon ⑤
Core team A. Tate—Body engrg., J. Smith—OC, R. James—Production, J. Jones—Maintenance ⑧

Process responsibility Body engrg./assembly operations ③
Key date 9X 03 01 ER 9X 08 26 Job #1 ⑥

FMEA number 1450 ①
Page 1 of 1
Prepared by J. Ford—X6521—Assy ops ④
FMEA date (orig.) 9X 05 17 (rev.) 9X 11 06 ⑦

Process function ⑨ Requirements	Potential failure mode ⑩	Potential effect(s) of failure ⑪	Sev ⑫	Class	Potential cause(s)/ mechanism(s) of failure ⑬ ⑭	Occur	Current process controls ⑮ ⑯	Detec ⑰	RPN	Recommended action(s) ⑱ ⑲	Responsibility and target completion date ⑳	Action results ㉒: Actions taken ㉑	Sev	Occ	Det	RPN
Manual application of wax inside door	Insufficient wax coverage over specified surface	Deteriorated life of door leading to: • Unsatisfactory appearance due to rust through paint over time • Impaired function of interior door hardware	7		Manually inserted spray head not inserted far enough	8	Visual check each hour—1/shift for film thickness (depth meter) and coverage	5	280	Add positive depth stop to sprayer	MFG Engrg 9X 10 15	Stop added, sprayer checked on line	7	2	5	70
										Automate spraying	Mfg Engrg 9X 12 15	Rejected due to complexity of different doors on same line				
To cover inner door, lower surfaces at minimum wax thickness to retard corrosion					Spray heads clogged • Viscosity too high • Temperature too low • Pressure too low	5	Test spray pattern at start-up and after idle periods, and preventive maintenance program to clean heads	3	105	Use Design of Experiments (DOE) on viscosity versus temperature versus pressure	Mfg Engrg 9X 10 01	Temp and press limits were determined and limit controls have been installed—control charts show process is in control Cpk=1.85	7	1	3	21
					Spray head deformed due to impact	2	Preventive maintenance programs to maintain head	2	28	None						
					Spray time insufficient	8	Operator instructions and lot sampling (10 doors / shift) to check for coverage of critical areas	7	392	Install spray timer	Maintenance 9X 09 15	Automatic spray timer installed—operator starts spray, timer controls shut-off, control charts show process is in control Cpk=2.05	7	1	7	49

SAMPLE

Figure 24.5 Process FMEA example.

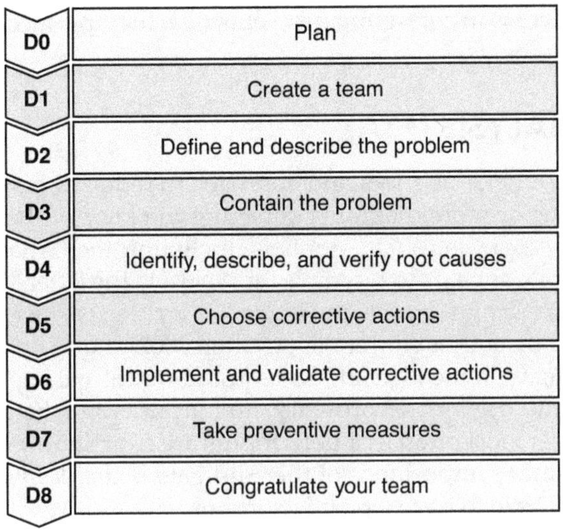

Figure 24.6 List of the eight discipline (8D).
source: Duffy (2013).

(graphical) tools to facilitate a more thorough understanding of process steps, performance characteristics, and paths to improvement.

The 8D methodology consists of the following components related to each other as presented in Figure 24.6.

5 WHYS

The *5 Whys* is a questioning methodology intended to help users understand the root cause(s) of a process failure or improvement opportunity. The underlying concept with the 5 Whys methodology is that members of a process improvement team cannot get to the point of understanding a process failure or process improvement opportunity until they can ask and answer the question of "Why" at least five times. The 5 Whys then helps users and team members develop a more thorough understanding of the process as well as process failures or improvement opportunities.

A second underlying concept with 5 Whys methodology is that making process adjustments or improvements without getting to the root cause(s) leads only to treating symptoms of the problems. What is critically important is that users or team members understand more than just symptoms by identifying the true root cause(s). A major concern for process owners is that premature interventions or treatments causes additional problems.

Just as with the 5 Whys methodology, the *5 Hows* methodology is intended to help users understand how an intervention or process improvement ultimately impacts the process operation or quality. Ultimately, asking "Why" at least five times (a process failure occurred) and asking "How" at least five times a proposed intervention may eliminate a root cause, users of the methodology will have a better chance of success in correcting problems or achieving success in implementing process improvements.

FAULT TREE ANALYSIS (FTA)

Fault Tree Analysis (FTA) is a tool and approach to reducing a system, even complex systems, into one or more diagrams depicting the logic of the system represented by two primary logic gates (i.e., symbols) including the "And" gate and the "Or" gate. The And gate appears to resemble most closely the letter "D" while the Or gate appears to resemble most closely the letter "C" written backward. Figure 24.7, originally developed by author Simha Pilot (March 2002) shows both the And as well as the Or gate along with the logic inputs required for the gate operation.

The And gate triggers, or provides, an output only when both the inputs are positive meaning most often that both inputs are functioning and are operational. A matter of primary importance of the And gate is that *both* inputs must be positive for the And gate to have a positive output.

The Or gate triggers, or provides, an output if either of the inputs are positive meaning that either or both of the inputs are functioning or are operational. A matter of primary importance of the Or gate is that *either or both* of the inputs can be positive for the Or gate to have a positive output.

Using a combination of the And/Or gates, it is possible to develop a graphic representation (i.e., fault tree) of even the most complex systems. Fault trees are particularly useful in analyzing how and where process failures may be expected to reveal, or manifest, themselves in system or process operation. FTA, originally developed at the Bell Laboratories, remains particularly useful in digital circuit analysis where systems operate on 5 volt logic and a signal is present or absent based on the presence or absence of a digital five volt signal.

The fault tree is said to be "top-down" because the diagrams are commonly drawn vertically from top to bottom. A more correct understanding of why fault trees are referred to as top-down is that constructing the diagram starts at the output and working "upstream" toward the inputs.

Fault trees are said to be "deductive" because they are based on reason and logic (i.e., represented by gates) generalized by a set of system or process characteristics (i.e., the presence or absence of signals representing system or process operation).

CORRECTIVE AND PREVENTIVE ACTION (CAPA)

Corrective and preventive action (CAPA) is intended to mitigate and prevent problems. It is a critically important quality tool. CAPA is a topic/tool new to the CQI BoK in 2025; however, it has been used for a very long time. We will address corrective actions first and then move into a discussion of preventive actions.

Corrective Action

Corrective action is defined in ANSI/ISO/ASQ Q9000-2015, "Quality management systems—Fundamentals and vocabulary" as follows: ". . . an action taken

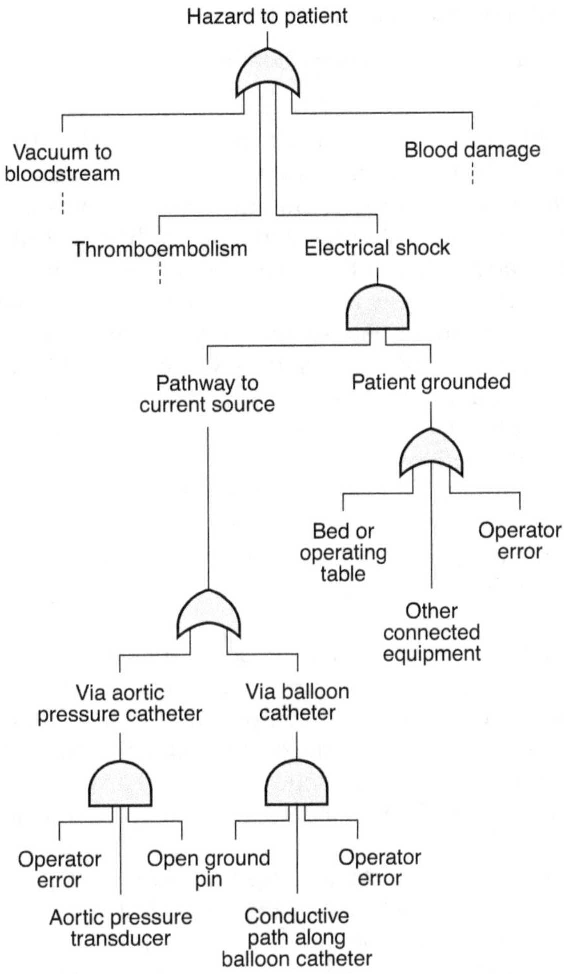

Figure 24.7 Fault tree depicting the root causes of hazard to patients during surgery.
Source: Pilot (2002).

to eliminate the cause of a detected non-conformity, which prevents the problem from recurring."

The BoK identifies corrective action as being composed of the following elements:

- Identify the problem.
- Contain the problem (interim action).
- Assign responsibility (personnel) to determine the causes of the problem and propose solutions to eliminate it or prevent its recurrence (permanent action).

- Verify that the solutions are implemented.
- Confirm their effectiveness (validation).

The corrective action process (CAP) begins with problem identification. Once one or more problems have been identified, steps are taken to contain the problem (ensure that the problem does not impact or affect process outputs or other processes). Problem containment is commonly referred to as *interim* action because the steps taken to ensure containment are generally disruptive to normal process operation, do not add value to the final product or service, do not permanently correct or prevent the problem, and are ultimately not sustainable. An example of problem containment might be re-inspecting and/or sorting good product from bad before that product reaches customers. Once the problem has been contained for the short term, a more in-depth analysis takes place; at that time, individuals involved with identifying the problem causes are in a position to propose long-term or permanent solutions.

The next step in a CAP is to verify that solutions have been implemented. Verifying solution implementation is a risk point in a CAP because the potential exists to implement proposed solutions that are (1) not needed (that is, do not address the real root cause of the problem), (2) implemented incorrectly, or (3) implemented incompletely.

Following solution implementation, it is imperative that solution effectiveness be validated. *Validation* is a process of ensuring that something has been completed, that whatever has been completed is in accordance with some set of instructions or specifications, that the completed action meets the intended goal(s) or outcomes(s), that the completed action does not harm or disrupt other elements of any given process or system, and that the completed action is sustained into future operations.

Some of the tools and techniques used to support a CAP are identified in the BoK, and those tools and techniques are covered at various other points in this book. Corrective action processes are used in conjunction with preventive action processes (PAPs), which is the subject of our next discussion.

Preventive Action

Preventive action is defined in ANSI/ISO/ASQ Q9000-2015, "Quality management systems—Fundamentals and vocabulary" as follows: ". . . an action taken to eliminate the cause of a potential non-conformity from occurring."

Preventive action is composed of the following elements:

- Identify potential failures, defects, or process deficiencies.
- Assign responsibility for process improvement.
- Evaluate lessons learned and conduct read across.
- Verify solution effectiveness.

The PAP begins with identification of potential failures, defects, or process deficiencies. As with the CAP, numerous tools and techniques are used to support the identification of potential failures, defects, or process deficiencies. Equally important to tools and techniques used in identification is the assignment of responsibility for process improvements. With responsibility comes accountability, and it is simply imperative that members of the leadership know who is responsible for implementation of process improvements, both as a source of accountability and so that they know where to flow resources and authorizations for process improvements.

While responsibility is normally assigned to an individual, leader, or manager, teams normally are assigned to implement the process improvements. Here, too, working in teams requires a set of knowledge, skills, and abilities.

Once the problem identification and process improvement responsibilities have been assigned and completed, the next step is verification of solution effectiveness. The steps in verification and validation in a CAP and PAP are the same.

When improvement efforts have been successful, it is often useful to document the information and consider whether the improvement can be applied to similar situations. For example, suppose a quality improvement team has learned that the occurrence of certain defects can be reduced if a tool is cleaned more frequently than specified. This "lesson learned" should be studied for applicability to similar processes. This type of study is called "read across" because the attempt is made to read across similar processes, procedures, and products.

PAPs are not used in isolation. Completion of each PAP element requires the use of one or more additional tools and/or techniques contained throughout this book. Corrective and preventive action processes are used as components of root cause analysis, and effective application of CAPA processes typically triggers use of other processes such as nonconforming material identification, handling, and disposition, material review boards, and so on.

Chapter 25

Resources

> 1. **Environmental and safety support.** Define and use various resources related to personal and environmental safety, such as safety data sheets (SDS), material data sheets (MDS), and personal protective equipment (PPE). (Apply)
> 2. **Reference documents.** Identify and use national and international standards (e.g., ISO, ANSI, ASTM, QS, and NIST) and customer requirements as authorities that support processes and procedures used to assure quality products. (Apply)
> 3. **Employees as resources.** Describe how employees can be empowered and the value they add to project teams or quality improvement teams. Describe typical team roles and responsibilities: facilitator, ground rules, project or team charter. Describe the stages of team development: forming, storming, norming, performing, and adjourning. (Remember)
> 4. **Quality Documentation:** Identify and apply basic quality documentation, including the correct form / revision for the process (e.g., ISO 9001, First Article Inspection Report, ISIR, PPAPs). Demonstrate proper usage of policy, procedure, work instructions, and forms, and proper documentation practices, such as document control, filling out forms completely, correcting misspellings, and initialing changes. (Apply).
>
> Body of Knowledge IV.G. 1–4

As we begin to conclude our discussion of the CQI BoK, it is important that we consider four remaining topics: environmental and safety support, reference documents, technical reports, and employees as resources. As we have done with all the other topics in this book, we will keep the scope of our discussion consistent with the ASQ CQI BoK. It is important to realize, however, that these four topics are the focus of three different career fields: environmental safety and health, professional communications, and human resources. Accordingly, there are several text and electronic resources available to provide additional depth and detail about these topics.

ENVIRONMENTAL AND SAFETY SUPPORT

Protection of employee safety and health while in the workplace is the obligation of employers. Provisions for employee safety and health are set forth in the many local, regional, and federal codes and statutes. The Occupational Safety and Health Administration (OSHA) is responsible for all safety and health regulation. The primary regulation for safety and health is the OSHA Code of Federal Regulations (CFR) Title 29. CFR Title 29 has multiple sections, each section addressing safety and health in select trades and industries.

Where safety and health are concerned, CFR Title 29 requires that protections be afforded to employees as they relate to the areas in the list below. Further, if some form of employee protection is required, CFR Title 29 requires that a safety program be developed and implemented wherein the safety program address hazards present; the selection, maintenance, and use of protections; training of employees; and program monitoring.

It cannot be overemphasized that employees do ultimately bear at least some responsibility in protecting their personal safety and health while at work. As a responsible professional, you should not take for granted that your safety and health are protected in the work environment.

There are at least six different areas of concern for employee safety and health:

- *Material safety data sheets (MSDSs).* MSDSs provide employees and emergency response personnel with documentation of hazardous or potentially hazardous substances. MSDSs must be made available to all employees, informing them of the type and nature of substances used in the workplace. Every employee is encouraged to review these MSDSs, as it may not be obvious what types of substances are in use. While there are several formats acceptable for MSDSs, they all contain the same basic information as it relates to physical data (melting point, boiling point, flash point, etc.), toxicity, health effects, first aid, reactivity, storage, disposal, protective equipment, and spill/leak procedures (ILPI 2012).
- *Personal protective equipment (PPE).* PPE is equipment used to ensure the safety of employees in the workplace. PPE is required by OSHA to "reduce employee exposure to hazards when engineering and administrative controls are not feasible or effective in reducing these exposures to acceptable levels. Employers are required to determine if PPE should be used to protect their workers" (OSHA 2012).

  PPE may include, but is not limited to, items such as safety glasses, gloves, footwear, ear plugs/muffs, respirators, and safety helmets. PPE may also include, but is not limited to, equipment such as air filters, gas detectors, fire alarms, and exit signs. As with MSDSs, employers are required to disclose PPE available for use.
- *Ergonomics.* Ergonomics refers to the study of the interface between humans and their workspaces, tools, and equipment. OSHA provides protections pertaining to ergonomics under CFR Title 29 (guidelines, enforcement, and outreach and assistance). Through application of basic ergonomic principles, many repetitive-motion injuries can be avoided;

hence, it is an employee's responsibility to identify any discomfort or pain arising from his or her working conditions. Once an employee identifies the need for assistance, it is the responsibility of the employer to meet the spirit of CFR Title 29 in. implementing a remedy.

- *Hearing conservation.* Excessive noise in the workplace can lead to hearing loss. Noise is measured on the decibel (db) scale, wherein exposure to 85 db or more for an extended period of time can lead to hearing loss. The db scale is *not* linear, which means the increase in noise magnitude from, say, 4 db to 5 db is not the same as the increase from 10 db to 11 db. Protection of hearing is provided by CFR Title 29 under the subject "Hearing Conservation."
- *Confined spaces.* A confined space is defined as any enclosure that limits entry or egress. Confined spaces typically have restricted airflow and tend to trap gases. In accordance with CFR Title 29, some confined spaces are required to limit entry by untrained and/or unequipped employees. Make no mistake about it, confined spaces are dangerous and should be avoided. Where access to a confined space is required, CFR Title 29 and many local regulations require specific procedures to be followed in terms of checking for hazardous gases, provision of breathable air, and so on. As employers are required to clearly mark and limit access to confined spaces, employees are strongly advised not to enter these spaces.
- *Lock-out/tag-out.* Lock-out/tag-out refers to removal of electrical power from tools, equipment, or other electrical circuits. It may be necessary to remove electrical power for safety, to facilitate maintenance, or to stop use of selected pieces of equipment or workspaces. When power is removed for any application, a locking device is affixed to wires and/or electrical connections to prevent reconnection of power—intentional or inadvertent. When power is removed and an appropriate lock-out device is installed, it is strictly forbidden under CFR Title 29 to reconnect power. Under CFR Title 29, the only person authorized to reconnect power is the person who disconnected the power and installed the lock-out device. Reconnecting power to equipment or workspaces where it has been removed puts at risk not only the life of the person reconnecting the power but also anyone who may potentially come in contact with the reconnected equipment or workspaces.

REFERENCE DOCUMENTS

As has been noted throughout this book, customer expectations help to define quality. An extension of customer expectations is a customer requirement wherein customers formalize their expectations in terms of the existence of selected quality systems, use of selected standards, and/or compliance with certain policies and procedures. These customer expectations may be set in the interest of standardization across multiple vendors or facilities. These customer expectations may also be set in the interest of communicating values or goals to customers as well as consumers.

Several sets of documents, reference documents, are used to specify customer expectations. While it is beyond the scope of this book to provide an exhaustive list of all reference standards, the standards can be accessed via the Internet at the addresses provided. These documents originate from organizations such as the following:

- *International Organization for Standardization (ISO).* ISO 9000 is a series of standards pertaining to research and development, manufacturing, and service delivery organizations. The ISO standards articulate requirements for the design, structure, implementation, operation, maintenance, and documentation of comprehensive quality systems (www.iso.org).
- *Quality Systems (QS).* QS 9000 is a set of standards for quality systems based on the ISO 9000 series of standards as tailored for the automotive industry. Information about the QS 9000 standards can be obtained from http://www.iso.org.

AS 9100 is a set of standards for quality systems based on the ISO 9000 series of standards (ISO 9001: 2000) as tailored for the aerospace industry. Information about the AS 9100 standards can be obtained from http://ts.nist.gov/Standards/Global/as9100.cfm.

- *U.S. Department of Defense (MIL-*).* The U.S. Department of Defense (DoD) established and maintains a set of reference documents that includes, but is not limited to, defense specifications, handbooks, and standards. Military Specifications (MIL-SPECs) "describe the physical and/or operational characteristics of a product," Military Standards (MIL-STDs) "detail the processes and materials to be used to make the product," and Military Handbooks (MIL-HDBKs) provide information and/or guidance (http://www.asq.org).
- *American National Standards Institute (ANSI).* ANSI is an organization that coordinates, maintains, archives, and distributes voluntary standards. These voluntary standards have applications in virtually every type of industry (http://www.ansi.org).
- *American Society for the Testing of Materials (ASTM).* ASTM serves as an international standards organization. ASTM develops procedures and protocols for the testing of all types of materials and applications (http://www.astm.org).
- *National Institute of Standards and Technology (NIST).* NIST maintains standards that facilitate and expedite international trade. NIST fosters compliance by evaluating conformity assessment accreditation bodies and ensuring adherence standards specified in international agreements (http://www.NIST.gov and https://www.nist.gov/standardsgov/what-we-do).

EMPLOYEES AS RESOURCES

Employees bring tremendous value to the workplace in terms of their knowledge, skills, abilities, and creativity. This value is so great that employees have come to be considered assets, in many cases of equal or greater value than buildings or equipment. Considering employees to be resources and assets is a relatively new phenomenon that has gained popularity particularly during the past two decades. When we manage the collective knowledge brought into the workplace by employees, we refer to this practice as "knowledge management" (Frappaolo 2006).

In addition to inventorying employee knowledge and allocating it to where it is most needed within an organization, other practices that facilitate employees making the best possible contributions in the work environment include empowering employees and employee involvement.

Empowering Employees

When we say we are "empowering employees," we mean that we are giving employees the latitude—in fact, the responsibility—to take ownership of certain aspects of their daily work life and to use their creativity in making certain types of decisions (Woods 2005). Empowering employees requires trust between leadership team members and staff/operators. That trust is developed over time (Blanchard et al. 2010) and requires significant effort in articulating and negotiating the boundaries within which employees are enabled to make decisions. A greater reliance on the input and participation of employees characterizes the work environment, particularly that of the CQI.

The CQI is well advised to learn as quickly as possible the amount of decision-making authority/empowerment he or she has as part of his or her normal work responsibilities. It is important to exercise the expected level of decision-making authority/empowerment but not to exceed it. As more workplace experience is gained, and as a history of making effective decisions is accumulated, additional empowerment will normally result, up to some point wherein leadership involvement is required. Where is that "point" where leadership involvement is required? That depends completely on the level of employee involvement in the decision-making process, for which the leadership team ultimately has final responsibility.

Employee Involvement

As noted in the preceding section, the amount of employee involvement in day-to-day business, production, or service delivery operations depends on a balance of employee competence and leadership authorization. Certainly, the trend is to enable employees to be more, rather than less, involved with decision-making processes.

There are a variety of ways to practice employee involvement (Ichniowski et al. 2001). It is likely that the CQI will, in fact, become involved with some form of employee involvement, and the form of involvement most likely for the CQI is serving on one or more types of teams. Common teams for the CQI are Six Sigma project teams, self-directed work teams, quality circles, and continuous improvement teams. It is beyond the scope of this book to explain the different types of teams; however, we will cover basic elements of team formation and stages of development, which is contained within the CQI BoK.

Team Roles and Responsibilities

Several characteristics are common to most team types:

- Teams have a clearly defined charter (assigned task or area of concern)
- Teams typically have a defined amount of time in which to complete their work
- Teams are provided with the resources needed (time, finances, access to equipment, access to information, etc.)
- Teams are normally provided with a list of expected outcomes
- Teams are composed of employees from as many stakeholder groups as possible
- Team size is kept to a minimum to accomplish assigned tasks (4–10 members is common)

In addition to these team characteristics, teams are composed of members who have different roles and responsibilities, such as the following:

- *Champion.* The team champion is an organization or company leader who has convened the team. The champion is responsible for ensuring that there is a clear path of communication to the company leadership, that the team functions effectively, and that the team has the resources it needs to complete its work.
- *Team leader.* The team leader is responsible for articulating strategy, formalizing and assigning work tasks, supervising day-to-day operation of the team, and documenting team activities. The team leader also typically tracks and monitors resources used and required by the team.
- *Facilitator.* The facilitator helps balance participation of team members, particularly during team meetings. The facilitator does not take an active role in the substance of team meeting agenda items or conversations; rather, the facilitator ensures meetings remain productive by enforcing team operating rules, practicing professional courtesy and collegiality, and effectively managing meeting time. The facilitator may also provide training in selected problem-solving approaches and tools/techniques that may be needed by members of the team.
- *Team members.* Team members are selected to serve on teams based on their ability to contribute to the solution of a problem or exploit an opportunity as well as their proximity to, or involvement with, a problem or opportunity. Team members are purposefully selected from as many areas as possible within or outside a work unit or company to ensure that all stakeholders are represented. A stakeholder is someone who is directly or indirectly affected by any given problem or opportunity.

No matter how experienced the members of a team may be within their areas of expertise, or in working on teams, team performance is very predictable based on stages of development. Each time a team is created, members of the team progress through these stages of development. The amount of time needed to progress through each stage depends on may factors; in fact, it is possible that a dysfunctional team may not be able to progress through these stages.

Stages of Team Development

The following are the stages of team development:

- *Forming.* During the forming stage, the team comes together; members learn the rules of operation, strategy, and expected outcomes. Team members also learn to work with one another; this stage commonly is a bit frustrating for team members.
- *Storming.* During the storming stage, team members come to grips with the project assignment. They begin to question the strategy, specific tasks, and expected outcomes. Team members may express many differences of opinion, and there is a tendency for some team members to withdraw from the work of the team while other team members begin to try to take charge. If the forming stage is a bit frustrating for team members, the storming stage can be downright difficult to manage—certainly more difficult than the forming stage (Scholtes et al. 2003).
- *Norming.* During the norming stage, team members begin to reach agreement on strategy, specific tasks, and expected outcomes. They begin to work collectively to accomplish common tasks. Once team development reaches the stage of norming, much of the frustration and difficulty experienced during the forming and storming stages of development subsides, and the day-to-day work atmosphere becomes much more congenial.
- *Performing.* During the performing stage, team members become efficient in completing tasks, they refine working relationships, and they work in unison to accomplish any remaining work. They gain a sense of pride and belonging with the team, which greatly helps prevent or minimize internal conflict.
- *Adjourning.* In the adjourning stage, most of the team's goals have been accomplished. The emphasis is on wrapping up final tasks and documenting the effort and results. As the work load is diminished, individual members may be reassigned to other teams, and the team disbands. There may be regret as the team ends; a ceremonial acknowledgement of the work and success of the team can be helpful. If the team is a standing committee with ongoing responsibility, members may be replaced by new people and the team can go back to a forming or storming stage and repeat the development process (https://courses.lumenlearning.com/suny-principlesmanagement/chapter/reading-the-five-stages-of-team-development).

Conflict Resolution

Regardless of how effective a team may be, and regardless of what stage of development a team may have obtained, conflicts will inevitably arise. A conflict involves a difference of opinion. Differences of opinion, while they have their merit and value, can be disruptive to any team. When working with teams, it is not a question of "if" a conflict will arise; it is a question of "when" a conflict will arise. Because the topic of conflict resolution is complex, it involves many experts and many approaches. While it is beyond the scope of this book to cover all the details of conflict management, we would be remiss if we did not provide at least some simple guidelines to help handle a conflict:

- All team members must be heard (one at a time).
- Team members must respect others' opinions and ideas.
- Rely on facts and data where possible. If the data aren't available, reach agreement on a method to obtain them.
- Bring in outside expertise if needed.
- If necessary, test more than one solution and collect data on each.

Consensus

Although it is sometimes necessary to put things to a team vote, it is usually advisable to reach agreement by continued discussion. Although this method, referred to as consensus building, takes more time than voting, it reduces the tendency for some team members to feel left out. Reaching a consensus means that people have to *give a little* in order for the team to make progress.

Brainstorming

In the early stages of problem solving, it is useful to get a large number of ideas. The brainstorming technique does this by asking each person to first express just one idea. This idea is written down so all can see it, and then the next person expresses one thought. After all team members have had a turn, each is asked for a second idea, and so on. One of the rules of a brainstorming session is that no idea is to be criticized or judged. Often members will piggyback on a previous idea and come up with a new or modified thought. The theory of brainstorming is that if all ideas are documented, it is likely that the best idea or solution is on the list somewhere. The next step is to compress the list. It may be possible to combine two or more ideas into one. Sometimes the ideas can be grouped into categories such as machining problems, supplier problems, painting problems, and so on. The team may elect to prioritize the items on the list, agreeing to study the highest-priority items first. Individuals may be assigned the task of pursuing individual ideas further and reporting on them at the next team meeting.

Meeting Management

The hours spent in team meetings are a very valuable resource. Well-managed meetings make the most of the time spent together. Barriers to effective use of meeting time include the following:

- Lack of a clear agenda
- Tendency to digress from the subject
- Feeling on the part of team members that the meeting is a waste of time or has a lower priority than other responsibilities
- Strong disagreement among team members
- Tendency for some members to dominate the discussion and others to withdraw participation

It is the responsibility of the team leader to minimize these and other barriers that may impede the team. Techniques that have proved useful include:

- Distribute the agenda in advance of the meeting. Begin the meeting by reviewing the agenda.
- Call the team back to the agenda when members stray too far or for too long.
- Keep the meeting moving. If an agenda item can't be adequately addressed, it may be best to postpone it for later consideration. Start and end on time. If the business of the meeting has been completed, end the meeting early.
- When conflicts among members arise, help them find a middle ground of agreement. Even an agreement on how to collect data is better than no agreement at all.
- Frequently go around the room and ask for input from each member to maintain an even level of participation.

Experience and training in meeting management will help the leader in using these techniques. When the team leader doesn't have the requisite skills, it may be useful to have a team facilitator assist with meeting management.

QUALITY DOCUMENTATION

There are two components to the quality documentation for purposes of this handbook—use of correct forms and revision control as well as the proper use of policy, procedure, work instructions, forms, and proper documentation practices.

The CQI is expected to apply concepts requiring that particular information be documented on particular forms, with particular expectations for revision control, as they relate to four primary quality systems and/or major quality processes including ISO 9001, First Article Inspection Reports, Initial Sample Inspection Reports (ISIR), and the Production Part Approval Process (PPAP). Key concepts relating to *proper use* include document control, filling out forms completely, correcting misspellings, and initialing changes.

ISO 9001

ISO is an acronym that stands for the International Standards Organization. ISO sponsors multiple quality systems of which, ISO 9001 is but one standard. As an organization, ISO has developed very comprehensive expectations for the documentation of quality systems in all of the standards they sponsor. It is important to note that documentation requirements are established by the ISO and are standardized in most applications which means that individuals and organizations implementing any given ISO standard generally are not at liberty to make choices about what information can or must be documented. That has been defined by the ISO and compliance with documentation requirements is a centerpiece of quality audits. Said another way, companies choose whether or not to become ISO certified, and if they do choose to become ISO certified, they must comply with the documentation requirements specified by the ISO.

General information about ISO 9001, and information about ISO 9001 documentation requirements in particular may be accessed electronically at https://committee.iso.org/sites/tc176sc2/home/page/iso-9001-auditing-practices-grou.html.

First Article Inspection Reports (FAI)

A *First Article Inspection (FAI)* is related to manufacturing, particularly for the assurance of quality prior to shifting from process setup to production.

> *"The purpose of the FAI is to provide objective evidence that all engineering design and specification requirements are properly understood, accounted for, verified, and documented . . . It is intended that the documentation generated will be a quality record of the supplier and customer for review of accountability and planning, for performing periodic surveillance and audits to verify conformance, for evaluating root cause and corrective action for any non-conformances, and for problem investigations" (AS9102: 2000).*

Differing from the ISO family of standards which apply in a standardized form and implementation to multiple industries, the FAI is generally guided by a standard produced for/by specific industries such as the aerospace and/or automotive industries. FAI standards most generally apply to industries that use production operations to produce parts (i.e., articles, pieces, etc.) in quantity. It is important to remember that the documentation requirements for FAI in one industry may not be the same as those in another industry. It is reasonable to expect commonalities in documentation, particularly with respect to a set of characteristics such as part numbers, quality requirements, engineering design specifications, part drawings, identification of specific manufacturing equipment required in production, etc.

Many vendors provide FAI software. Querying the term "First Article Inspection" on the Internet will produce multiple FAI products and vendors. Examples of FAI templates are readily available via the Internet as well. Below is an example set of working papers for FAI as utilized by the MOOG corporation. A thorough review of the example, and comparison with other FAI software, templates, and examples will reveal the commonalities and differences noted above.

Moog	First Article Inspection Report Approval Form	Report No: FAI

Supplier Name & Address		**Supplier Ref. No./ Cert of Conformity**	
		Supplier Code	
Part Number/ Scheme Number		**Drawing/Scheme Issue/Revision**	
Part Description		**Previous FAIR No.**	N/A
Classification	Unclassified Critical Sensitive	**Previous FAIR Date**	

Component Weight		**Any Special Ref.**		**Key Characteristics (CCF's) Specified**	Yes	No

Type of FAIR	Full FAIR		Repeat FAIR		Partial FAIR	
Reason for FAIR	New Part/Drawing/Scheme		Change of Source or Location		Change of Method or Sequence	
	Drawing issue change		Lapse in Production (> 24 months)		Request by MOOG	
	Description of Reason	NEW PART				

Supplier Declaration: This is to certify that:

- This FAIR demonstrates conformance of the product to all requirements and that the information is a true record of the production method.
- Where conformance control features (key characteristics) are specified, stage 2 data collection has commenced.

Name:		Date:		Signature:	
Phone No.		Fax No.		E-Mail	

FAIR Stage 1 Approval	Satisfactory	Conditional	Reject	Name (printed)	Date
Final Approval by *Authorized Supplier*					
Dimensional Report					
Materials Function e.g. Lab./Eng. etc.					
Final Approval					

Comments

FAIR Stage 2 Approval	Decision Signature Satisfactory Reject		Name (printed)	Date
Engineering				
Final Approval				

Comments

Decision Signature

AS9102 Rev A: First Article Inspection Sheet 3 of 6
Form 1: Part Number Accountability

1. Part Number	**2. Part Name**	***3. Serial Number***	**4. FAI Report Number** FAI
5. Part Revision Level	***6. Drawing Number***	***7. Drawing revision level***	***8. Additional Changes***
9. Manufacturing Process Reference	**10. Organization Name**	**11. Supplier Code**	**12. P.O. Number**
13. Detail FAI **Detail FAI** **Assembly FAI**	**14. Full FAI** **Partial FAI**	***Baseline Part Number including revision level***	
	Reason for Partial FAI:		

a) if above part number is a detail part only, go to Field 19
b) if above part number is an assembly, go to the "INDEX" section below.

INDEX of part numbers or sub-assembly numbers required to make the assembly noted above.

15. Part Number	***16. Part Name***	***17. Part Serial Number***	**18. FAI Report Number**

1) Signature indicates that all characteristics are accounted for; meet drawing requirements or are properly documented for disposition.

2) Also indicate if the FAI is complete per Section 5.4: FAI complete FAI not Complete

19. Signature	**20. Date**
21. Reviewed By	**22. Date**
23. Customer Approval	**24. Date**

AS9102 Form 1 Moog
C:\Temp\notesD30550\AS9102_Blank_forms.doc

The example above was accessed on September 3, 2018 at: http://www.moog.com/literature/Corporate/Suppliers/First_Article_Inspection-_Aerospace_Forms.pdf. It should be noted that the example above is illustrative of comprehensive FAI documentation based on the AS9102 standard as applied to the aerospace industry by the MOOG Corporation.

Initial Sample Inspection Reports (ISIR)

The *Initial Sample Inspection Report (ISIR)* differs from the FAI in the scope of documentation provided. The FAI provides more detail and support information than does the ISIR. There are many differences of opinion that are communicated as fact that are not necessarily true implying that the FAI and the ISIR are the same thing. There are many examples in online forums where dialogue participants interchange FAI and ISIR. It should be noted that the FAI is more comprehensive than the ISIR in that, among other things, the FAI contains more information/documentation about the production process and is typically more comprehensive than the ISIR, which is frequently, and correctly, referring to part layout and specification information/documentation separate from production information/documentation.

It was noted in the discussion about FAI that the FAI is generally guided by a standard produced for/by specific industries such as the aerospace and/or automotive industries. FAI standards most generally apply to industries that use production operations to produce parts (i.e., articles, pieces, etc.) in quantity. ISIR standards, content, and procedures also differ by industry and company. It is common to specify the details of what is, and what is not, included in an ISIR when ordering parts and generating purchase contracts. In this case, as was the case with the FAI, what one industry or company may stipulate is included in an ISIR may not be the same as the documentation requirements in another industry or company. It is reasonable to expect commonalities in documentation among different ISIR documents.

As noted in the FAI discussion, many vendors provide ISIR software. Querying the term "Initial Sample Inspection Report" on the Internet will produce multiple ISIR products and vendors.

A final comment on the FAI as compared to the ISIR—These reports do differ in the amount of information contained. The FAI is the more comprehensive document. Both the FAI and the ISIR are used in different industries and within different companies. Inclusion or exclusion of an FAI or an ISIR should be addressed in a purchase contract. It would be prudent practice to specify the elements or pieces of information included in an FAI or ISIR in a purchase contract as well.

Production Part Approval Process (PPAP)

Production Part Approval Process (PPAP) was designed as part of a wider and nationwide effort within the United States to regain manufacturing competitiveness. The PPAP is applied when producing large volumes of complex products. The PPAP process, accordingly, is used primarily within the automotive industry, not confined to the United States but worldwide.

The automotive industry professional organization titled the Automotive Industry Action Group (AIAG) publishes several important standards and documents that the CQI is encouraged to know and understand. These standards and documents include topics such as Measurement Systems Analysis, FMEA, Advanced Product Quality Planning and Control Plans, and Statistical Process Control, to name a few. These resources can be accessed electronically at http://www.aiag.org/store/quality/publications.

When considering PPAP relative to ISO 9001, FAI, and ISIR (as was discussed in this section of the CQI Handbook and the CQI BoK), PPAP would be considered in a scale of complexity and cost to implement in the following descending order: ISO 9001, PPAP, FAI, ISIR.

A primary intent of the PPAP is to ensure that there is a sufficiently robust quality system and documentation effort to ensure that product designs can be produced at a large scale. It is one thing to consider, and ensure, overall product quality in smaller scale production, ensuring that a quality system exists and is fully functional during large scale production operations, often at multiple plants and facilities, often spanning multiple countries, requires a significant amount of support infrastructure and documentation.

PPAP is composed of 19 components as follows:

1. Design Documentation
2. Engineering Change Documentation
3. Engineering Approvals
4. Design Failure Modes and Effects Analyses (DFMEA)
5. Process Flow Diagrams
6. Process Failure Modes and Effects Analyses (PFMEA)
7. Control Plans
8. Measurement Systems Analyses
9. Dimensional Results
10. Records of Material/Performance Tests
11. Initial Sample Inspection Reports
12. Initial Process Studies
13. Qualified Laboratory Documentation
14. Appearance Approval Reports
15. Sample Production Parts
16. Master Production Samples
17. Checking Aides
18. Customer-Specific Requirements
19. Part Submission Warrants

The PPAP components provided above were accessed at: https://www.automotiveengineeringhq.com/production-part-approval-process-ppap/.

A final comparison of FAI, ISIR, and PPAP should confirm the relationship discussed earlier in this chapter. Ranging from more complex to less complex is PPAP, then FAI, and then ISIR. In particular, it should now be clear that, while sometimes confused when used interchangeably, that FAI and ISIR are commonly components of the more detailed and robust PPAP. The CQI should remember, however, that the FAI encompasses more content and information than does the ISIR.

BASIC QUALITY DOCUMENTATION

The second and last major section of the 2018 CQI BoK, section Vi.G.4, relates to *proper use* which includes, as intended in the BoK: proper use of policy, procedure, work instructions and forms, and proper documentation practices (such a filling out forms completely, correcting misspellings, initialing changes, etc.).

There has now been an in-depth discussion of quality documentation. Several points may be summarized as follows:

1. Producing high-quality parts or services does not occur by happenstance.
2. The more complex the parts produced, or the services delivered, the more robust the quality systems must be to ensure quality.
3. There are four particularly important approaches to quality: ISO 9001, FAI, ISIR, and PPAP.
4. Each approach to quality is heavily dependent on documentation.
5. There are clear expectations related to how, why, when, where, and under what circumstances appropriate documentation is generated, used, stored, and revised.
6. Quality documentation serves not only as historical records, but also as the basis of communications, contracts, and performance expectations.
7. As historical records, quality documentation serves as a permanent record of your professional performance and the ASQ Code of Ethics calls for your professional practice as a CQI to produce a high-quality work effort as reflected in your generation and use of quality documentation.

Appendix A

Certified Quality Inspector Body Of Knowledge 2025

The topics in this body of knowledge include additional detail in the form of subtext explanations and the cognitive level at which the questions will be written. This information will provide useful guidance for both the Exam Development Committee and the candidate preparing to take the exam. The subtext is not intended to limit the subject matter or be all-inclusive of what might be covered in an exam. It is meant to clarify the type of content to be included in the exam. The descriptor in parentheses at the end of each line of subtext refers to the maximum cognitive level at which the topic will be tested. A complete description of cognitive levels is provided at the end of this document.
Note: Approximately 20% of the questions in each exam will require calculation.

I. TECHNICAL MATHEMATICS (19 QUESTIONS)

A. Basic Shop Math

Solve basic shop math problems using addition, subtraction, multiplication, division of fractions and decimals, squares, and square roots. Use methods such as truncating and rounding to obtain significant digits for positive and negative numbers. Demonstrate when and how to use % change calculation (new value—original value/original value x 100). (Apply)

B. Basic Algebra

Solve or simplify first-degree and single-variable equations. (Apply)

C. Basic Geometry

Calculate general parameters for basic geometric shapes, such as area, circumference, diameter, radius, perimeter, and volume. Calculate complementary and supplementary angles. Select the minimum number of coordinate points for shapes (e.g., lines and circles). (Apply)

D. Basic Trigonometry

Solve for angles and lengths using trigonometric functions, such as sine, cosine, tangent, and the Pythagorean Theorem. (Apply)

E. Measurement Systems

Convert units within and between English and metric measurement systems (SI), such as inch to microinch and meter to millimeter. (Apply)

F. Numeric Conversions

Use various numbering methods, such as scientific notation, decimals, and fractions, and convert values between these methods. (Apply)

G. Basic Statistics and Applications

1. Measures of central tendency
Calculate mean, median, and mode. (Apply)

2. Measures of dispersion
Calculate range, standard deviation, and variance. (Apply)

3. Measures of proportion
Calculate percentage and ratio measures for various data sets. (Apply)

4. Graphical displays
Define, interpret, and use scatter diagrams, tally sheets, and bar charts to effectively display data in various situations. (Apply)

5. Normal distribution
Explain various characteristics of a normal distribution, such as symmetry, bell curve, and central tendency. (Apply)

II. METROLOGY (26 QUESTIONS)

A. Common Gauges and Measurement Instruments

1. Variable gauges
Identify and use variable gauges, such as micrometers, calipers, dial indicators, and Coordinate Measuring Machines (CMMs). Understand linear scales, such as steel rule and gauge blocks. Use borescopes, thermometers, and temperature probes. (Apply)

2. Attribute gauges
Identify and use attribute gauges, such as thread plugs, progressive gauges, ring gauges, flush pins, pin gauges, and radius gauges. (Apply)

3. Transfer gauges
Identify and use transfer gauges, such as small-hole gauges, telescoping gauges, and spring calipers. (Apply)

4. Measurement scales
Identify and use measurement scales, such as dial, digital, and vernier scales. (Apply)

B. Special Gauges and Applications

Identify and describe the following tools and components. (Remember)

1. Electronic gauging tools: oscilloscopes and multimeters
2. Automatic gauging components: machine vision, ultrasonic, X-ray, and laser
3. Pneumatic gauging components: probes and rings
4. Force gauging: torque wrenches and load cells
5. Environment instrumentation: chart recorders and data loggers

C. Gauge Selection, Handling, and Use

1. 10:1 rule
Apply the 10:1 rule: inspection measurements require better than the tolerance of a dimension by a factor of 10, and calibration standards require better than the inspection measurements by a factor of 10. (Apply)

2. Gauge selection
Select gauges according to the feature or characteristic to be measured, the applicable tolerance, and the accuracy, environment, resolution, and capability of the test instrument. Determine whether the type of measurement should be direct, differential, or transfer. (Apply)

3. Gauge handling, preservation, and storage
Identify and apply various methods of cleaning, handling, and storing gauges. (Apply)

4. Gauge correlation
Identify and apply methods for establishing the correlation between measurement instruments, such as gauge-to-gauge or manual-to-automated process. (Apply)

D. Surface Plate Tools and Techniques

1. Surface plate equipment
Select and use height gauges, V-blocks, and other indicators to measure various types of features. Understand the care, cleaning, calibration, and lapping of a surface plate. (Apply)

2. Angle measurement instruments
Identify and use protractors, sine bars, and angle blocks. (Apply)

E. Specialized Inspection Equipment

1. Measuring mass
Describe and apply weights, balances, and scales. (Apply)

2. Measuring finish
Describe and apply profilometers (e.g., optical and stylus). (Apply)

3. Measuring shape and profile
Understand and describe mechanical comparators, roundness testers, precision spindles, and profile tracers. (Understand)

4. Optical equipment
Describe and apply optical comparators and microscopes. (Apply)

5. Software-based measurement systems
Define and use digital cameras, vision inspection systems (white light/blue light), and other digital systems for product inspection. Recognize software limitations with regard to locating functional datums, target points and areas, hole positions, and the basic operation of the x, y, and z axes. (Apply)

6. Measuring inclination
Define and describe the measurement of the slope or slant of various equipment (mechanical/laser). (Understand)

F. Calibration

1. Calibration systems
Describe the principles and purpose of a calibration system, including the importance of establishing calibration intervals and uncertainty. Identify and use basic tracking and identification methods, such as logs, stickers, radio frequency identifications (RFID), barcodes, and other identification codes to control calibration equipment. (Apply)

2. Calibration standards and equipment traceability
Understand and describe the hierarchy of standards, from working standards through international standards, and the documentation process of a measurement device traceable to international standards. Recognize measurement results before and after an adjustment or repair is made. (Understand)

3. Gauge calibration environment
Understand the effects that environmental conditions have on the calibration process, such as temperature, humidity, vibration, and cleanliness of the gauge. (Understand)

4. Out-of-calibration effects
Describe the effects that out-of-calibration instruments can have on product acceptance and the actions to take in response to this situation. (Apply)

G. Measurement System Analysis (MSA)

Define and understand the following elements of MSA. (Understand)

1. Bias

2. Stability

3. Precision

4. Accuracy

5. Linearity

6. Repeatability and reproducibility (R&R) studies

III. INSPECTION AND TEST (32 QUESTIONS)

A. Blueprints, Drawings, Geometric Dimensioning & Tolerancing (GD&T), and Model Based Definitions

1. Blueprints, engineering drawings, and model based definitions
Define and interpret various sections of technical drawings, such as title blocks, tolerances, and change or revision blocks, as well as notes, scale, and size details. Use this information to conduct inspections. (Apply)

2. Terminology and symbols
Define and interpret drawing views and details for product specifications or other controlling documents. Define and use various terms and symbols from the ASME Y14.5 Standard. (Analyze)

3. Position and bonus tolerances
Calculate position and bonus tolerances from various drawings. (Analyze)

4. Part alignment and datum structure
Determine part alignment and setup using the datum structure. (Analyze)

B. Sampling

Define and utilize the following concepts related to sampling. (Apply)

1. **Acceptance quality limit (AQL)**
2. **Random sampling**
3. **Lot and sample size**
4. **Acceptance number**
5. **Sampling plans**

C. Inspection Planning and Processes

1. Inspection types
Define and distinguish between inspection types, such as incoming material, first-article (first-piece), in-process, and final. (Analyze)

2. Inspection errors
Identify potential inspection errors and their sources, such as bias, fatigue, flinching, distraction, poor time management, setup, and cosine error when using indicators. (Apply)

3. Product traceability
Identify methods to trace products and materials, such as age control, shelf life, first-in first-out (FIFO), barcoding, date codes, and lot and part numbering. (Apply)

4. Identification of nonconforming material
Describe and apply various methods of identifying nonconforming material, such as tagging, labeling, and segregating. (Apply)

5. Levels of severity
Define and describe levels of severity (i.e., critical, major, and minor) and apply them to product features and defects. (Apply)

6. Disposition of nonconforming material
Describe disposition methods, such as rework, reprocess, reinspect, scrap, and customer waiver as determined by a material review board (MRB) or other authority. (Understand)

D. Testing Methods

Define and use the following methods in various situations.

1. Nondestructive testing: X-ray, eddy current, ultrasonic, dye penetrant, magnetic particle, optical, visual, and profile (Understand)

2. Destructive testing: tensile, force testing, and drop test (Apply)

3. Functionality testing: tension, torque, leak testing, and compression (Apply)

4. Hardness testing: Brinell, Rockwell, durometer, and micro-hardness scales (Apply)

E. Software for test equipment

Identify and describe basic tools (e.g., safeguarding, functional checks, comparison of test results, and identification of attributes and parameters) used to ensure that the software for test equipment adequately and correctly performs its intended functions. Recognize various software validation methods. (Remember)

IV. QUALITY ASSURANCE AND IMPROVEMENT (23 QUESTIONS)

A. Quality Improvement

1. Terms and concepts
Define basic quality improvement concepts, such as defect detection and prevention, the cost of poor quality, total quality management (TQM), and the importance of customer satisfaction. (Understand)

2. Products and processes
Define and distinguish between products and processes. Describe the interrelationships of product design, materials used, manufacturing processes, and final output, and how individual steps in a process can affect the final product or the system as a whole. (Understand)

B. Quality Audits

1. Types of audits
Define and describe various types of audits, such as internal, external, system, product, and process. (Understand)

2. Audit process
Explain various stages of the audit process (planning, performance, and closure), such as audit scope and purpose, resources needed, audit schedule, opening meeting, interviewing, data gathering, document and record review, analysis of results, closing meeting, audit documentation (reporting), recordkeeping, and verification of corrective actions. (Understand)

3. Audit tools
Define and describe the purpose of checklists, log sheets, sampling plans, record reviews, document reviews, and forward-and backward-tracing. (Understand)

4. Communication tools and techniques
Define and describe the use of graphs, charts, diagrams, and other aids for written and oral presentations, including interview techniques and listening skills. (Understand)

5. Corrective action requests (CARs)
Describe how CARs from audits can support quality improvement. (Understand)

C. Quality Tools and Techniques

Apply the following quality tools and techniques. (Apply)

1. Pareto charts

2. Cause and effect diagrams

3. Flowcharts

4. Check sheets

5. Scatter diagrams

6. Histograms

D. Statistical Process Control (SPC)

1. Common and special cause variation
Explain the difference between these causes of variation. Illustrate whether a process is in statistical control by monitoring data patterns (e.g., runs, trends, and hugging), and identify what actions should be taken in response. (Apply)

2. Control limits and specification limits
Define, describe, and illustrate the differences between these limits as used in SPC. (Apply)

3. Variables charts
Identify characteristics of and apply $\bar{X} - R$ and $\bar{X} - s$ charts. (Apply)

4. Attributes charts
Identify characteristics of and apply p, np, c, and u charts. (Apply)

5. Process capability analysis
Define and describe the differences between C_p, C_{pk}, P_p, and P_{pk} studies, and identify their application to various types of data. (Understand)

E. Problem-solving Tools and Continuous Improvement Techniques

Define and describe the following tools and techniques in various situations.

1. Plan-do-check-act (PDCA) or plan-do-study-act (PDSA) cycles (Understand)

2. Lean tools for eliminating waste (e.g., error-proofing and value-stream mapping) and lean concepts (e.g., kaizen, flow, and pull) (Understand)

3. 5S/6S (Apply)

4. Six sigma phases: define, measure, analyze, improve, control (DMAIC) (Understand)

5. Failure mode and effects analysis (FMEA) (Understand)

6. 8D Methodology (Understand)

7. 5 Whys (Understand)

8. Fault tree analysis (Understand)

9. Corrective and preventive action (CAPA) (Understand)

R. Resources

1. Environmental and safety support
Define and use various resources related to personal and environmental safety, such as safety data sheets (SDS), material data sheets (MDS), and personal protective equipment (PPE). (Apply)

2. Reference documents
Identify and use national and international standards (e.g., ISO, ANSI, ASTM, QS, and NIST) and customer requirements as authorities that support processes and procedures used to assure quality products. (Apply)

3. Employees as resources
Describe how employees can be empowered and the value they add to project teams or quality improvement teams. Describe typical team roles and responsibilities, such as facilitator, ground rules, and project or team charter. Describe the stages of team development: forming, storming, norming, performing, and adjourning. (Remember)

4. Quality documentation
Identify and apply basic quality documentation, including the correct form/revision for the process (e.g., ISO 9001, First Article Inspection Report, ISIR, PPAPs). Demonstrate proper usage of policy, procedure, work instructions, and forms, and proper documentation practices, such as document control, filling out forms completely, correcting misspellings, and initialing changes. (Apply)

Appendix B

Levels of Cognition

BASED ON BLOOM'S TAXONOMY—REVISED (2001)

In addition to **content** specifics, the subtext for each topic in this BoK also indicates the intended **complexity level** of the test questions for that topic. These levels are based on "Levels of Cognition" (from Bloom's Taxonomy—Revised, 2001) and are presented below in rank order, from least complex to most complex.

REMEMBER

Recall or recognize terms, definitions, facts, ideas, materials, patterns, sequences, methods, principles, etc.

UNDERSTAND

Read and understand descriptions, communications, reports, tables, diagrams, directions, regulations, etc.

APPLY

Know when and how to use ideas, procedures, methods, formulas, principles, theories, etc.

ANALYZE

Break down information into its constituent parts and recognize their relationship to one another and how they are organized; identify sublevel factors or salient data from a complex scenario.

EVALUATE

Make judgments about the value of proposed ideas, solutions, etc., by comparing the proposal to specific criteria or standards.

CREATE

Put parts or elements together in such a way as to reveal a pattern or structure not clearly there before; identify which data or information from a complex set is appropriate to examine further or from which supported conclusions can be drawn.

Appendix C

General Tables of Units of Measurement

These tables have been prepared for the benefit of those requiring tables of units for occasional ready reference. In Section 4 of this Appendix, the tables are carried out to a large number of decimal places and exact values are indicated by underlining. In most of the other tables, only a limited number of decimal places are given, therefore making the tables better adopted by the average user.

1. TABLES OF METRIC UNITS OF MEASUREMENT

In the metric system of measurement, designations of multiples and subdivisions of any unit may be arrived at by combining with the name of the unit the prefixes deka, hecto, and kilo, meaning, respectively, 10, 100, and 1000, and deci, centi, and milli, meaning, respectively, one-tenth, one-hundredth, and one-thousandth. In some of the following metric tables, some such multiples and subdivisions have not been included for the reason that these have little, if any, currency in actual usage.

In certain cases, particularly in scientific usage, it becomes convenient to provide for multiples larger than 1000 and for subdivisions smaller than one-thousandth. Accordingly, the following prefixes have been introduced and these are now generally recognized:

yotta	(Y)	meaning 10^{24}	deci	(d)	meaning 10^{-1}
zetta	(Z)	meaning 10^{21}	centi	(c)	meaning 10^{-2}
exa	(E)	meaning 10^{18}	milli	(m)	meaning 10^{-3}
peta	(P)	meaning 10^{15}	micro	(μ)	meaning 10^{-6}
tera	(T)	meaning 10^{12}	nano	(n)	meaning 10^{-9}
giga	(G)	meaning 10^{9}	pico	(p)	meaning 10^{-12}
mega	(M)	meaning 10^{6}	femto	(f)	meaning 10^{-15}
kilo	(k)	meaning 10^{3}	atto	(a)	meaning 10^{-18}
hecto	(h)	meaning 10^{2}	zepto	(z)	meaning 10^{-21}
deka	(da)	meaning 10^{1}	yocto	(y)	meaning 10^{-24}

Thus a kilometer is 1000 meters and a millimeter is 0.001 meter.

* Reproduced by permission of NIST. http://ts.nist.gov/WeightsAndMeasures/Publications/upload/h4402_appenc.pdf.

Units of Length

10 millimeters (mm)	= 1 centimeter (cm)
10 centimeters	= 1 decimeter (dm) = 100 millimeters
10 decimeters	= 1 meter (m) = 1000 millimeters
10 meters	= 1 dekameter (dam)
10 dekameters	= 1 hectometer (hm) = 100 meters
10 hectometers	= 1 kilometer (km) = 1000 meters

Units of Area

100 square millimeters (mm^2)	= 1 square centimeter (cm^2)
100 square centimeters	= 1 square decimeter (dm^2)
100 square decimeters	= 1 square meter (m^2)
100 square meters	= 1 square dekameter (dam^2) = 1 are
100 square dekameters	= 1 square hectometer (hm^2) = 1 hectare (ha)
100 square hectometers	= 1 square kilometer (km^2)

Units of Liquid Volume

10 milliliters (mL)	= 1 centiliter (cL)
10 centiliters	= 1 deciliter (dL) = 100 milliliters
10 deciliters	= 1 liter[1] = 1000 milliliters
10 liters	= 1 dekaliter (daL)
10 dekaliters	= 1 hectoliter (hL) = 100 liters
10 hectoliters	= 1 kiloliter (kL) = 1000 liters

Units of Volume

1000 cubic millimeters (mm^3)	= 1 cubic centimeter (cm^3)
1000 cubic centimeters	= 1 cubic decimeter (dm^3) = 1,000,000 cubic millimeters
1000 cubic decimeters	= 1 cubic meter (m^3) = 1,000,000 cubic centimeters = 1,000,000,000 cubic millimeters

[1] By action of the 12th General Conference on Weights and Measures (1964) the liter is a special name for the cubic decimeter.

Units of Mass

10 milligrams (mg)	= 1 centigram (cg)
10 centigrams	= 1 decigram (dg) = 100 milligrams
10 decigrams	= 1 gram (g) = 1000 milligrams
10 grams	= 1 dekagram (dag)
10 dekagrams	= 1 hectogram (hg) = 100 grams
10 hectograms	= 1 kilogram (kg) = 1000 grams
1000 kilograms	= 1 megagram (Mg) or 1 metric ton (t)

2. TABLES OF U.S. UNITS OF MEASUREMENT[2]

In these tables where foot or mile is underlined, it is survey foot or U.S. statute mile rather than international foot or mile that is meant.

Units of Length

12 inches (in)	= 1 foot (ft)
3 feet	= 1 yard (yd)
16-1/2 feet	= 1 rod (rd), pole, or perch
40 rods	= 1 furlong (fur) = 660 feet
8 furlongs	= 1 U.S. statute mile (mi) = 5280 feet
1852 meters	= 6076.11549 feet (approximately) = 1 international nautical mile

Units of Area[3]

144 square inches (in^2)	= 1 square foot (ft^2)
9 square feet	= 1 square yard (yd) = 1296 square inches
272-1/4 square feet	= 1 square rod (sq rd)
160 square rods	= 1 acre = 43,560 square feet
640 acres	= 1 square mile (mi^2)
1 mile square	= 1 section of land
6 miles square	= 1 township = 36 sections = 36 square miles

[2] This section lists units of measurement that have traditionally been used in the United States. In keeping with the Omnibus Trade and Competitiveness Act of 1988, the ultimate objective is to make the International System of Units the primary measurement system used in the United States.

[3] Squares and cubes of customary but not of metric units are sometimes expressed by the use of abbreviations rather than symbols. For example, sq ft means square foot, and cu ft means cubic foot.

Units of Volume[4]

1728 cubic inches (in^3)	= 1 cubic foot (ft^3)
27 cubic feet	= 1 cubic yard (yd^3)

Gunter's or Surveyor's Chain Units of Measurement

0.66 foot (ft)	= 1 link (li)
100 links	= 1 chain (ch) = 4 rods = 66 feet
80 chains	= 1 U.S. statute mile (mi) = 320 rods = 5280 feet

Units of Liquid Volume[5]

4 gills (gi)	= 1 pint (pt) = 28.875 cubic inches
2 pints	= 1 quart (qt) = 57.75 cubic inches
4 quarts	= 1 gallon (gal) = 231 cubic inches = 8 pints = 32 gills

Apothecaries Units of Liquid Volume

60 minims (min or ♏)	= 1 fluid dram (fl dr or ƒʒ) = 0.2256 cubic inch
8 fluid drams	= 1 fluid ounce (fl oz or ƒ℥) = 1.8047 cubic inches
16 fluid ounces	= 1 pint (pt or O) = 28.875 cubic inches = 128 fluid drams
2 pints	= 1 quart (qt) = 57.75 cubic inches = 32 fluid ounces = 256 fluid drams
4 quarts	= 1 gallon (gal) = 231 cubic inches = 128 fluid ounces = 1024 fluid drams

Units of Dry Volume[6]

2 pints (pt)	= 1 quart (qt) = 67.200 6 cubic inches
8 quarts	= 1 peck (pk) = 537.605 cubic inches = 16 pints
4 pecks	= 1 bushel (bu) = 2150.42 cubic inches = 32 quarts

[4] Squares and cubes of customary but not of metric units are sometimes expressed by the use of abbreviations rather than symbols. For example, sq ft means square foot, and cu ft means cubic foot.

[5] When necessary to distinguish the *liquid* pint or quart from the *dry* pint or quart, the word "liquid" or the abbreviation "liq" should be used in combination with the name or abbreviation of the *liquid* unit.

[6] When necessary to distinguish *dry* pint or quart from the *liquid* pint or quart, the word "dry" should be used in combination with the name or abbreviation of the dry unit.

Avoirdupois Units of Mass[7]

The "grain" is the same in avoirdupois, troy, and apothecaries units of mass.

27-11/32 grains	= 1 dram (dr)
16 drams	= 1 ounce (oz) = 437-1/2 grains
16 ounces	= 1 pound (lb) = 256 drams = 7000 grains
100 pounds	= 1 hundredweight (cwt)[8]
20 hundredweights	= 1 ton = 2000 pounds[8]

In "gross" or "long" measure, the following values are recognized:

112 pounds	= 1 gross or long hundredweight[8]
20 gross or long hundredweights	= 1 gross or long ton = 2240 pounds[8]

Troy Units of Mass

The "grain" is the same in avoirdupois, troy, and apothecaries units of mass.

24 grains	= 1 pennyweight (dwt)
20 pennyweights	= 1 ounce troy (oz t) = 480 grains
12 ounces troy	= 1 pound troy (lb t) = 240 pennyweights = 5760 grains

Apothecaries Units of Mass

The "grain" is the same in avoirdupois, troy, and apothecaries units of mass.

20 grains	= 1 scruple (s ap or ℈)
3 scruples	= 1 dram apothecaries (dr ap or ʒ) = 60 grains
8 drams apothecaries	= 1 ounce apothecaries (oz ap or ℥) = 24 scruples = 480 grains
12 ounces apothecaries	= 1 pound apothecaries (lb ap) = 96 drams apothecaries = 288 scruples = 5760 grains

[7] When necessary to distinguish the *avoirdupois* dram from the *apothecaries* dram, or to distinguish the *avoirdupois* dram or ounce from the *fluid* dram or ounce, or to distinguish the *avoirdupois* ounce or pound from the *troy* or *apothecaries* ounce or pound, the word "avoirdupois" or the abbreviation "avdp" should be used in combination with the name or abbreviation of the *avoirdupois* unit.

[8] When the terms "hundredweight" and "ton" are used unmodified, they are commonly understood to mean the 100-pound hundredweight and the 2000-pound ton, respectively; these units may be designated "net" or "short" when necessary to distinguish them from the corresponding units in *gross* or *long* measure.

3. NOTES ON BRITISH UNITS OF MEASUREMENT

In Great Britain, the yard, the avoirdupois pound, the troy pound, and the apothecaries pound are identical with the units of the same names used in the United States. The tables of British linear measure, troy mass, and apothecaries mass are the same as the corresponding United States tables, except for the British spelling "drachm" in the table of apothecaries mass. The table of British avoirdupois mass is the same as the United States table up to one pound; above that point the table reads:

14 pounds	= 1 stone
2 stones	= 1 quarter = 28 pounds
4 quarters	= 1 hundredweight = 112 pounds
20 hundredweight	= 1 ton = 2240 pounds

The present British gallon and bushel—known as the "Imperial gallon" and "Imperial bushel"—are, respectively, about 20 percent and 3 percent larger than the United States gallon and bushel. The Imperial gallon is defined as the volume of 10 avoirdupois pounds of water under specified conditions, and the Imperial bushel is defined as 8 Imperial gallons. Also, the subdivision of the Imperial gallon as presented in the table of British apothecaries fluid measure differs in two important respects from the corresponding United States subdivision, in that the Imperial gallon is divided into 160 fluid ounces (whereas the United States gallon is divided into 128 fluid ounces), and a "fluid scruple" is included. The full table of British measures of capacity (which are used alike for liquid and for dry commodities) is as follows:

4 gills	= 1 pint
2 pints	= 1 quart
4 quarts	= 1 gallon
2 gallons	= 1 peck
8 gallons (4 pecks)	= 1 bushel
8 bushels	= 1 quarter

The full table of British apothecaries measure is as follows:

20 minims	= 1 fluid scruple
3 fluid scruples	= 1 fluid drachm = 60 minims
8 fluid drachms	= 1 fluid ounce
20 fluid ounces	= 1 pint
8 pints	= 1 gallon (160 fluid ounces)

4. TABLES OF UNITS OF MEASUREMENT

All underlined figures are exact.

Units of Length—International Measure[9]

Units		Inches	Feet	Yards	Miles	Centimeters	Meters
1 inch	=	1	0.08333333	0.02777778	0.00001578283	2.54	0.0254
1 foot	=	12	1	0.3333333	0.0001893939	30.48	0.3048
1 yard	=	36	3	1	0.0005681818	91.44	0.9144
1 mile	=	63,360	5280	1760	1	160,934.4	1609.344
1 centimeter	=	0.3937008	0.03280840	0.01093613	0.000006213712	1	0.01
1 meter	=	39.37008	3.280840	1.093613	0.0006213712	100	1

Units of Length—Survey Measure[9]

Units		Links	Feet	Rods	Chains	Miles	Meters
1 link	=	1	0.66	0.04	0.01	0.000125	0.2011684
1 foot	=	1.515152	1	0.06060606	0.01515152	0.0001893939	0.3048006
1 rod	=	25	16.5	1	0.25	0.003125	5.029210
1 chain	=	100	66	4	1	0.0125	20.11684
1 mile	=	8000	5280	320	80	1	1609.347
1 meter	=	4.970960	3.280833	0.1988384	0.04970960	0.0006213699	1

Units of Area—International Measure[10]

All underlined figures are exact.

Units		Square inches	Square feet	Square yards
1 square inch	=	1	0.006944444	0.0007716049
1 square foot	=	144	1	0.1111111
1 square yard	=	1296	9	1
1 square mile	=	4,014,489,600	27,878,400	3,097,600
1 square centimeter	=	0.1550003	0.001076391	0.0001195990
1 square meter	=	1550.003	10.76391	1.195990

---

[9] One international foot = 0.999998 survey foot (exactly)
One international mile = 0.999998 survey mile (exactly)

Note: 1 survey foot = 1200/3937 meter (exactly)
1 international foot = 12 × 0.0254 meter (exactly)
1 international foot = 0.0254 × 39.37 survey foot (exactly)

[10] One square survey foot = 1.000004 square international feet
One square survey mile = 1.000004 square international miles

Units		Square miles	Square centimeters	Square meters
1 square inch	=	0.0000000002490977	6.4516	0.00064516
1 square foot	=	0.00000003587006	929.0304	0.09290304
1 square yard	=	0.0000003228306	8361.2736	0.83612736
1 square mile	=	1	25,899,881,103.36	2,589,988.110336
1 square centimeter	=	0.00000000003861022	1	0.0001
1 square meter	=	0.0000003861022	10,000	1

Units of Area—Survey Measure[11]

Units		Square feet	Square rods	Square chains	Acres
1 square foot	=	1	0.003673095	0.0002295684	0.00002295684
1 square rod	=	272.25	1	0.0625	0.00625
1 square chain	=	4356	16	1	0.1
1 acre	=	43,560	160	10	1
1 square mile	=	27,878,400	102,400	6400	640
1 square meter	=	10.76387	0.03953670	0.002471044	0.0002471044
1 hectare	=	107,638.7	395.3670	24.71044	2.471044

Units		Square miles	Square meters	Hectares
1 square foot	=	0.00000003587006	0.09290341	0.000009290341
1 square rod	=	0.000009765625	25.29295	0.002529295
1 square chain	=	0.00015625	404.6873	0.04046873
1 acre	=	0.0015625	4046.873	0.4046873
1 square mile	=	1	2,589,998	258.9998
1 square meter	=	0.0000003861006	1	0.0001
1 hectare	=	0.003861006	10,000	1

Units of Volume

All underlined figures are exact.

Units		Cubic inches	Cubic feet	Cubic yards
1 cubic inch	=	1	0.0005787037	0.00002143347
1 cubic foot	=	1728	1	0.03703704
1 cubic yard	=	46,656	27	1
1 cubic centimeter	=	0.06102374	0.00003531467	0.000001307951
1 cubic decimeter	=	61.02374	0.03531467	0.001307951
1 cubic meter	=	61,023.74	35.31467	1.307951

[11] One square survey foot = 1.000004 square international feet
One square survey mile = 1.000004 square international miles

Units		Milliliters (cubic centimeters)	Liters (cubic decimeters)	Cubic meters
1 cubic inch	=	16.387064	0.016387064	0.000016387064
1 cubic foot	=	28,316.846592	28.316846592	0.028316846592
1 cubic yard	=	764,554.857984	764.554857984	0.764554857984
1 cubic centimeter	=	1	0.001	0.000001
1 cubic decimeter	=	1000	1	0.001
1 cubic meter	=	1,000,000	1000	1

Units of Capacity or Volume—Dry Volume Measure

All underlined figures are exact.

Units		Dry pints	Dry quarts	Pecks	Bushels
1 dry pint	=	1	0.5	0.0625	0.015625
1 dry quart	=	2	1	0.125	0.03125
1 peck	=	16	8	1	0.25
1 bushel	=	64	32	4	1
1 cubic inch	=	0.0297616	0.0148808	0.00186010	0.000465025
1 cubic foot	=	51.42809	25.71405	3.214256	0.80356395
1 liter	=	1.816166	0.9080830	0.1135104	0.02837759
1 cubic meter	=	1816.166	908.0830	113.5104	28.37759

Units		Cubic inches	Cubic feet	Liters	Cubic meters
1 dry pint	=	33.6003125	0.01944463	0.5506105	0.0005506105
1 dry quart	=	67.200625	0.03888925	1.101221	0.001101221
1 peck	=	537.605	0.311114	8.809768	0.008809768
1 bushel	=	2150.42	1.244456	35.23907	0.03523907
1 cubic inch	=	1	0.0005787037	0.01638706	0.00001638706
1 cubic foot	=	1728	1	28.31685	0.02831685
1 liter	=	61.02374	0.03531467	1	0.001
1 cubic meter	=	61,023.74	35.31467	1000	1

Units of Capacity or Volume—Liquid Volume Measure

All underlined figures are exact.

Units		Minims	Fluid drams	Fluid ounces	Gills
1 minim	=	1	0.01666667	0.002083333	0.0005208333
1 fluid dram	=	60	1	0.125	0.03125
1 fluid ounce	=	480	8	1	0.25
1 gill	=	1920	32	4	1
1 liquid pint	=	7680	128	16	4
1 liquid quart	=	15,360	256	32	8
1 gallon	=	61,440	1024	128	32
1 cubic inch	=	265.9740	4.432900	0.5541126	0.1385281
1 cubic foot	=	459,603.1	7660.052	957.5065	239.3766
1 milliliter	=	16.23073	0.2705122	0.03381402	0.008453506
1 liter	=	16,230.73	270.5122	33.81402	8.453506

Units		Liquid pints	Liquid quarts	Gallons	Cubic inches
1 minim	=	0.0001302083	0.00006510417	0.00001627604	0.003759766
1 fluid dram	=	0.0078125	0.00390625	0.0009765625	0.22558594
1 fluid ounce	=	0.0625	0.03125	0.0078125	1.8046875
1 gill	=	0.25	0.125	0.03125	7.21875
1 liquid pint	=	1	0.5	0.125	28.875
1 liquid quart	=	2	1	0.25	57.75
1 gallon	=	8	4	1	231
1 cubic inch	=	0.03463203	0.01731602	0.004329004	1
1 cubic foot	=	59.84416	29.92208	7.480519	1728
1 milliliter	=	0.002113376	0.001056688	0.0002641721	0.06102374
1 liter	=	2.113376	1.056688	0.2641721	61.02374

Units		Cubic feet	Milliliters	Liters
1 minim	=	0.000002175790	0.06161152	0.00006161152
1 fluid dram	=	0.0001305474	3.696691	0.003696691
1 fluid ounce	=	0.001044379	29.57353	0.02957353
1 gill	=	0.004177517	118.2941	0.1182941
1 liquid pint	=	0.01671007	473.1765	0.4731765
1 liquid quart	=	0.03342014	946.3529	0.9463529
1 gallon	=	0.1336806	3785.412	3.785412
1 cubic inch	=	0.0005787037	16.38706	0.01638706
1 cubic foot	=	1	28,316.85	28.31685
1 milliliter	=	0.00003531467	1	0.001
1 liter	=	0.03531467	1000	1

Units of Mass Not Less Than Avoirdupois Ounces

All underlined figures are exact.

Units		Avoirdupois ounces	Avoirdupois pounds	Short hundredweights	Short tons
1 avoirdupois ounce	=	1	0.0625	0.000625	0.00003125
1 avoirdupois pound	=	16	1	0.01	0.0005
1 short hundredweight	=	1600	100	1	0.05
1 short ton	=	32,000	2000	20	1
1 long ton	=	35,840	2240	22.4	1.12
1 kilogram	=	35.27396	2.204623	0.02204623	0.001102311
1 metric ton	=	35,273.96	2204.623	22.04623	1.102311

Units		Long tons	Kilograms	Metric tons
1 avoirdupois ounce	=	0.00002790179	0.028349523125	0.000028349523125
1 avoirdupois pound	=	0.0004464286	0.45359237	0.00045359237
1 short hundredweight	=	0.04464286	45.359237	0.045359237
1 short ton	=	0.8928571	907.18474	0.90718474
1 long ton	=	1	1016.0469088	1.0160469088
1 kilogram	=	0.0009842065	1	0.001
1 metric ton	=	0.9842065	1000	1

Units of Mass Not Greater Than Pounds and Kilograms

All underlined figures are exact.

Units		Grains	Apothecaries scruples	Pennyweights	Avoirdupois drams
1 grain	=	1	0.05	0.04166667	0.03657143
1 apoth. scruple	=	20	1	0.8333333	0.7314286
1 pennyweight	=	24	1.2	1	0.8777143
1 avdp. dram	=	27.34375	1.3671875	1.139323	1
1 apoth. dram	=	60	3	2.5	2.194286
1 avdp. ounce	=	437.5	21.875	18.22917	16
1 apoth. or troy oz.	=	480	24	20	17.55429
1 apoth. or troy pound	=	5760	288	240	210.6514
1 avdp. pound	=	7000	350	291.6667	256
1 milligram	=	0.01543236	0.0007716179	0.0006430149	0.0005643834
1 gram	=	15.43236	0.7716179	0.6430149	0.5643834
1 kilogram	=	15,432.36	771.6179	643.0149	564.3834

Units		Apothecaries drams	Avoirdupois ounces	Apothecaries or troy ounces	Apothecaries or troy pounds
1 grain	=	0.01666667	0.002285714	0.002083333	0.0001736111
1 apoth. scruple	=	0.3333333	0.04571429	0.04166667	0.003472222
1 pennyweight	=	0.4	0.05485714	0.05	0.004166667
1 avdp. dram	=	0.4557292	0.0625	0.5696615	0.004747179
1 apoth. dram	=	1	0.1371429	0.125	0.01041667
1 avdp. ounce	=	7.291667	1	0.9114583	0.07595486
1 apoth. or troy ounce	=	8	1.097143	1	0.083333333
1 apoth. or troy pound	=	96	13.16571	12	1
1 avdp. pound	=	116.6667	16	14.58333	1.215278
1 milligram	=	0.0002572060	0.00003527396	0.00003215075	0.000002679229
1 gram	=	0.2572060	0.03527396	0.03215075	0.002679229
1 kilogram	=	257.2060	35.27396	32.15075	2.679229

Units		Avoirdupois pounds	Milligrams	Grams	Kilograms
1 grain	=	0.0001428571	64.79891	0.06479891	0.00006479891
1 apoth. scruple	=	0.002857143	1295.9782	1.2959782	0.0012959782
1 pennyweight	=	0.003428571	1555.17384	1.55517384	0.00155517384
1 avdp. dram	=	0.00390625	1771.8451953125	1.7718451953125	0.0017718451953125
1 apoth. dram	=	0.008571429	3887.9346	3.8879346	0.0038879346
1 avdp. ounce	=	0.0625	28,349.523125	28.349523125	0.028349523125
1 apoth. or troy ounce	=	0.06857143	31,103.4768	31.1034768	0.0311034768
1 apoth. or troy pound	=	0.8228571	373,241.7216	373.2417216	0.3732417216
1 avdp. pound	=	1	453,592.37	453.59237	0.45359237
1 milligram	=	0.000002204623	1	0.001	0.000001
1 gram	=	0.002204623	1000	1	0.001
1 kilogram	=	2.204623	1,000,000	1000	1

5. TABLES OF EQUIVALENTS

In these tables it is necessary to differentiate between the "international foot" and the "survey foot"—the survey foot is underlined.

When the name of a unit is enclosed in brackets (thus, [1 hand] . . .), this indicates (1) that the unit is not in general current use in the United States, or (2) that the unit is believed to be based on "custom and usage" rather than on formal authoritative definition.

Equivalents involving decimals are, in most instances, rounded off to the third decimal place except where they are exact, in which cases these exact equivalents are so designated. The equivalents of the imprecise units "tablespoon" and "teaspoon" are rounded to the nearest milliliter.

Units of Length

angstrom (Å)[12]	0.1 nanometer (exactly). 0.0001 micrometer (exactly). 0.0000001 millimeter (exactly). 0.000000004 inch.
1 cable's length	120 fathoms (exactly). 720 feet (exactly). 219 meters.
1 centimeter (cm)	0.3937 inch.
1 chain (ch) (Gunter's or surveyor's)	66 feet (exactly). 20.1168 meters.
1 decimeter (dm)	3.937 inches.
1 dekameter (dam)	32.808 feet.
1 fathom	6 feet (exactly). 1.8288 meters.
1 foot (ft)	0.3048 meter (exactly).
1 furlong (fur)	10 chains (surveyor's) (exactly). 660 feet (exactly). 1/8 U.S. statute mile (exactly). 201.168 meters.
[1 hand]	4 inches.
1 inch (in)	2.54 centimeters (exactly).
1 kilometer (km)	0.621 mile.
1 league (land)	3 U.S. statute miles (exactly). 4.828 kilometers.
1 link (li) (Gunter's or surveyor's)	0.66 foot (exactly). 0.201168 meter.
1 meter (m)	39.37 inches. 1.094 yards.
1 micrometer	0.001 millimeter (exactly). 0.00003937 inch.
1 mil	0.001 inch (exactly). 0.0254 millimeter (exactly).
1 mile (mi) (U.S. statute)[13]	5280 feet survey (exactly). 1.609 kilometers.
1 mile (mi) (international)	5280 feet international (exactly).

[12] The angstrom is basically defined as 10^{-10} meter.

[13] The term "statute mile" originated with Queen Elizabeth I who changed the definition of the mile from the Roman mile of 5000 feet to the statute mile of 5280 feet. The international mile and the U.S. statute mile differ by about 3 millimeters although both are defined as being equal to 5280 feet. The international mile is based on the international foot (0.3048 meter) whereas the U.S. statute mile is based on the survey foot (1200/3937 meter).

1 mile (mi) (international nautical)[14]	1.852 kilometers (exactly). 1.151 survey miles.
1 millimeter (mm)	0.03937 inch.
1 nanometer (nm)	0.001 micrometer (exactly). 0.00000003937 inch.
1 Point (typography)	0.013837 inch (exactly). 1/72 inch (approximately). 0.351 millimeter.
1 rod (rd), pole, or perch	16-1/2 feet (exactly). 5.0292 meters.
1 yard (yd)	0.9144 meter (exactly).

Units of Area

1 acre[15]	43,560 square feet (exactly). 0.405 hectare.
1 are	119.599 square yards. 0.025 acre.
1 hectare	2.471 acres.
[1 square (building)]	100 square feet.
1 square centimeter (cm2)	0.155 square inch.
1 square decimeter (dm^2)	15.500 square inches.
1 square foot (ft^2)	929.030 square centimeters.
1 square inch (in^2)	6.4516 square centimeters (exactly).
1 square kilometer (km^2)	247.104 acres. 0.386 square mile.
1 square meter (m^2)	1.196 square yards. 10.764 square feet.
1 square mile (mi^2)	258.999 hectares.
1 square millimeter (mm^2)	0.002 square inch.
1 square rod (rd^2), sq pole, or sq perch	25.293 square meters.
1 square yard (yd^2)	0.836 square meter.

[14] The international nautical mile of 1852 meters (6076.11549 . . . feet) was adopted effective July 1, 1954, for use in the United States. The value formerly used in the United States was 6080.20 feet = 1 nautical (geographical or sea) mile.

[15] The question is often asked as to the length of a side of an acre of ground. An acre is a unit of area containing 43,560 square *feet*. It is not necessarily square, or even rectangular. But, if it is square, then the length of a side is equal to $\sqrt{43{,}560} = 208.710+$ *feet*.

Units of Capacity or Volume

1 barrel (bbl), liquid	31 to 42 gallons.[16]
1 barrel (bbl), standard for fruits, vegetables, and other dry commodities, except cranberries	7056 cubic inches. 105 dry quarts. 3.281 bushels, struck measure.
1 barrel (bbl), standard, cranberry	5826 cubic inches. 86-45/64 dry quarts. 2.709 bushels, struck measure.
1 bushel (bu) (U.S.) struck measure	2150.42 cubic inches (exactly). 35.239 liters.
[1 bushel, heaped (U.S.)]	2747.715 cubic inches. 1.278 bushels, struck measure.[17]
[1 bushel (bu) (British Imperial) 1.032 U.S. bushels, struck measure (struck measure)]	2219.36 cubic inches.
1 cord (cd) (firewood)	128 cubic feet (exactly).
1 cubic centimeter (cm^3)	0.061 cubic inch.
1 cubic decimeter (dm^3)	61.024 cubic inches.
1 cubic foot (ft^3)	7.481 gallons. 28.316 cubic decimeters.
1 cubic inch (in^3)	0.554 fluid ounce. 4.433 fluid drams. 16.387 cubic centimeters.
1 cubic meter (m^3)	1.308 cubic yards.
1 cubic yard (yd^3)	0.765 cubic meter.
1 cup, measuring	8 fluid ounces (exactly). 237 milliliters. 1/2 liquid pint (exactly).
1 dekaliter (daL)	2.642 gallons. 1.135 pecks.
1 dram, fluid (or liquid) (fl dr or f℥) (U.S.)	1/8 fluid ounce (exactly). 0.226 cubic inch. 3.697 milliliters. 1.041 British fluid drachms.
[1 drachm, fluid (fl dr) (British)]	0.961 U.S. fluid dram. 0.217 cubic inch. 3.552 milliliters.

16 There are a variety of "barrels" established by law or usage. For example, Federal taxes on fermented liquors are based on a barrel of 31 gallons; many State laws fix the "barrel for liquids" as 31-1/2 gallons; one State fixes a 36-gallon barrel for cistern measurement; Federal law recognizes a 40-gallon barrel for "proof spirits"; by custom, 42 gallons comprise a barrel of crude oil or petroleum products for statistical purposes, and this equivalent is recognized "for liquids" by four States.

17 Frequently recognized as 1-1/4 bushels, struck measure.

1 gallon (gal) (U.S.)	231 cubic inches (exactly). 3.785 liters. 0.833 British gallon. 128 U.S. fluid ounces (exactly).
[1 gallon (gal) (British Imperial)]	277.42 cubic inches. 1.201 U.S. gallons. 4.546 liters. 160 British fluid ounces (exactly).
1 gill (gi)	7.219 cubic inches. 4 fluid ounces (exactly). 0.118 liter.
1 hectoliter (hL)	26.418 gallons. 2.838 bushels.
1 liter (1 cubic decimeter exactly)	1.057 liquid quarts. 0.908 dry quart. 61.025 cubic inches.
1 milliliter (mL)	0.271 fluid dram. 16.231 minims. 0.061 cubic inch.
1 ounce, fluid (or liquid) (fl oz or f℥)(U.S.)	1.805 cubic inches. 29.573 milliliters. 1.041 British fluid ounces.
[1 ounce, fluid (fl oz) (British)]	0.961 U.S. fluid ounce. 1.734 cubic inches. 28.412 milliliters.
1 peck (pk)	8.810 liters.
1 pint (pt), dry	33.600 cubic inches. 0.551 liter.
1 pint (pt), liquid	28.875 cubic inches (exactly). 0.473 liter.
1 quart (qt), dry (U.S.)	67.201 cubic inches. 1.101 liters. 0.969 British quart.
1 quart (qt), liquid (U.S.)	57.75 cubic inches (exactly). 0.946 liter. 0.833 British quart.
[1 quart (qt) (British)]	69.354 cubic inches. 1.032 U.S. dry quarts. 1.201 U.S. liquid quarts.
1 tablespoon, measuring	3 teaspoons (exactly). 15 milliliters. 4 fluid drams. 1/2 fluid ounce (exactly).
1 teaspoon, measuring	1/3 tablespoon (exactly). 5 milliliters. 1-1/3 fluid drams.[18]
1 water ton (English)	270.91 U.S. gallons. 224 British Imperial gallons (exactly).

[18] The equivalent "1 teaspoon = 1-1/3 fluid drams" has been found by the Bureau to correspond more closely with the actual capacities of "measuring" and silver teaspoons than the equivalent "1 teaspoon = 1 fluid dram," which is given by a number of dictionaries.

Units of Mass

1 assay ton[19] (AT)	29.167 grams.
1 carat (c)	200 milligrams (exactly). 3.086 grains.
1 dram apothecaries (dr ap or ʒ)	60 grains (exactly). 3.888 grams.
1 dram avoirdupois (dr avdp)	27-11/32 (= 27.344) grains. 1.777 grams.
1 gamma (γ)	1 microgram (exactly).
1 grain	64.79891 milligrams (exactly).
1 gram (g)	15.432 grains. 0.035 ounce, avoirdupois.
1 hundredweight, gross or long[20] (gross cwt)	112 pounds (exactly). 50.802 kilograms.
1 hundredweight, gross or short (cwt or net cwt)	100 pounds (exactly). 45.359 kilograms.
1 kilogram (kg)	2.205 pounds.
1 microgram (μg) [the Greek letter mu in combination with the letter g]	0.000001 gram (exactly).
1 milligram (mg)	0.015 grain.
1 ounce, avoirdupois (oz avdp)	437.5 grains (exactly). 0.911 troy or apothecaries ounce. 28.350 grams.
1 ounce, troy or apothecaries (oz t or oz ap or ℥)	480 grains (exactly). 1.097 avoirdupois ounces. 31.103 grams.
1 pennyweight (dwt)	1.555 grams.
1 point	0.01 carat. 2 milligrams.
1 pound, avoirdupois (lb avdp)	7000 grains (exactly). 1.215 troy or apothecaries pounds. 453.59237 grams (exactly).
1 pound, troy or apothecaries (lb t or lb ap)	5760 grains (exactly). 0.823 avoirdupois pound. 373.242 grams.
1 scruple (s ap or ℈)	20 grains (exactly). 1.296 grams.

19 Used in assaying. The assay ton bears the same relation to the milligram that a ton of 2000 pounds avoirdupois bears to the ounce troy; hence the mass in milligrams of precious metal obtained from one assay ton of ore gives directly the number of troy ounces to the net ton.

20 The gross or long ton and hundredweight are used commercially in the United States to only a very limited extent, usually in restricted industrial fields. The units are the same as the British "ton" and "hundredweight."

1 ton, gross or long[21]	2240 pounds (exactly). 1.12 net tons (exactly). 1.016 metric tons.
1 ton, metric (t)	2204.623 pounds. 0.984 gross ton. 1.102 net tons.
1 ton, net or short	2000 pounds (exactly). 0.893 gross ton. 0.907 metric ton.

Appendix D

Cumulative Standard Normal Distribution Table

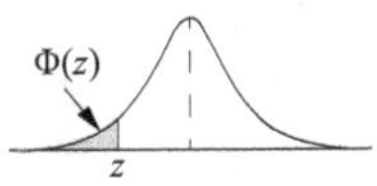

Standard Normal Z Table

Table values give ϕ (z) = Pr(Z < z) for negative values of z

z	0.0	0.01	0.02	0.03	0.04	0.05	0.06	0.07	0.08	0.09
–4.0	0.0000	0.0000	0.0000	0.0000	0.0000	0.0000	0.0000	0.0000	0.0000	0.0000
–3.9	0.0000	0.0000	0.0000	0.0000	0.0000	0.0000	0.0000	0.0000	0.0000	0.0000
–3.8	0.0001	0.0001	0.0001	0.0001	0.0001	0.0001	0.0001	0.0001	0.0001	0.0001
–3.7	0.0001	0.0001	0.0001	0.0001	0.0001	0.0001	0.0001	0.0001	0.0001	0.0001
–3.6	0.0002	0.0002	0.0001	0.0001	0.0001	0.0001	0.0001	0.0001	0.0001	0.0001
–3.5	0.0002	0.0002	0.0002	0.0002	0.0002	0.0002	0.0000	0.0002	0.0002	0.0002
–3.4	0.0003	0.0003	0.0003	0.0003	0.0003	0.0003	0.0003	0.0003	0.0003	0.0002
–3.3	0.0005	0.0005	0.0005	0.0004	0.0004	0.0004	0.0004	0.0004	0.0004	0.0003
–3.2	0.0007	0.0007	0.0006	0.0006	0.0006	0.0006	0.0006	0.0005	0.0005	0.0005
–3.1	0.0010	0.0009	0.0009	0.0009	0.0008	0.0008	0.0008	0.0008	0.0007	0.0007
–3.0	0.0013	0.0013	0.0013	0.0012	0.0012	0.0011	0.0011	0.0011	0.0010	0.0010
–2.9	0.0019	0.0018	0.0018	0.0017	0.0016	0.0016	0.0015	0.0015	0.0014	0.0014
–2.8	0.0026	0.0025	0.0024	0.0023	0.0023	0.0022	0.0021	0.0021	0.0020	0.0019
–2.7	0.0035	0.0034	0.0033	0.0032	0.0031	0.0030	0.0029	0.0028	0.0027	0.0006
–2.6	0.0047	0.0045	0.0044	0.0043	0.0041	0.0040	0.0039	0.0038	0.0037	0.0036
–2.5	0.0062	0.0060	0.0059	0.0057	0.0055	0.0054	0.0052	0.0051	0.0049	0.0048
–2.4	0.0082	0.0080	0.0078	0.0075	0.0073	0.0071	0.0069	0.0068	0.0066	0.0064
–2.3	0.0107	0.0104	0.0102	0.0099	0.0096	0.0094	0.0091	0.0089	0.0087	0.0084
–2.2	0.0139	0.0136	0.0132	0.0129	0.0125	0.0122	0.0119	0.0116	0.0113	0.0110
–2.1	0.0179	0.0174	0.0170	0.0166	0.0162	0.0158	0.0154	0.0150	0.0146	0.0143
–2.0	0.0228	0.0222	0.0217	0.0212	0.0207	0.0202	0.0197	0.0192	0.0188	0.0183
–1.9	0.0287	0.0281	0.0274	0.0268	0.0262	0.0256	0.0250	0.0244	0.0239	0.0233
–1.8	0.0359	0.0351	0.0344	0.0356	0.0329	0.0322	0.0314	0.0307	0.0301	0.0294
–1.7	0.0446	0.0436	0.0427	0.0418	0.0409	0.0401	0.0392	0.0384	0.0375	0.0367
–1.6	0.0548	0.0537	0.0526	0.0516	0.0505	0.0495	0.0485	0.0475	0.0465	0.0455
–1.5	0.0668	0.0655	0.0643	0.0630	0.0618	0.0606	0.0694	0.0582	0.0572	0.0559
–1.4	0.0808	0.0793	0.0778	0.0764	0.0749	0.0735	0.0721	0.0708	0.0694	0.0681
–1.3	0.0968	0.0951	0.0934	0.0918	0.0901	0.0885	0.0869	0.0853	0.0838	0.0823
–1.2	0.1151	0.1131	0.1112	0.1093	0.1075	0.1056	0.1038	0.1020	0.1003	0.0985
–1.1	0.1357	0.1335	0.1314	0.1292	0.1271	0.1251	0.1230	0.1210	0.1190	0.1170
–1.0	0.1587	0.1562	0.1539	0.1515	0.1492	0.1469	0.1446	0.1423	0.1401	0.1379
–0.9	0.1841	0.1814	0.1788	0.1762	0.1736	0.1711	0.1685	0.1660	0.1639	0.1611
–0.8	0.2119	0.2090	0.2061	0.2033	0.2005	0.1977	0.1949	0.1922	0.1894	0.1867
–0.7	0.2420	0.2389	0.2358	0.2527	0.2296	0.2266	0.2236	0.2206	0.2177	0.2148
–0.6	0.2743	0.2709	0.2676	0.2643	0.2611	0.2578	0.2546	0.2514	0.2483	0.2451
–0.5	0.3085	0.3050	0.3015	0.2581	0.2946	0.2912	0.2877	0.2843	0.2810	0.2776
–0.4	0.3446	0.3409	0.3372	0.3336	0.3300	0.3264	0.3228	0.3192	0.3156	0.3121
–0.3	0.3321	0.3783	0.3745	0.3707	0.3669	0.3632	0.3594	0.3557	0.3520	0.3483
–0.2	0.4207	0.4168	0.4129	0.4090	0.4052	0.4013	0.3974	0.3936	0.3897	0.3659
–0.1	0.4602	0.4562	0.4522	0.4483	0.4443	0.4404	0.4364	0.4325	0.4286	0.4245
–0	0.5000	0.4960	0.4920	0.4880	0.4840	0.4801	0.4761	0.4721	0.4681	0.4641

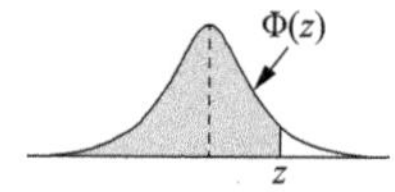

Standard Normal Z Table

Table values give ϕ (z) = Pr(Z < z) for positive values of z

z	0.0	0.01	0.02	0.03	0.04	0.05	0.06	0.07	0.08	0.09
0.0	0.5000	0.5040	0.5080	0.5120	0.5160	0.5199	0.5239	0.5279	0.5319	0.5359
0.1	0.5398	0.5438	0.5478	0.5517	0.5557	0.5596	0.5636	0.5675	0.5714	0.5753
0.2	0.5793	0.5832	0.5871	0.5910	0.5948	0.5987	0.6026	0.6064	0.6103	0.6141
0.3	0.6179	0.6217	0.6255	0.6293	0.6331	0.6368	0.6406	0.6443	0.6480	0.6517
0.4	0.6554	0.6591	0.6628	0.6664	0.6700	0.6736	0.6772	0.6808	0.6844	0.6879
0.5	0.6915	0.6950	0.6985	0.7019	0.7054	0.7088	0.7123	0.7157	0.7190	0.7224
0.6	0.7257	0.7291	0.7324	0.7357	0.7389	0.7422	0.7454	0.7486	0.7517	0.7549
0.7	0.7580	0.7611	0.7642	0.7673	0.7704	0.7734	0.7764	0.7794	0.7823	0.7852
0.8	0.7881	0.7910	0.7939	0.7967	0.7995	0.8023	0.8051	0.8078	0.8106	0.8133
0.9	0.8159	0.8186	0.8212	0.8238	0.8264	0.8289	0.8315	0.8340	0.8365	0.8389
1.0	0.8413	0.8438	0.8461	0.8485	0.8508	0.8531	0.8554	0.8577	0.8599	0.8621
1.1	0.8643	0.8665	0.8686	0.8708	0.8729	0.8749	0.8770	0.8790	0.8810	0.8830
1.2	0.8849	0.8869	0.8888	0.8907	0.8925	0.8944	0.8962	0.8980	0.8997	0.9015
1.3	0.9032	0.9049	0.9066	0.9082	0.9099	0.9115	0.9131	0.9147	0.9162	0.9177
1.4	0.9192	0.9207	0.9222	0.9236	0.9251	0.9265	0.9279	0.9292	0.9306	0.9319
1.5	0.9332	0.9345	0.9357	0.9370	0.9382	0.9394	0.9406	0.9418	0.9429	0.9441
1.6	0.9452	0.9463	0.9474	0.9484	0.9495	0.9505	0.9515	0.9525	0.9535	0.9545
1.7	0.9554	0.9564	0.9573	0.9582	0.9591	0.9599	0.9608	0.9616	0.9625	0.9633
1.8	0.9641	0.9649	0.9656	0.9664	0.9671	0.9678	0.9686	0.9693	0.9699	0.9706
1.9	0.9713	0.9719	0.9726	0.9732	0.9738	0.9744	0.9750	0.9756	0.9761	0.9767
2.0	0.9772	0.9778	0.9783	0.9788	0.9793	0.9798	0.9803	0.9808	0.9812	0.9817
2.1	0.9821	0.9826	0.9830	0.9834	0.9838	0.9842	0.9846	0.9850	0.9854	0.9857
2.2	0.9861	0.9864	0.9868	0.9871	0.9875	0.9878	0.9881	0.9884	0.9887	0.9890
2.3	0.9893	0.9896	0.9898	0.9901	0.9904	0.9906	0.9909	0.9911	0.9913	0.9916
2.4	0.9918	0.9920	0.9922	0.9925	0.9927	0.9929	0.9931	0.9932	0.9934	0.9936
2.5	0.9938	0.9940	0.9941	0.9943	0.9945	0.9946	0.9948	0.9949	0.9951	0.9952
2.6	0.9953	0.9955	0.9956	0.9957	0.9959	0.9960	0.9961	0.9962	0.9963	0.9964
2.7	0.9965	0.9966	0.9967	0.9968	0.9969	0.9970	0.9971	0.9972	0.9973	0.9974
2.8	0.9974	0.9975	0.9976	0.9977	0.9977	0.9978	0.9979	0.9979	0.9980	0.9981
2.9	0.9981	0.9982	0.9982	0.9983	0.9984	0.9984	0.9985	0.9985	0.9986	0.9986
3.0	0.9987	0.9987	0.9987	0.9988	0.9988	0.9989	0.9989	0.9989	0.9990	0.9990
3.1	0.9990	0.9991	0.9991	0.9991	0.9992	0.9992	0.9992	0.9992	0.9993	0.9993
3.2	0.9993	0.9993	0.9994	0.9994	0.9994	0.9994	0.9994	0.9995	0.9995	0.9995
3.3	0.9995	0.9995	0.9995	0.9996	0.9996	0.9996	0.9996	0.9996	0.9996	0.9997
3.4	0.9997	0.9997	0.9997	0.9997	0.9997	0.9997	0.9997	0.9997	0.9997	0.9998
3.5	0.9998	0.9998	0.9998	0.9998	0.9998	0.9998	0.9998	0.9998	0.9998	0.9998
3.6	0.9998	0.9998	0.9999	0.9999	0.9999	0.9999	0.9999	0.9999	0.9999	0.9999
3.7	0.9999	0.9999	0.9999	0.9999	0.9999	0.9999	0.9999	0.9999	0.9999	0.9999
3.8	0.9999	0.9999	0.9999	0.9999	0.9999	0.9999	0.9999	0.9999	0.9999	0.9999
3.9	1.0000	1.0000	1.0000	1.0000	1.0000	1.0000	1.0000	1.0000	1.0000	1.0000
4.0	1.0000	1.0000	1.0000	1.0000	1.0000	1.0000	1.0000	1.0000	1.0000	1.0000

Appendix E

Factors Helpful in Constructing Control Charts for Variables

n	A	A_2	A_3	C_4	$1/C_4$	B_3	B_4	B_5
2	2.12130	1.88060	2.65870	0.79788	1.25332	0.00000	3.26657	0.00000
3	1.73205	1.02307	1.95440	0.88623	1.12838	0.00000	2.56814	0.00000
4	1.50000	0.72851	1.62810	0.92132	1.08540	0.00000	2.26603	0.00000
5	1.34164	0.57680	1.42729	0.93999	1.06384	0.00000	2.08895	0.00000
6	1.22474	0.48332	1.28713	0.95153	1.05094	0.03033	1.96967	0.02886
7	1.13389	0.41934	1.18191	0.95937	1.04235	0.11770	1.88230	0.11292
8	1.06066	0.37255	1.09910	0.96503	1.03624	0.18508	1.81492	0.17861
9	1.00000	0.33670	1.03166	0.96931	1.03166	0.23912	1.76088	0.23179
10	0.94868	0.30821	0.97535	0.97266	1.02811	0.28372	1.71628	0.27596
11	0.90453	0.28507	0.92739	0.97535	1.02527	0.32128	1.67872	0.31336
12	0.86603	0.26582	0.88591	0.97756	1.02296	0.35352	1.64648	0.34559
13	0.83205	0.24942	0.84954	0.97941	1.02102	0.38162	1.61838	0.37377
14	0.80178	0.23533	0.81734	0.98097	1.01940	0.40622	1.59378	0.39849
15	0.77460	0.22310	0.78854	0.98232	1.01800	0.42826	1.57174	0.42069
16	0.75000	0.21234	0.76260	0.98348	1.01680	0.44783	1.55217	0.44043
17	0.72761	0.20279	0.73905	0.98451	1.01573	0.46574	1.53426	0.45852
18	0.70711	0.19426	0.71758	0.98541	1.01481	0.48185	1.51815	0.47482
19	0.68825	0.18657	0.69787	0.98621	1.01398	0.49656	1.50344	0.48971
20	0.67082	0.17960	0.67970	0.98693	1.01324	0.51015	1.48985	0.50348
21	0.65465	0.17328	0.66289	0.98758	1.01258	0.52272	1.47728	0.51623
22	0.63960	0.16748	0.64726	0.98817	1.01197	0.53440	1.46560	0.52808
23	0.62554	0.16214	0.63269	0.98870	1.01143	0.54514	1.45486	0.53898
24	0.61237	0.15869	0.61906	0.98919	1.01093	0.55527	1.44473	0.54927
25	0.60000	0.15263	0.60628	0.98964	1.01047	0.56478	1.43522	0.55893

Continued

Continued

n	B_6	d_2	$1/d_2$	d_3	D_1	D_2	D_3	D_4
2	2.60633	1.128	0.88652	0.853	0.000	3.687	0.0000	3.26862
3	2.27597	1.693	0.59067	0.888	0.000	4.357	0.0000	2.57354
4	2.08774	2.059	0.48567	0.880	0.000	4.699	0.0000	2.28218
5	1.96360	2.326	0.42992	0.864	0.000	4.918	0.0000	2.11436
6	1.87420	2.534	0.39463	0.848	0.000	5.078	0.0000	2.00395
7	1.80582	2.704	0.36982	0.833	0.205	5.203	0.0758	1.92419
8	1.75145	2.847	0.35125	0.820	0.387	5.307	0.1359	1.86407
9	1.70683	2.970	0.33670	0.808	0.546	5.394	0.1838	1.81616
10	1.66936	3.078	0.32489	0.797	0.687	5.469	0.2232	1.77680
11	1.63734	3.173	0.31516	0.787	0.812	5.534	0.2559	1.74409
12	1.60953	3.258	0.30694	0.778	0.924	5.592	0.2836	1.71639
13	1.58505	3.336	0.29976	0.770	1.026	5.646	0.3076	1.69245
14	1.56345	3.407	0.29351	0.763	1.118	5.696	0.3282	1.67185
15	1.54395	3.472	0.28802	0.756	1.204	5.740	0.3468	1.65323
16	1.52653	3.532	0.28313	0.750	1.282	5.782	0.3630	1.63703
17	1.51050	3.588	0.27871	0.744	1.356	5.820	0.3779	1.62207
18	1.49600	3.640	0.27473	0.739	1.423	5.857	0.3909	1.60907
19	1.48271	3.689	0.27108	0.734	1.487	5.891	0.4031	1.59691
20	1.47038	3.735	0.26774	0.729	1.548	5.922	0.4145	1.58554
21	1.45893	3.778	0.26469	0.724	1.606	5.950	0.4251	1.57491
22	1.44826	3.819	0.26185	0.720	1.659	5.979	0.4344	1.56559
23	1.43842	3.858	0.25920	0.716	1.710	6.006	0.4432	1.55677
24	1.42911	3.859	0.25913	0.712	1.723	5.995	0.4465	1.55351
25	1.42035	3.931	0.25439	0.708	1.807	6.055	0.4597	1.54032

Appendix F

Values of K1 for Computing Repeatability Using the Range Method

	Number of trials						
n	**2**	**3**	**4**	**5**	**6**	**7**	**8**
1	3.65	2.70	2.30	2.08	1.93	1.82	1.74
2	4.02	2.85	2.40	2.15	1.98	1.86	1.77
3	4.19	2.91	2.43	2.16	2.00	1.87	1.78
4	4.26	2.94	2.44	2.17	2.00	1.88	1.79
5	4.33	2.96	2.45	2.18	2.01	1.89	1.79
6	4.36	2.98	2.46	2.19	2.01	1.89	1.79
7	4.40	2.98	2.46	2.19	2.02	1.89	1.79
8	4.44	2.99	2.48	2.19	2.02	1.89	1.79
9	4.44	2.99	2.48	2.20	2.02	1.89	1.80
10	4.44	2.99	2.48	2.20	2.02	1.89	1.80
11	4.44	3.01	2.48	2.20	2.02	1.89	1.80
12	4.48	3.01	2.49	2.20	2.02	1.89	1.81
13	4.48	3.01	2.49	2.20	2.02	1.90	1.81
14	4.48	3.01	2.49	2.20	2.03	1.90	1.81
15	4.48	3.01	2.49	2.20	2.03	1.90	1.81
***n* ≥ 16**	4.56	3.05	2.50	2.21	2.04	1.91	1.81

n = (# of parts [samples]) × (# of operators)

Continued

Continued

	Number of trials						
n	9	10	11	12	13	14	15
1	1.67	1.62	1.57	1.54	1.51	1.48	1.47
2	1.71	1.65	1.60	1.56	1.52	1.49	1.47
3	1.71	1.66	1.60	1.57	1.53	1.50	1.47
4	1.72	1.66	1.61	1.57	1.53	1.50	1.48
5	1.72	1.66	1.61	1.57	1.54	1.51	1.48
6	1.72	1.66	1.61	1.57	1.54	1.51	1.48
7	1.72	1.66	1.61	1.57	1.54	1.51	1.48
8	1.73	1.67	1.61	1.57	1.54	1.51	1.48
9	1.73	1.67	1.62	1.57	1.54	1.51	1.48
10	1.73	1.67	1.62	1.57	1.54	1.51	1.48
11	1.73	1.67	1.62	1.57	1.54	1.51	1.48
12	1.73	1.67	1.62	1.57	1.54	1.51	1.48
13	1.73	1.67	1.62	1.57	1.54	1.51	1.48
14	1.73	1.67	1.62	1.57	1.54	1.51	1.48
15	1.73	1.67	1.62	1.58	1.54	1.51	1.48
$n \geq 16$	1.73	1.67	1.62	1.58	1.54	1.51	1.48

n = (# of parts [samples]) × (# of operators)

Appendix G

Values of K2 for Computing Reproducibility Using the Range Method

Number of operators												
3	4	5	6	7	8	9	10	11	12	13	14	15
2.70	2.30	2.08	1.93	1.82	1.74	1.67	1.62	1.57	1.54	1.51	1.48	1.47

References

2GIG Engineering. n.d. "MEMS Inclinometer." Accessed July 26, 2018. https://www.2gig-eng.com/mems-inclinometer.

Achatz, Thomas. 2006. *Technical Shop Mathematics*. 3rd ed. New York: Industrial Press.

ADMET Material Testing. 2006. "Tensile Property Testing of Plastics." Accessed May 2012. http://asia.matweb.com/reference/tensilestrength.asp.

Advancing Standards Transforming Markets (ASTM International), Subcommittee E11.50. 2024. *Standard Guide for Measurement Systems Analysis (MSA)*. West Conshohocken, PA.

Ahearn, G. 2015. "Digital Inspection: A Camera Perspective." *Quality Magazine*, May 1. Accessed November 27, 2018. https://www.qualitymag.com/articles/ 92585-digital-inspection-a-camera-perspective.

Aircraft NDT. 2005. "NDT Terminology at Level III NDT." Accessed May 2012. http:// www.aircraftndt.com/ndt_terms.htm.

American National Standards Institute (ANSI). *ANSI Z540-1-1994 (Revised 2002). American National Standard for Calibration: Calibration Laboratories and Measuring and Test Equipment (General Requirements)*. Boulder, CO: ANSI.

American Society of Mechanical Engineers (ASME). 1985. *ANSI/ASME B89.1.12M-1985 Methods for Performance Evaluation of Coordinate Measuring Machines*. New York: ASME.

———. *ANSI/ASME 1986. B46.1-1095 Surface Texture—Surface Roughness, Waviness, and Lay.* New York: ASME.

ANSI/ASQ (American National Standards Institute/American Society for Quality). 2008a. *Z1.4-2008 Sampling Procedures and Tables for Inspection by Attributes*. Milwaukee: ASQ Quality Press.

———. 2008b. *Z1.9-2008 Sampling Procedures and Tables for Inspection by Variables*. Milwaukee: ASQ Quality Press.

Anderson, J. 2015. "Select the Best Identification Technology." *Quality Magazine*, March 3. Accessed November 27, 2018. https://www.qualitymag.com/articles/ 92479-select-the-best-identification-technology.

Anleitner, M. 2011. *The Power of Deduction: Failure Modes and Effects Analysis for Design*. Milwaukee, WI: ASQ Quality Press.

Arter, D., Charles Cianfrani, and Jack West. 2012. *How to Audit the Process-Based Quality Management System*. 2nd ed. Milwaukee, WI: ASQ Quality Press

Arter, D. 2003. *Quality Audits for Improved Performance*. 3rd ed. Milwaukee, WI: ASQ Quality Press.

Aslam, M., Aamir Saghir, & Liaquat Ahmad. 2021. *Introduction to Statistical Process Control*. Hoboken, NY: John Wiley & Sons Publishing.

ASM International. 2005. *Metallography and Microstructures*. Vol. 9 of ASM Handbook. Materials Park, OH: ASM International.

ASME (American Society of Mechanical Engineers). 1985. *ANSI/ASME B89.1.12M-1985 Methods for Performance Evaluation of Coordinate Measuring Machines*. New York: ASME.

ASME (American Society of Mechanical Engineers). 1994. *Y14.5M Geometric Dimensioning and Tolerancing (GD&T).* New York: ASME.

———. 2003. *Y14.3-2003 Multiview and Sectional View Drawings.* New York: ASME.

ASNT (American Society for Nondestructive Testing). 2012. Accessed May 2012. http://www.asnt.org/ndt/primer1.htm.

ASQ Statistics Division. 2005. *Glossary and Tables for Statistical Quality Control.* 4th ed. Milwaukee: ASQ Quality Press.

ASQ. "Mistake Proofing." Accessed July, 24, 2025. https://asq.org/quality-resources/quality-glossary.

ASQ. "Quality Glossary," s.v. "flow." http://asq.org/glossary/f.html.

ASQ. "Quality Glossary," s.v. "kaizen." Accessed July, 24, 2025. https://asq.org/quality-resources/quality-glossary.

ASQ/ANSI/ISO 19011:2018. 2018. *Guidelines for Auditing Management Systems.* Milwaukee, WI: ASQ Quality Press.

ASTM (American Society for Testing and Materials). 1996. *D4356 Standard Practice for Establishing Consistent Test Method Tolerances.* West Conshohocken, PA: ASTM.

Axelrod, N. 2010. "Machine Vision Adds Value." *Quality Magazine,* February 25.

Bond, T. P. 1983. "Basics of an MRB." *Quality,* November, 48.

Bosch, J. A. 1984. *66 Centuries of Measurement.* Dayton, OH: Sheffield Measurement Division.

Briggs, D., and T. Hendrix. 2003. *Recommended Practice Guide for Geometric Tolerances.* Chicago: Boeing Commercial Airplanes.

Bucher, J. 2004. *The Metrology Handbook.* Milwaukee: ASQ Quality Press.

———. 2006. *The Quality Calibration Handbook: Developing and Managing a Calibration Program.* Milwaukee: ASQ Quality Press.

Bucher, J. 2004. *The Metrology Handbook.* Milwaukee: ASQ Quality Press.

Burgess, G. K. 2006. *Pyrometers: The Measurement of High Temperatures.* Palm Springs, CA: Wexford College Press.

Burr, W. 1976. *Statistical Quality Control Methods.* New York: Marcel Dekker.

Busch, T. 1966. *Fundamentals of Dimensional Metrology.* Albany, NY: Delmar.

Cañas, A. J., and J. D. Novak, eds. 2006. "Geometric Tolerancing Explained by the Means of Concept Maps." *Proceedings of the Second International Conference on Concept Mapping.* San José, Costa Rica.

CALCE and the University of Maryland. 2001. "Material Hardness." Accessed May 2012. http://www.calce.umd.edu/general/Facilities/Hardness_ad_.htm#rf11.

Chrysler, Ford, and General Motors Corporations. 2012. "A Potential Failure Modes And Effects Analysis." *The American Society for Quality and the Automotive Industry Action Group.* 4th ed. Milwaukee, WI.

Clark, R. 2012. "Bonus Tolerances for GD&T: A True Balancing Act." *HighBeam Research.* February 1. Accessed May 15, 2018. http://www.highbeam.com/ doc/1G1–142386165.html.

Coleman, L. B. 2023. *Advanced Quality Auditing.* Milwaukee, WI: ASQ Quality Press.

———. 2023. *The Certified Quality Auditor Handbook.* 5th ed. Milwaukee, WI: ASQ Quality Press.

Cox, H. 2010. "Other Dimensions: Working with Measurement Uncertainty." *Quality Magazine,* January 28.

Curtis, M. 2010. *Handbook of Dimensional Measurement.* 4th ed. New York: Industrial Press.

Darmody, W. J. 1967. "Elements of a Generalized Measuring System." In *Handbook of Industrial Metrology.* Englewood Cliffs, NJ: Prentice-Hall.

Davies, E. R. 2005. *Machine Vision: Theory, Algorithms, Practicalities.* 3rd ed. San Francisco: Morgan Kaufmann/Elsevier.

Davis, J. R. 2004. *Tensile Testing.* 2nd ed. Materials Park, OH: ASM International.

De Silva, G. M. S. 2002. *Basic Metrology for ISO 9000 Certification.* Woburn, MA: Butterworth-Heinemann.

Defeo, J. 2017. *Juran's Quality Handbook: The Complete Guide to Performance Excellence.* 7th ed. New York: McGraw-Hill.

Deleryd, M. 1996. "Process capability studies in theory and practice." Licentiate Thesis, Division of Quality Technology and Statistics, Luleå University of Technology.

Deming, W. Edwards. 1951. *Elementary Principles of the Statistical Control of Quality.* 2nd ed. Tokyo: Nippon Kagaku Gijutsu Remmei.

———. 1982. *Quality, Productivity, and Competitive Position.* Cambridge, MA: MIT, Center for Advanced Engineering Study.

Deming, W. E. 1986. *Out of the Crisis.* Cambridge, MA: MIT Center for Advanced Engineering.

Department of Defense. 1980. *Procedures for Performing a Failure Mode, Effects, and Criticality Analysis.* Washington, DC: General Accounting Office Technical Library.

Dettmers, D. 2004. "NDT Methods for Mechanical Integrity." *IRC. Best Practices Workshop,* November.

Devor, R., T. Chang, and J. Sutherland. 2006. *Statistical Quality Design and Control.* 2nd ed. Upper Saddle River, NJ: Prentice Hall.

Digital Design. 2012. "Drawing Sheet Layout." Accessed May 12, 2012, at http://www.metrication.com/drafting/layout.htm.

Dirks, A. L. 1995–2000. "Introduction to Technical Drawing." Accessed May 10, 2012. http://webhost.bridgew.edu/adirks/ald/courses/design/desdraft.htm.

Dorris, A. L., and B. L. Foote. 1978. "Inspection Errors and Statistical Quality Control: A Survey." *AIIE Transactions* 10, no. 2 (1978): 184–92.

Dotson, C. 2006. *Fundamentals of Dimensional Metrology.* 5th ed. New York: Delmar Cengage Learning.

Dovich, R. 1992. "Acceptance Sampling." In *Quality Engineering Handbook,* edited by T. Pyzdek and R. Berger. Milwaukee: ASQC Quality Press; New York: Marcel Dekker.

Drews, W. E. 1978. "How to Measure Roundness." *Tooling and Production,* June, 156–60.

Durivage, M. 2015. *Practical Measurement and Variable Measurement Systems Analysis.* Milwaukee, WI: ASQ Press.

Duffy, G., 2014. *Modern Kaizen: Continuous and Breakthrough Improvement.* ASQ Press, Milwaukee, WI.

Duncan, A. J. 1974. *Quality Control and Industrial Statistics.* 4th ed. Homewood, IL: Richard D. Irwin.

Effective Training. 2012. "GD&T Glossary." Accessed May 15, 2012. http://www. etinews .com/gdt_glossary.html.

Elshennawy, A. K. and G. S. Weheba. 2015. *Manufacturing Processes and Materials.* 4th ed. Dearborn, MI: Society of Manufacturing Engineers.

Elshennawy, A. K., I. Ham, and P. H. Cohen. 1988. "Evaluating the Performance of Coordinate Measuring Machines." *ASQC Quality Progress,* January, 59–65.

England, G. 2012. "Hardness Testing." *Surface Engineering Forum.* Accessed May 2012. Available at http://www.gordonengland.co.uk/hardness/.

Fabel, G. 1997. "Machine Vision Systems Looking Better All the Time." *Quality Digest,* October. Accessed November 27, 2018. http://www.qualitydigest.com/ oct97/html /machvis.html.

Farago, F. T., and M. Curtis. 1994. *Handbook of Dimensional Measurement.* 3rd ed. New York: Industrial Press.

Feigenbaum, A. V. 1991. *Total Quality Control.* 3rd ed., revised. New York: McGraw-Hill.

Forsgren, R. 2021. *Lean Knowledge Management: How NASA Implemented A Practical KM Program.* New York: Business Expert Press.

Fox, D. 2012. "Professional Skills." Department of Civil Engineering, University of Portsmouth. Accessed May 15, 2012. http://www.scribd.com/doc/93761941/ 22248861-Professional-Skills-v1-1.

Gaylor, J. F. W., and C. R. Shotbolt. 1964. *Metrology for Engineers*. London: Cassell & Company.

General FTA Image. Accessed September 2, 2018. https://www.bing.com/images/search?view=detailV2&ccid=6Q5ABuhe&id=B9994B84EAD31D278152B8E3E6EE3ED77C26D75A&thid=OIP.6Q5ABuheyNEPBbluf7p7mQHaGR&mediaurl=http%3a%2f%2fwww.conceptdraw.com%2fHow-To-Guide%2fpicture%2fEngineering-Fault-Tree-Analysis-Diagrams-Accident-Analytic-Tree-FTA-Diagram.png&exph=725&expw=855&q=Fault+Tree+Example+for+Manufacturer&simid=608034532623188989&selectedIndex=104&ajaxhist=0

Gerson, S., and S. Gerson. 2011. *Technical Writing: Process & Product*. 8th ed. New York: Prentice Hall.

Gordon, D. K. 2002. "Where Does Quality Begin?" ASQ *Quality Progress*, March.

Griffith, Gary G. 1986. *Quality Technician's Handbook*. Eaglewood Cliffs, New Jersey: Prentice Hall.

Groover, Mikell P. 2002. *Fundamentals of Modern Manufacturing*. New York: John Wiley & Sons.

Gryna, F. M., R. C. H. Chua, and J. A. Defeo. 2007. *Juran's Quality Planning and Analysis for Enterprise Quality*. 5th ed. New York: McGraw-Hill.

Gryna, F. 2001. *Quality Planning & Analysis: From Product Development Through Use*. 4th ed. New York: McGraw-Hill.

Gryna, F. M. 1988. "Manufacturing Planning." In *Juran's Quality Control Handbook*, edited by J. M. Juran. 4th ed. New York: McGraw-Hill.

———. 2001. *Quality Planning and Analysis*. 4th ed. New York: McGraw-Hill.

Gunderson, D. S., 1994. *Blueprint Reading and Sketching*, NAVEDTRA 14040. NAVSUP Logistics Tracking Number 0504-LP-026-7150. Naval Education and Training Professional Development and Technology Center.

Gupta, Bhisham C. and Irwin Guttman. 2013. *Statistics and Probability with Applications for Engineers and Scientists*. New York: John Wiley & Sons.

Gupta, Bhisham C., and H. Fred Walker. 2007a. *Applied Statistics for the Six Sigma Green Belt*. Revised ed. Milwaukee, WI: ASQ Quality Press.

———. 2007b. *Statistical Quality Control for the Six Sigma Green Belt*. Milwaukee, WI: ASQ Quality Press.

Heil, P., and R. Daigle. 2004. "Vision Systems Look to the Future." *Quality Digest*, March. Accessed November 27, 2018. https://www.qualitydigest.com/mar04/ articles/01_article.shtml.

Hickman, I. 2001. *Oscilloscopes*. 5th ed. Burlington, MA: Newnes/Elsevier.

Horton, Holbrook L. 1999. *Mathematics at Work*. 4th ed. New York: Industrial Press.

IEEE, 2017. *IEEE/ISO/IEC 24765 Systems and Software Engineering—Vocabulary*. Geneva.

Instron. 2012. "Tensile Testing." *Material Testing Solutions*. Accessed May 2012. Available at http://www.instron.us/wa/applications/test_types/tension/default.aspx.

Integrated Publishing. 2012. "Destructive Testing." Accessed May 2012. Available at http://www.tpub.com/content/construction/14250/css/14250_169.htm.

ISO (International Organization for Standardization). 2002. *ISO Standards Handbook, Technical Drawings, Volume 1: Technical Drawings in General*. Geneva: ISO.

ISO (International Organization for Standardization). 2006. ISO 80000–4:2006 *Quantities and units—Part 4: Mechanics*. Geneva: ISO.

ISO TR 11462–3:2020.2020 Guidelines for implementation of statistical process control (SPC)—Part 3, Digital Document. Geneva, Switzerland.

ISO (International Organization for Standardization). 2007. *ISO Guide 99: International Vocabulary of Metrology—Basic and General Concepts and Associated Terms (VIM)*. Geneva: ISO.

ISO (International Organization for Standardization). 2005. *ISO/IEC 17025:2005 General requirements for the competence of testing and calibration laboratories.* Geneva: ISO.

ISO (International Organization for Standardization). 2007. *ISO/IEC Guide 99: International vocabulary of metrology—Basic and general concepts and associated terms (VIM).* Geneva: ISO.

Jenkins, J. D. 2003. "The Uncertain World of Measurement Credibility." *Quality Digest,* October. Accessed November 28, 2018. http://www.qualitydigest.com/ oct03/articles/03_article.shtml.

Juran, J. and Blanton Godfrey. 1999. *Juran's Quality Control Handbook.* 5th ed. New York, NY: McGraw-Hill Publishing.

Juran, J., Frank Gryna, and R. Bingham. 1974. *Juran's Quality Control Handbook.* 3rd ed. New York, NY: McGraw-Hill Higher Education.

Juran, J. M., and F. N. Gryna Jr. 2005. *Juran's Quality Planning and Analysis for Enterprise Quality.* New York: McGraw-Hill.

Juran, J., and J. De Feo. 2016. *Juran's Quality Handbook.* 7th ed. New York: McGraw-Hill.

Kausek, J. 2006. *The Management System Auditor's Handbook.* Milwaukee, WI: ASQ Quality Press.

Kennedy, C. W., E. G. Hoffman, and S. D. Bond. 1987. *Inspection and Gaging.* New York: Industrial Press.

Kenton, W. "Lean Six Sigma: Definition, Principles, and Benefits." Accessed July, 24, 2025. https://www.investopedia.com/terms/l/lean-six-sigma.asp.

Kimothi, S. K. 2002. *The Uncertainty of Measurements: Physical and Chemical Metrology and Analysis.* Milwaukee: ASQ Quality Press.

Klepp, P. 1998. "The Future of Optical Gaging Equipment." *Quality Digest,* July. Accessed November 27, 2018. http://www.qualitydigest.com/july98/html/ optgage.html.

Knoche, S. 2006. "Embrace Model Based Definition." *Quality Magazine,* July 1. Accessed November 27, 2018. https://www.qualitymag.com/articles/84810-embrace-model-based-definition.

Konz, S., G. Peterson, and A. Joshi. 1981. "Reducing Inspection Errors." *ASQC Quality Progress* 7 (1981): 24–26.

Kotz, S., and Cynthia R. Lovelace. 1998. *Process Capability Indices in Theory and Practice.* London, New York, Sydney, Auckland: Arnold.

Kubiak, T. M., and Donald. W. Benbow. 2012. *The Certified Six Sigma Black Belt Handbook.* 2nd ed. Milwaukee, WI: ASQ Quality Press.

Lamit, L. G., and K. L. Kitto. 1994. *Principles of Engineering Drawing.* Minneapolis: West Publishing.

Larson, B. 2000. PowerPoint Presentation. Center for NDE, Iowa State University.

Leach, R., ed. 2011. *Optical Measurement of Surface Topography.* Berlin/Heidelberg: Springer-Verlag.

Lee, Q., and Brad Snyder. 2008. *Value Stream and Process Mapping.* Bellingham, WA: Enna Publishing.

Manos, A., and Chad Vincent. 2012. *The Lean Handbook: A Guide to the Bronze Certification Body of Knowledge.* Milwaukee, WI: ASQ Quality Press.

Martin, C. 2005. *The Surface Texture Book.* London: Thomas and Hudson.

Martin, J. 2015. "Tracking Progress: Re-evaluating Laser Tracker Technology." *Quality Magazine,* January 1. Accessed November 27, 2018. https://www.quality mag.com /articles/92339-tracking-progress-re-evaluating-laser-tracker-technology.

Martin, P. 2023. *Technical Writing for Engineers Who Don't Like to Write: A Straightforward Guide Written by an Engineer for Engineers.* Lititz, PA: Kitzi Press.

Mazur, G. 2009. *Digital Millimeter Principles.* 4th ed. Orland Park, IL: American Technical Publishers.

McKenzie, R. M. 1958. "On the Accuracy of Inspectors." *Ergonomics* 1 (1958): 258–72.

McKinley, J. 2006. "Re: PDSA or PDCA?" Accessed January 2008. http://deming.eng.clemson.edu/den/archive/2006.07/msg00055.html.

McMaster, R. C., and S. A. Wenk. 1951. *A Basic Guide for Management's Choice of Nondestructive Tests*. Special Technical Publication No. 112. Philadelphia, PA: American Society for Testing and Materials.

McShane Vaughn, Mary. 2016. *The Probability Handbook*. Milwaukee, WI: ASQ Quality Press.

———. 2022. *The ASQ Certified Six Sigma Black Belt Handbook*, 4th ed. Milwaukee, WI: ASQ Quality Press.

Megaw, E. D. 1979. "Factors Affecting Visual Inspection Accuracy." *Applied Ergonomics* 10 (1979): 27–32.

Meyers, M. A., and K. K. Chawla. 1999. "Section 3.8." In *Mechanical Behavior of Materials*. Upper Saddle River, NJ: Prentice Hall.

Micropyretics Heaters International. 2012. Accessed May 2012. Available at http://www.mhi-inc.com/PG4/tensile.html.

Mil-Std 1629A, 1980. Military Standard: Procedures for Performing a Failure Mode, Effects, and Criticality Analysis. *EverySpec*. http://www.everyspec.com/MIL-STD/MIL-STD-1600-1699/MIL_STD_1629A_1556/.

Montgomery, D. 2005. *Design and Analysis of Experiments*. 6th ed. New York, NY: John Wiley and Sons Publishing.

Montgomery, D. C. 2005. *Introduction to Statistical Quality Control*. 5th ed. New York: John Wiley & Sons.

Moog Corporation. 2018. "First Article Inspection Documentation." Accessed September 3. http://www.moog.com/literature/Corporate/Suppliers/First_Article_Inspection-_Aerospace_Forms.pdf

Moyer, Robert and Murray R Spiegel. 2005. *Schaum's Outline of College Algebra*. 3rd ed. New York: McGraw-Hill.

MTAG (Manufacturing Technology Advisory Group). 2001. "Core Competencies." Accessed May 2, 2012. http://learningconnections.org/mtag/mtag_files/bio.htm.

Munns, I. 2004. "Liquid Penetrant Inspection." TWI World Centre for Materials Joining Technology.

Myler, H. 1999. *Fundamentals of Machine Vision*. Bellingham, WA: Society of Photo-Optical Instrumentation Engineers.

NASA (National Aeronautics and Space Administration). 2010. *Measurement Uncertainty Analysis: Principles and Methods, NASA Measurement Quality Assurance Handbook*. Washington, DC: NASA.

NASA (National Aeronautics and Space Administration). 2010. *Measuring and Test Equipment Specifications, NASA Measurement Quality Assurance Handbook*. Washington, DC: NASA, 2010.

National Conference of Standards Laboratories (NCSL). 1996. *NCSL-RP-1 Establishment and Adjustment of Calibration Intervals*. Boulder, CO: NCSL.

National Institute of Standards and Technology (NIST). 2003. *Specifications, Tolerances, and Other Technical Requirements for Weighing and Measuring Devices*. NIST Handbook 44. Gaithersburg, MD: NIST.

NDT-ED.org. 2012a. "Basic Principles of Ultrasonic Testing." Education Resources. Accessed May 2012. http://www.ndt-ed.org/EducationResources/CommunityCollege/Ultra sonics/Introduction/description.htm.

———. 2012b. "Introduction to Magnetic Particle Inspection." Education Resources. Accessed May 2012. http://www.ndt-ed.org/EducationResources/CommunityCollege/MagParticle/Introduction/introduction.htm.

———. 2012c. "Present State of Radiology." Education Resources. Accessed May 2012. http://www.ndt-ed.org/EducationResources/CommunityCollege/ Radiography/Introduction/presentstate.htm.

Nevado, G. 2007. "Quality 101: CMM Part Alignment." *Quality Magazine,* October 30. Accessed November 27, 2018. https://www.qualitymag.com/articles/85176-quality-101-cmm-part-alignment.

nResults. 2008. "Functionality Testing." Accessed May 2012. http://nresult.com/quality-assurance/functionality-testing/.

Ohno, T. 1988. *Toyota Production System: Beyond Large Scale Production.* Portland, OR: Productivity Press.

Okes, D. 2019. *Root Cause Analysis: The Core of Problem Solving and Corrective Action.* 2nd ed. Milwaukee, WI: ASQ Quality Press.

Peled, Doron A. 2010. *Software Reliability Methods.* New York: Springer-Verlag.

Pennella, R. 2004. *Managing the Metrology System.* Milwaukee: ASQ Quality Press.

Phillips, A.W. 2015. *ISO 9001:2015 Internal Audits Made Easy.* 4th ed. Milwaukee, WI: ASQ Quality Press.

Pilot, S. 2002. "Back to Basics: What Is a Fault Tree Analysis? Use a general conclusion to determine specific causes of a system failure" ASQ *Quality Progress,* March.

Pullum, Laura L. 2001. *Software Fault Tolerance Techniques and Implementation.* Boston: Artech House Publishers.

Pyzdek, T., and R. Berger, eds. 2003. *Quality Engineering Handbook.* Milwaukee: ASQ Quality Press; New York: Marcel Dekker.

Pzydek, Thomas. 2003. *The Six Sigma Handbook: A Complete Guide for Green Belt, Black Belt, and Managers at All Levels: Revised and Expanded.* New York: McGraw-Hill.

Quality Magazine. 2016. "A Perfect Quality Inspector." *Quality Magazine,* September 1.

Qualitybooker. "Production Part Approval Process." Accessed March 25, 2025. https://www.qualitybook.org/production-part-approval-process-ppap.

Rademaker, G. 2005. "X-Ray Inspection." *Quality Digest,* April. Accessed November 27, 2018. http://www.qualitydigest.com/april05/articles/01_article.shtml

Rashed, A. F., and A. M. Hamouda. 1974. *Technology for Real Quality.* Alexandria, Egypt: Egyptian University House.

Raz, T. 1992. "Inspection." In *Quality Engineering Handbook,* edited by T. Pyzdek and R. Berger. Milwaukee, WI: ASQC Quality Press; New York: Marcel Dekker.

Reason, R. E. 1960. *The Measurement of Surface Texture.* London: Cleaver Hume Press.

Rice, G. 1976. "Measurement Systems and the Standards Laboratory." *Conference on the Management of Laboratory Instruments.* Cairo, Egypt.

Rich, Barnett and Philip Schmidt. 2004. *Schaum's Outline of Elementary Algebra.* 3rd ed. New York: McGraw-Hill.

Roymech, UK. 2007. "Drawing Title Blocks." Accessed May 2012. http://www.roymech.co.uk/Useful_Tables/Drawing/Title_blocks.html.

Russell, J. P. 2005. *Process Auditing Techniques.* Milwaukee, WI: ASQ Press.

Russell, J. P. 2015. *The Quality Auditing Handbook.* 4th ed. Milwaukee, WI: ASQ Quality Press.

Scholtes, P., B. Joiner, and B. Streibel. 2003. *The Team Handbook.* 3rd ed. Madison, WI: Oriel.

Schrader, G. F., and A. K. Elshennawy. 2000. *Manufacturing Processes and Materials.* 4th ed. Dearborn, MI: Society of Manufacturing Engineers.

Serway, R. A., and J. W. Jewett Jr. 2003. *Physics for Scientists and Engineers.* 6th ed. Belmont, CA: Brooks Cole.

Simpson, J. A. 1981. "Foundations of Metrology." *Journal of Research of the National Bureau of Standards* 86, no. 3 (1981): 36–42.

Smith, J. B., 2012. *Auditing Beyond Compliance.* 3rd ed. Milwaukee, WI: ASQ Quality Press.

Soleimani, M., W. R. B. Lionheart, A. J. Peyton, X. Ma, and S. Higson. 2006. "A 3D Inverse Finite Element Method Applied to the Experimental Eddy Current Imaging Data." IEEE *Trans Magnetics* 42, no. 5: 1560–67.

Sommerville, Ian.2006. *Software Engineering.* 8th ed. New York: Addison-Wesley.

SPS Electronics GmBH. " Software for measuring and testing devices—efficiency and precision in the testing environment" Accessed April 14, 2025. https://spselectronic.com/products-services/measuring-testing-equipment/accessories/software/.

Stamatis, D. 2003. *Failure Mode and Effect Analysis: FMEA Theory To Execution*. 2nd ed. Milwaukee, WI: ASQ Quality Press.

Stojanovic, S. "List of mandatory documents required by ISO 9001:2015." Accessed September 3, 2018. https://advisera.com/9001academy/knowledgebase/list-of-mandatory-documents-required-by-iso-90012015/.

Synergy Global Sourcing. "Initial Sample Inspection Report." Accessed September 3, 2018. https://www.synergyglobal.in%2fimages%2fquality_inspection_docs%2fsynergy_inspection_report.jpg&exph=1616&expw=1392&q=Initial+Sample+Inspection+Report+Template&simid=607997707581262268&selectedIndex=29&ajaxhist=0.

Tague, Nancy. 2023. *The Quality Toolbox*, 2nd ed. Milwaukee, WI: ASQ Quality Press.

Taylor, B. N. 1995. *Guide for the Use of the International System of Units (SI).* NIST Special Publication 881. Gaithersburg, MD: NIST.

Tec-Ease. 2012. "GD&T Training and Materials." Accessed May 12, 2012. http://www. tec-ease.com.

Triola, Mario F. 2006. *Elementary Statistics*,10th ed. New York: Pearson Education.

Vogt, T. 1980. *Optimizing Calibration Recall Intervals and Algorithms.* NBS-GCR-80-283. Washington, DC: NIST Press.

Wald, A. 1973. *Sequential Analysis.* New York: Dover Publications.

Wallace, Dolores R.., and RogerU. Fujii. 1989. "Software Verification and Validation: An Overview." *IEEE Software*, May, 10–17.

Ware, M. 1999. *Cyanotype: The History, Science and Art of Photographic Printing in Prussian Blue*. London: NMSI Trading.

Wenk, S. A., and R. C. McMaster. 1987. *Choosing NDT: Applications, Costs and Benefits of Nondestructive Testing in Your Quality Assurance Program*. Columbus, OH: American Society for Nondestructive Testing.

Western Electric. 1956. *Statistical Quality Control Handbook.* Indianapolis, IN: Western Electric Corporation.

Westfall, Linda. 2024. *The ASQ Certified Software Engineer Handbook*. Milwaukee, WI: ASQ Quality Press.

Wunchell, W. 1996. *Inspection and Measurement in Manufacturing.* Dearborn, MI: Society of Manufacturing Engineers.

Zipin, R. B. 1971. "Dimensional Measurements and Standards in Manufacturing." *Bendix Technical Journal* 1, no. 4 (1971): 15–19.

About the Editors

Dr. H. Fred Walker is a practicing Principal Quality Engineer in the nuclear power industry. Over the past 30 years Dr. Walker has been led quality initiatives in nearly 40 national and international organizations. Dr. Walker is a co-author and project lead of *The Quality Engineer Handbook* (1st edition), *The Quality Technician Handbook* (four editions), *The Quality Inspector Handbook* (four editions), *and Applied Statistics for the Six Sigma Green Belt.* Fred earned his Ph.D. in Industrial Education and Technology and the Master of Systems Engineering degrees from Iowa State University and an MBA and BS in Industrial Technology from California State University—Fresno. Dr. Walker holds the following professional certifications from ASQ: CMQ/OE, CQE, CRE, CSQE, SSBB, SSGB, CQA, CBA, HACCP/Critical Control Point Auditor, CQI, CCT, CQT, CQPA, CQIA.

Dr. Ahmad K. Elshennawy, PhD is Professor and Executive Director of The UCF Quality Institute in the Department of Industrial Engineering and Management Systems at the University of Central Florida (UCF). Dr. Elshennawy combines over 30 years of international experience in the areas of quality management and strategic improvement, as a researcher, academician and a consultant in the United States and different parts of the world. Dr. Elshennawy holds Ph.D. and M.Eng. degrees in Industrial Engineering from Penn State University and BS and MS degrees in Production Engineering from Alexandria University. Dr. Elshennawy is a Certified Lean Six Sigma Master Black Belt (LSSMBB), an ASQ Fellow, Certified Quality Engineer (CQE), and Certified Reliability Engineer (CRE).

Dr. Mary McShane-Vaughn is the founder of University Training Partners, a company that develops and delivers Lean Six Sigma training for corporations and universities. Over the past fifteen years, she has trained over 1600 Yellow, Green, and Black Belts.

She is author of *The Probability Handbook, The Probability Workbook,* and *Lean Six Sigma Leadership,* and is the editor of the newest editions of *The ASQ Certified Reliability Engineer Handbook, The ASQ Certified Six Sigma Black Belt Handbook* and *The ASQ Six Sigma Black Belt Study Guide.* Mary earned her PhD in Industrial Engineering and an MS in Statistics from the Georgia Institute of Technology, and a BS in Industrial Engineering from Kettering University in Michigan. Dr. McShane-Vaughn holds ASQ certifications as a Six Sigma Black Belt, Quality Engineer, and Reliability Engineer.

Index

A

B

C

D

G

H

I

N

O

P

Q

R

S

T

www.ingramcontent.com/pod-product-compliance
Lightning Source LLC
LaVergne TN
LVHW080644210926
848485LV00003B/323